OPTICAL FIBER AND WIRELESS COMMUNICATIONS

Edited by **Rastislav Róka**

Optical Fiber and Wireless Communications
http://dx.doi.org/10.5772/66006
Edited by Rastislav Róka

Contributors

Yongli Zhao, Vjaceslavs Bobrovs, Sandis Spolitis, Sergejs Olonkins, Jurgis Porins, Girts Ivanovs, Xuelin Yang, Adnan Hajomer, Weisheng Hu, Konstantinos Manousakis, Georgios Ellinas, Panayiotis Kolios, Zaichen Zhang, Jian Dang, Liang Wu, Dmitry Svistunov, Jin-Yuan Wang, Jun-Bo Wang, Yongjin Wang, Mohammad Faiz Liew Abdullah, Rahmat Talib, Boris I. Lembrikov, Yossef Ben-Ezra, Danilo Bórquez-Paredes, Alejandra Beghelli, Ariel Leiva, Tianhua Xu, Mohammed Alresheedi, J M H Elmirghani, Ahmed Taha Hussein, Mitsuru Kihara, Khadijah Ismail, Lisawati Sainawi, Joao Pedro, Bodhisattwa Gangopadhyay, Stefan Spaelter, Joao Batista Rosolem, Waqas Ahmed Imtiaz

Published by InTech
Janeza Trdine 9, 51000 Rijeka, Croatia

Publishing Process Manager Martina Blecic
Technical Editor SPi Global
Cover InTech Design team

First published June, 2017
Printed in Croatia
Legal deposit, Croatia: National and University Library in Zagreb

Additional hard copies can be obtained from orders@intechopen.com

Optical Fiber and Wireless Communications, Edited by Rastislav Róka
p. cm.
Print ISBN 978-953-51-3275-2
Online ISBN 978-953-51-3276-9

For EU product safety concerns:
IN TECH d.o.o., Prolaz Marije Krucifikse Kozulić 3, 51000 Rijeka, Croatia,
info@intechopen.com or visit our website at intechopen.com.

Meet the editor

Rastislav Róka was born in Šaľa, Slovakia, on January 27, 1972. He received his MSc and PhD degrees in Telecommunications from the Slovak University of Technology, Bratislava, in 1995 and 2002, respectively. Since 1997, he has been working as a senior lecturer at the Institute of Multimedia Information and Communication Technologies, Faculty of Electrical Engineering and Information Technology, Slovak University of Technology in Bratislava. Since 2009, he is working as an associate professor at this institute. His teaching and educational activities are realized in areas of fixed transmission media, designing and planning of telecommunication networks, and optocommunication transmission systems. At present, his research activity is focused on the signal transmission through optical transport, metropolitan, and access networks using advanced optical signal processing including various multiplexing, modulation, and encoding techniques. His main effort is dedicated to effective utilization of the optical fiber's transmission capacity of the broadband optical networks by means of dynamic bandwidth and wavelength allocation algorithms applied in various advanced hybrid passive optical network infrastructures.

Contents

Preface

The book *Optical Fiber and Wireless Communications* provides a platform for practicing researchers, academics, PhD students, and other scientists to review, plan, design, analyze, evaluate, intend, process, and implement diversiform issues of optical fiber and wireless systems and networks, optical technology components, optical signal processing, and security. The 17 chapters of the book demonstrate the capabilities and potentialities of optical communication to solve scientific and engineering problems with varied degrees of complexity.

The first six chapters, related to optical fiber systems and networking, provide details of software-defined optical networking, network virtualization over elastic optical networks, and multiperiod attack-aware optical network planning. Also, a future electrical multiplexing technique, the OFDM technique in cooperation with intensity modulation and direct detection, and a technique of selective mode excitation are taken into consideration for advanced optical fiber systems.

The second three chapters, associated with optical wireless systems, demonstrate possibilities of holograms in optical wireless communications, visible light communication with input-dependent noise, and receiver performance in radio-over-fiber network transmissions.

The next five chapters are concerned with components for optical technologies, especially with parametric and hybrid optical amplifiers, universal traffic switching and flexible optical transponders, novel single-mode multifiber connectors, power-over-fiber, and ultrafast all-optical memory applications.

The last three chapters deal with the optical signal processing and security represented by DPS techniques for compensating and equalizing transmission impairments, physical-layer encryption based on digital chaos, and protection architecture feasibility for passive optical networks.

I hope that beginners and professionals in the field would benefit by going through the details given in the chapters of this book.

Rastislav Róka
Slovak University of Technology,
Institute of MICT,
FEI STU Bratislava,
Slovakia

Section 1

Optical Fiber Systems and Networking

Chapter 1

Software-Defined Optical Networking (SDON): Principles and Applications

Yongli Zhao, Yuqiao Wang, Wei Wang and Xiaosong Yu

Additional information is available at the end of the chapter

http://dx.doi.org/10.5772/intechopen.68294

Abstract

Featured by the advantages of high capacity, long transmission distance, and low energy consumption, optical network has been deployed widely as the most important infrastructure for backbone transport network. With the development of Internet, datacenter has become the popular infrastructure for cloud computing, which needs to be connected with high bitrate transport network to support heterogeneous applications. In this case, optical network also becomes a promising option for intra and inter-datacenter networking. In the networking field, software-defined networking (SDN) has gained a lot of attention from both academic and industry, and it aims to provide a flexible and programmable control plane. SDN is applicable to optical network, and the optical network integrated with SDN, namely software-defined optical network (SDON), are expected as the future transport solutions, which can provide both high bitrate connectivity and flexible network applications. The principles and applications of SDON are introduced in this chapter.

Keywords: optical network, SDON, Datacenter, key technologies, applications

1. Introduction

The continuous growth in the number of large-scale datacenters (DC) poses new challenges to the capacity and elasticity of current optical network. Datacenter applications, such as scheduled high-speed data replication, require the substrate network to provide ultra-high bandwidth dynamically [1]. This is quite different from the service model of traditional optical network, which arises the need for that the datacenter network (DCN) architecture to be redesigned [2]. Recently, elastic optical network (EON) with software-defined networking (SDN) architecture has been introduced into datacenters to improve the performance of service provisioning. This can jointly consider both IT resources and network resources and enable network to be programmed like general hardware [3–5]. The optical network integrated with SDN, namely software-defined optical network (SDON), can provide high bandwidth with IT resources for multiple tenants [6]. It separates the control plane from the data plane and turns the control manner into a centralized and flexible one. In addition, it enables datacenter operators to control and program network functions such as bandwidth provisioning with Quality of service (QoS) guarantees. Resources of datacenter are provisioned by virtualizing resources of distributed datacenters and operator's optical network in a coordinating manner [7]. A hierarchical control mechanism is proposed to control multi-domain optical network [8], which can support connections between datacenters in multi-domain scenario. On the basis of above studies, the objective of this chapter is to find what applications can be deployed on SDON.

2. Software-defined optical network architecture

A case of SDON integrated with datacenters is schematically depicted in **Figure 1** [9]. There are datacenters (DC) geographically distributed in different locations. Datacenter network typically consists of multi-domain and multi-technology network, such as Ethernet access network and optical transport network. For each network domain, an SDN controller is deployed with extended protocols. The Control Virtual Network Interface and the corresponding OpenFlow (OF) agents (there are some other potential protocols, such as Path Computation Element Communication Protocol (PCEP) and Simple Network Management Protocol (SNMP)) are implemented for optical devices and equipment. The control plane is designed as a hierarchical architecture, in which a father controller is deployed for cross-domain orchestration and each domain controller is deployed for local domain resource management.

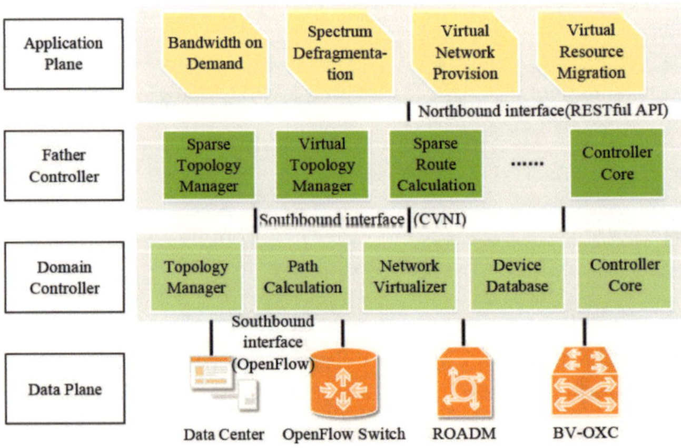

Figure 1. SDON architecture.

2.1. Southbound and northbound interfaces

The southbound interface (SBI) is mainly used to synchronize control information between controllers and network devices and datacenters. The control information contains topology graph, label resources, statistics data, alarm events, and kinds of flow tables.

In December 2013, a series of suggestions about SDON southbound interface were given from the white paper of Open Networking Foundation (ONF), namely control data plane interface (CDPI), which has defined the programming protocols between SDN controllers and optical transport network devices [10].

Here are some basic functions that controller southbound interface should satisfy:

(1) Controller can get local topology and resource occupation status via southbound interface.

(2) Controller can configure local network connection and control functions related to service, including connection creating, deleting, and adjusting via southbound interface. The configuration parameters include the information about QoS, Operation-Administration-Maintenance (OAM), and so on.

(3) Controller can get local network's alarming information via southbound interface asynchronously.

The northbound interface (NBI) is designed for communications between controllers and network customers. Via the NBI, controllers can virtualize the heterogeneous features of physical devices and provide abstracted network information for customers to deploy their applications.

According to a research about OpenDayLight project, which is demonstrated on Open Networking Conference 2014 in California, the development of SDN northbound interface functions is pushed forward swiftly. It matches the purpose of SDN research, which means that services can be accepted flexibly and clients can take control of devices and resources in network by using the software programmable technique. Consequently, it is obvious that there will be a good perspective of developing SDN northbound interfaces.

Northbound interface should support basic functions as follows.

(1) NBI should support upper layer controller to get all or parts of the network topology, including top layer subnet, as well as all nodes, links, ports it contains.

(2) NBI should support service management, which is mainly about:

    (a) Support creating, deleting, and adjusting point-to-point, point-to-multi-point, and multi-point-to-multi-point services.

    (b) Support getting running status of all or selected services.

(3) NBI should support point-to-point connection control function.

(4) According to flow distribution between nodes on client side, NBI should support creating, deleting, and adjusting virtual network services.

(5) NBI can optionally support path computation function when the controller is controlled by other upper layer orchestrators.

   (a) Support posting routing computation request and offering path computation function according to demand of service strategy.

   (b) Support posting reconfiguring path request, adjusting path connected, in order to satisfy new requirement.

(6) NBI should support post controller resource data updating request. When network resource data is updating directly, instead of based on network notifications, new request will be needed through northbound interface to update data. And resource should be released after connections are deleted.

(7) Northbound should support triggering the function of resynchronizing controller resource data.

2.2. Protocols

There are several protocols that can be used to implement SBI:

(1) OpenFlow protocol: OpenFlow protocol for network devices should correspond to the requirement of ONF TS-024 (OpenFlow Switch Specification 1.4.1) and ONF TS-022 (Optical Transport Protocol Extensions Ver1.0) or higher version. OpenFlow protocol supports resource reporting and network device configuring.

(2) OF-Config protocol: As a supplementary of OpenFlow protocol, it is used for configuring OpenFlow network devices.

(3) Traditional management protocols: Qx, SNMP, TL1 protocols, and so on.

(4) ASON/GMPLS control panel protocol: Signaling protocol should correspond to GB-T 21645.4-2010(ASON) and Request For Comments (RFC) 3471; routing protocol should correspond to GB/T 216545.8-2012 and RFC4203. Automatic discovering protocol should correspond to GB/T 21645.7-2010 and RFC4204.

(5) Path computation element protocol (PCEP): Corresponding to the requirement of Internet Engineering Task Force (IETF) RFC 5440 (PCE communication protocol) and other derived RFC documents.

SDON NBI can be implemented by SDN-specific protocols (RESTCONF, PCEP, etc.) or other popular protocols used in current software engineering (Restful, Protobuf, etc.).

In some cases, notification function is mandatory for NBI to push update information to customers. RESTCONF is one option, which supports notification event defined in YANG model. RESTCONF clients receive notifications by means of subscribing URL of notification message's resource. NBI protocol can follow RESTCONF protocol defined in IETF draft-ietf-netconf-restconf-07 to realize functions above.

It is JSON or XML format that content encoding of northbound interface communication should use, which corresponds to RFC7159 and RFC3032, respectively.

3. Key technologies in SDON

SDON is stretching optical network intelligence. It represents the fact that the control panel of optical network changes from switching intelligence to a comprehensive automatic one, where services diversity and management automation are also considered. In order to adapt to this revolution, SDON needs to make breakthroughs on key technologies like services assignment strategies, heterogeneous networks, and programmable optical transmit devices.

3.1. Sliceable transponder

Sliceable transponder is a kind of photonic, which can be divided into several sub-transponders, and each sub-transponder can be used to construct an independent lightpath without "O-E-O" switching on transit nodes. From the perspective of optical layer, multiple optical flows can be multiplexed into one optical transponder. And the de-multiplexing procedure for each optical flow will be operated in the receiver at destination node of that flow. **Figure 2** shows a sliceable flexible optical transponder.

A physical transponder can generate different ligthpaths with different spectrum segments. Multi flow optical transponder has been proposed and proved by experiment [11]. In addition, multi-spectrum sliceable and bandwidth-extendable coherence optical transponder based on any waveform optical generator, which can also generate multiple optical paths, has been proposed and proved [12].

The most significant difference between sliceable transponder and unsliceable transponder is that there are multiple send-receive interfaces in sliceable transponder, which means multiple clients' signals can be handled by one transponder. Consequently, there should be a programmable switch matrix between interface card and sliceable transponder, in order to routing

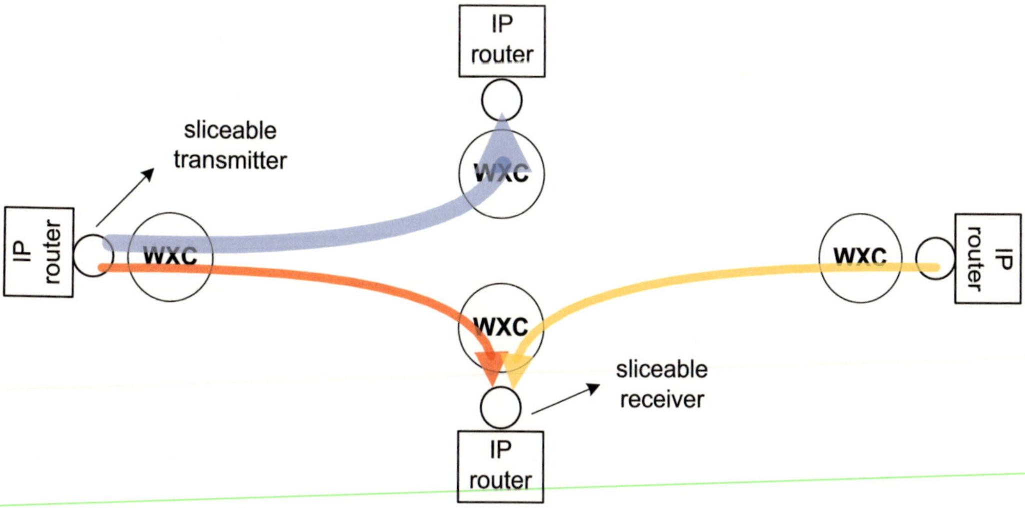

Figure 2. Sliceable optical transponder schematic diagram.

signals between them. In this case, the number of interface cards M and transponder slices are the key factor, which decide the network capability and networking cost.

3.2. Wavelength-selective switching (WSS)

The advent of all-optical switches based on wavelength-selective switching (WSS) has a significant impact on the development of extensibility and flexibility on optical networking. It makes the structure of all-optical switches change unexpectedly. This kind of WSS lets each wavelength in an input Wavelength-Division Multiplexing (WDM) signal transmit to anyone of the output ports, while the N×1 WSS has the contrary function.

Each WSS chooses an output previously configured wavelength from each input optical fiber. The number of WSS will grow according to the expansion of its dimensions. As to interfaces on client side, several WSS can be deployed as multiplexing/de-multiplexing role, instead of deploying one WSS for each add/drop fiber.

Figure 3 shows an example flexible switching node, which can realize high-spectral efficient multi-grid grade switching.

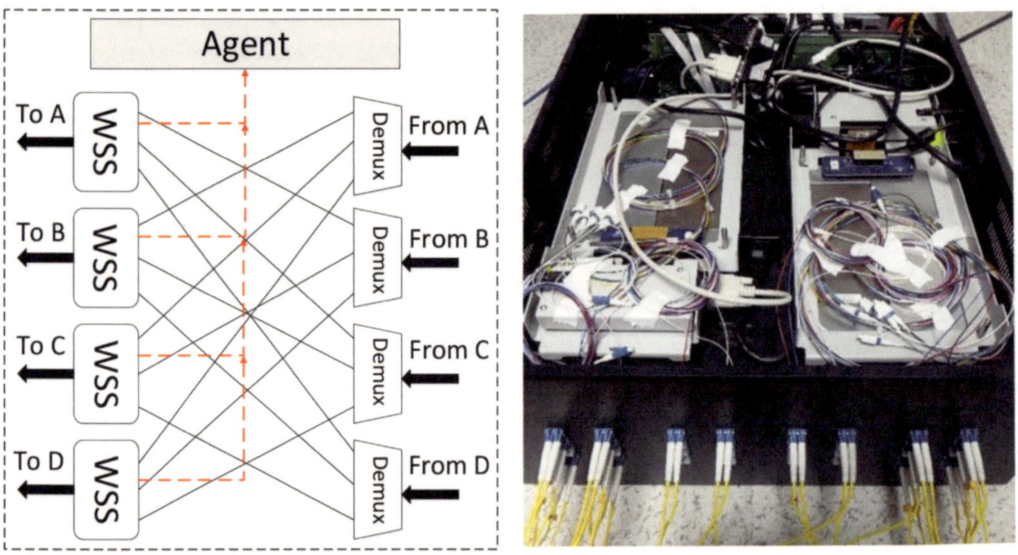

Figure 3. Software-defined flexible switching node.

3.3. Bandwidth-variable optical cross-connect

Bandwidth-variable optical cross-connect (BV-OXC) is an important kind of switching node in flexible spectrum optical network. A multi-dimension BV-OXC is composed of some splitters and some bandwidth-variable wavelength-selective switches (BV-WSS) as shown in **Figure 4**.

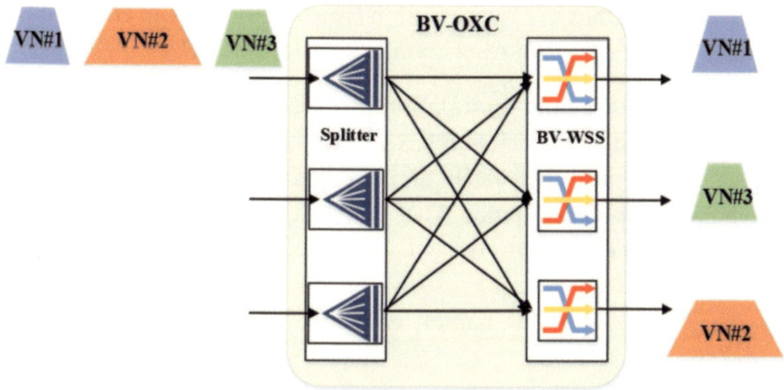

Figure 4. Bandwidth-variable optical switching node schematic diagram.

Bandwidth-variable optical cross-connect (BV-OXC) changes the switching granularity from wavelength to sub-carrier level. The basic unit bandwidth of sub-carrier is 12.5 GHz, while the interval is 6.25 GHz, both of which are much less than 50 GHz—the International Telegraph Union Telecommunication Standardization Sector (ITU-T) standard wavelength.

In addition, each wavelength in OXC is isolated from the others. Even though there is a blocking interval between two adjacent wavelengths, this feature makes it impossible for a bandwidth that is more than a wavelength to pass inviolately. However, BV-OXC eliminates the slot between sub-carriers and makes it possible for multiple sub-carriers to bind together and become a broadband. BV-OXC makes it possible to switch a signal as a super-channel.

Thus, BV-OXC has not only better switching granularity but also makes switching capability rise dramatically. Subsequently, this technique can be used in elastic optical network (EON), for the reason that the core idea of EON is to provide necessary spectrum resource with better granularity for users' different connection requirements.

3.4. Routing and spectrum assignment

In elastic optical network, high-spectral efficiency and expansibility have been already realized. Meanwhile, the flexibility introduces new challenges in network research. For instance, taking route and spectrum assignment into account, it is not suitable for traditional routing and wavelength assignment (RWA) algorithm being used in flexible spectrum network any longer. One of the reasons is that there lies a new feature—constraint condition of spectrum continuity in flexible spectrum network, which means that network needs to assign a slot of continuing and available spectrum resource for each lightpath. In addition, the flexibility of grid makes the network model more complicated, resulting in higher difficulties of optimization.

In elastic optical network, spectrum is further divided into a series of sub-carriers that has lower data rate and finer granularity than those in WDM network [13]. Compared with RWA problem in traditional network schedule, the primary problem in elastic network is to allocate contiguous spectrum resources to provision high bandwidth demand connections. This problem is called Routing and Spectrum Allocation (RSA) [14].

According to required bitrates and modulation format, lightpath is constructed from source node to destination node in frequency domain by means of allocating continuous sub-carriers on all links. In order to demodulate easily on receivers, one or more sub-carriers are needed as gap between adjacent spectrum paths. There are three kinds of constraints that we should obey when operating RSA computation:

(1) Spectrum consistency constraint, namely each link in the path needs to allocate the same spectrum resource.

(2) Spectrum continuity constraint, only to make sure that each sub-carrier in the path must connect with the others.

(3) Spectrum conflict constraint, allocated spectrum in different paths on the same fiber cannot be overlapped [15–17].

When connections are deleted, the allocated spectrum resource will be released and can be provided for other coming services. However, due to the fact that network being usually used to serve multiple applications, lightpaths may be constructed and deleted randomly, which leads to the appearance of spectrum fragments inevitably. When contiguous spectrum is divided into fragmentary segments, new coming requests might be blocked for lacking wider contiguous spectrum segment, though there are still lots of unused spectrum slots. This means that the spectrum is not used efficiently. Some heuristic algorithms have been proposed for RSA problems [18–20] and defragmentation problems [21–23].

3.5. Virtual optical network mapping

Virtual optical network (VON) is a new kind of transport service, which aims to allocate spectrum resource to customers at topology level. Compared with RSA, VON mapping is more complex because it needs not only spectrum allocation but also mapping virtual topology to physical substrates. **Figure 5** shows a use case of virtual network mapping. The left is abstracted topology and the right is the physical topology. If node mapping method is adopted, node 1 on the abstracted topology is mapped on node 2 on physical topology. Similarly, nodes 2 and 3 are mapped on nodes 4 and 1 on physical topology. If link mapping method is adopted, link 1–2 on abstracted topology is mapped on link 2-3-4 on physical topology, and link 1–3 is mapped on link 2–1 on physical topology. The procedure of virtual network mapping can be divided into two phases:

(1) Node mapping: Map virtual nodes to different real physical nodes and satisfy virtual nodes' resources demand.

(2) Link mapping: Map virtual links between virtual nodes to real physical links.

From the perspective of VON mapping, all of the mapping algorithms can be summarized as:

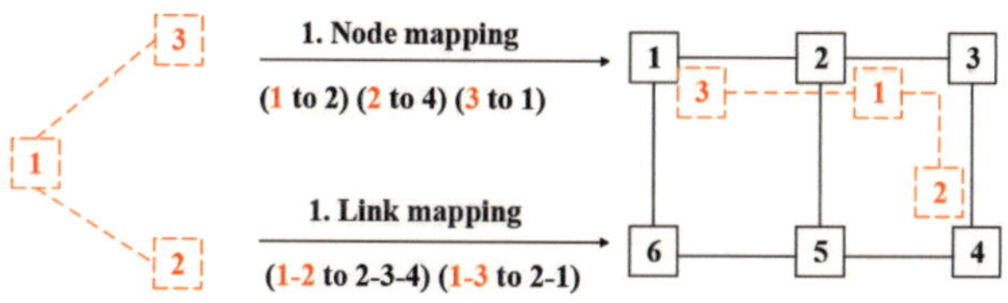

Figure 5. An example of virtual optical network mapping.

(1) Use a kind of greedy algorithm when mapping nodes, for example, begin from the physical node which owes the most physical resources [24, 25].

(2) If the network does not support distribution technique when mapping links, Shortest Path First algorithm should be considered [24]. Else, if it supports, multi-path algorithm should be considered [25, 26].

(3) When mapping nodes and links at the same time, this method can find a fairly good solution. However, the algorithm complexity is high and expansibility is not good.

As for mapping algorithm, concepts like sliceable ability of link, optical transponder, and switching nodes are proposed to describe resource abstraction method in flexible spectrum network. And based on that above, LS-based virtual optical network mapping strategy and NS-based virtual optical network mapping strategy are proposed [27]. Result from experiment shows that LS-based virtual optical network mapping and NS-based virtual optical network mapping, compared to baseline virtual optical network mapping, get lower blocking probability and higher profit.

3.6. Cross stratum optimization algorithm

In datacenter network, network-based services are provisioned by both IT and network resource. In order to achieve the optimization of IT and network resource, cross stratum optimization (CSO) is proposed [28], which can enable a joint optimization of IT and network resource and take optical as a service (OaaS).

The SDN-based control architecture with CSO between optical network and application stratum resource in inter-datacenter network is proposed to partially meet the QoS requirement.

Unified control architecture is proposed as depicted in **Figure 6(a)**, which especially emphasizes on the cooperation between application controller (AC) and service controller (SC) based on OpenFlow to realize the CSO of application and network resource. According to the requirement, AC customizes appropriate virtual resource through related SC with extended OpenFlow protocol. In addition, dynamic global load balancing strategy is implemented based on both application and network resource, while SC can trigger service-aware Path Computation Element algorithm based on the result of dynamic global load balancing strategy. SC interacts with each other for the security and virtual network information. OpenFlow-enabled routers and optical transport nodes are realized by extending match domain in extended OpenFlow protocol. The responsibilities and interactions among entities are provided as shown in **Figure 6(b)** and **(c)**.

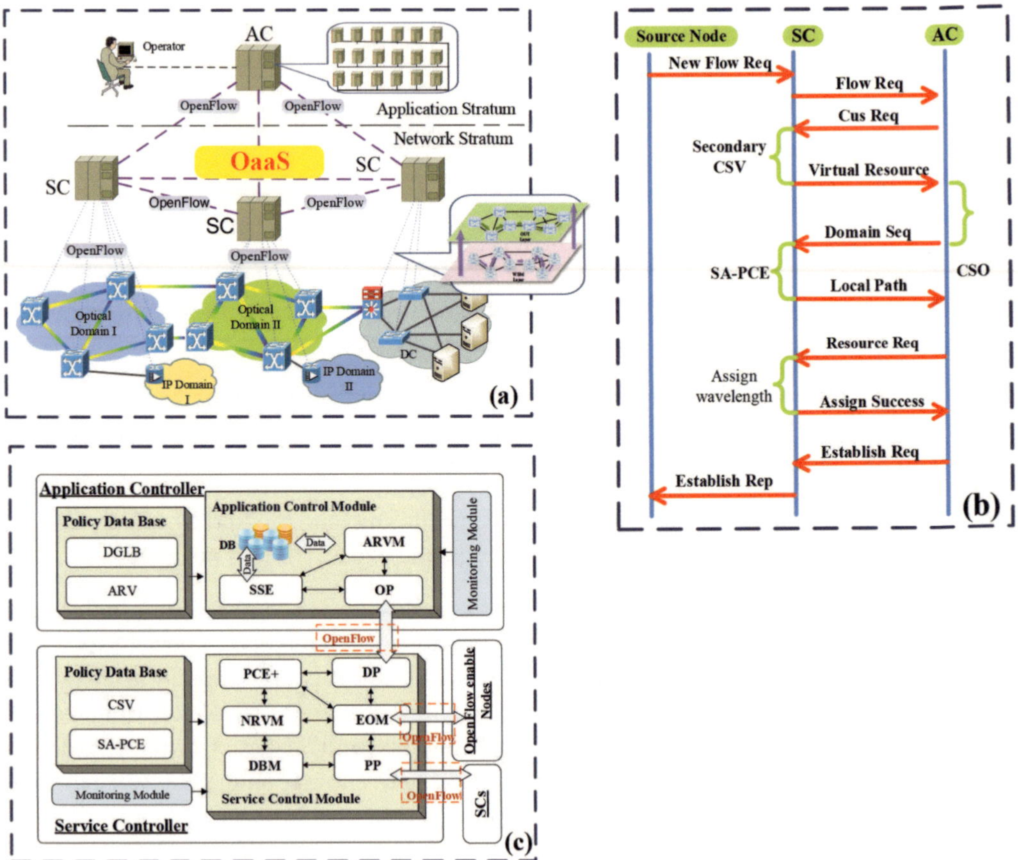

Figure 6. Architecture and procedure of OaaS: (a) architecture, (b) CSO procedures, (c) module diagram.

4. Novel applications

SDON is a promising solution for next-generation high-speed optical transport network, which embraces a wide application perspective. Here are some sample applications, which can make optical network being used in a more flexible manner. The sample applications include bandwidth on demand [29], Virtual Optical Network (VON) services [30], and some emerging applications [31, 32].

4.1. Bandwidth on demand

Based on the centralized control plane of SDON, providers are able to predict resource occupation status in time dimension, and accordingly arrange and accommodate users' requests to avoid spectrum conflicts and fragmentations. One application, which is called time-aware bandwidth on demand, has already been implemented and demonstrated to do such things. Bandwidth on demand can improve the utilization of spectrum resource in SDON. The overall feasibility and efficiency of this time-aware Bandwidth on demand architecture is experimentally verified on a testbed with real OpenFlow-enabled tunable optical modules in terms of blocking probability and resource occupation rate, as shown in **Figure 7**. First, the users' requests are sent to the controller, and then the controller will schedule these requests to make the spectrum resource utilization optimized, finally the controller will send messages to optical nodes to set up lightpaths. As one of the basic types of services in intelligent optical network, bandwidth on demand can prove the integrity and availability of networking by offering bandwidth resource directly. Also, bandwidth on demand can be a basic measurement index of network automation.

4.2. Virtual optical network

Virtual optical network (VON), which aims to provide multiple dedicated virtual network over shared network infrastructure, has gained a lot of attention by network providers. By VON, service providers can request a dedicated network for each application on a per-need basis and have full control ability over that network. Network virtualization technologies can partition/composite of network infrastructure (i.e., the physical optical nodes and links) into independent virtual resource, adopting the same functionality as the physical resource. Also, the composition of these virtual resource (i.e., virtual optical nodes and links) allows deploying multiple VON. A VON must be composed of not only a virtual transport plane but also a virtual control plane, with the purpose of providing the required independent, and full control functionalities. In datacenter optical network, since the computing resources are highly distributed, it is impossible to build a dedicated optical network for a special service. So joint considering computing and networking resource is very important if they are under the control of same provider. Network virtualization can eradicate the ossification of the network, stimulate innovation of new network architectures and applications, and enable network operators to operate different VON that share a common physical infrastructure. **Figure 8** shows the schematic of SDN-based VON provisioning.

There are many challenges in VON provisioning, such as VON design and VON resource management. VON design refers to the procedures to find a subset of the underlying physical

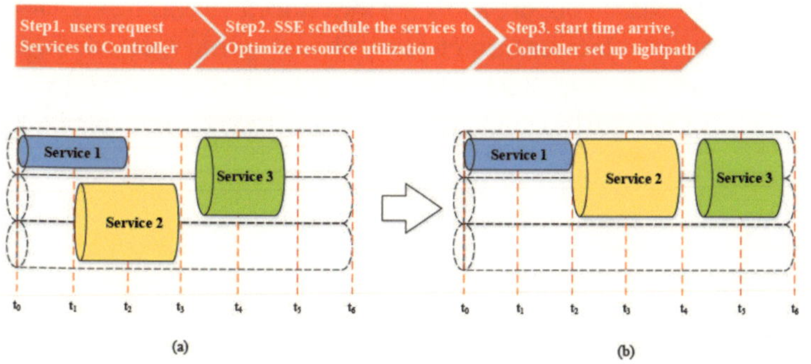

Figure 7. Time-aware bandwidth on demand: (a) without SS, three slots occupied; (b) with SS, two slots occupied, one slot resource is saved.

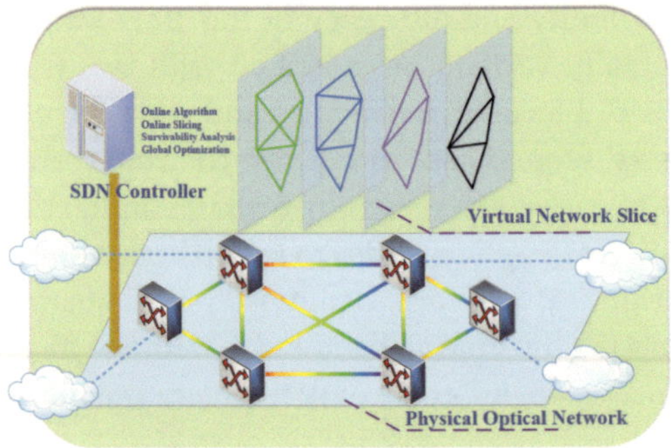

Figure 8. VON provisioning.

network topology connecting the computing resources with virtual nodes hosted on the physical nodes and virtual links spanning over the physical links. The VON design problem is related to the traffic demand model and the entire physical network topology information. VON resource management includes two schemes, that is, dedicated scheme and shared scheme. The dedicated VON assumes that the link resources are exclusively assigned to a VON user, so each VON topology runs independently, and its reserved link resources cannot be occupied by other VON. The dedicated VON is relatively easier to achieve, but it has a drawback that it increasingly decreases the efficiency of resource utilization. Note that the sharing VON scheme can be fully or partially shared, and partially shared VON manner is more profitable. On each physical link belonging to a VON, a minimum guaranteed capacity must be assigned to it, while residual capacity is shared by all the VON on that link. The challenge in employing the partially shared VON approach is that the service provider should treat the dedicated and shared resources with different policies, which inevitably increases implementation complexity and difficulty.

5. Conclusions

In this chapter, we summarize the basic principles and key technologies of SDON. First, the architecture of SDON is described with southbound and northbound interfaces and protocols. Then, several key technologies of SDON are introduced, including sliceable transponder, WSS, BV-OXC, routing and spectrum assignment, virtual optical network mapping, and CSO algorithms. Finally, two novel applications are prospected, such as bandwidth on demand and VON. More studies need to be conducted to promote the evolution of SDON in the future.

Author details

Yongli Zhao*, Yuqiao Wang, Wei Wang and Xiaosong Yu

*Address all correspondence to: yonglizhao@bupt.edu.cn

State Key Laboratory of Information Photonics and Optical Communications, Beijing University of Posts and Telecommunications, Beijing, China

References

[1] Velasco L, Paolucci F, Gifre Ll, et al. First Experimental Demonstration of ABNO-driven In-Operation Flex-grid Network Re-Optimization. San Francisco, USA: OFC; 2014. DOI: 10.1364/OFC.2014.Th3K.3

[2] Marc Makkes X, Reginald C, Ana-Maria O, et al. Scaling Bottlenecks in Data Center Networks. San Francisco, USA: OFC; 2014. DOI: 10.1364/OFC.2014.Tu2I.1

[3] Zhao Y, Zhang J, Yang H, et al. Data center optical networks (DCON) with OpenFlow based Software Defined Networking (SDN). 2013. Guilin, China. DOI: 10.1109/China Com.2013.6694698

[4] Zhang J, Zhao Y, Yang H, et al. First Demonstration of enhanced Software Defined Networking (eSDN) over elastic Grid Optical Networks (eGrid) for Data Center Service Migration. Anaheim, CA, USA: OFC 2013, Post deadline paper; 2013. DOI: 10.1364/OFC.2013.PDP5B.1

[5] Jinno M, Sone Y, Takara H, et al. Multi-flow optical transponder for efficient multilayer optical networking. IEEE Communications Magazine. 2012;**50**(5):56. DOI: 10.1109/MCOM.2012.6194383

[6] Zhang J, Yang H, Zhao Y, et al. Experimental demonstration of elastic optical networks based on enhanced software defined networking (eSDN) for data center application. Optics Express. 2013;**21**(22):26990–27002. DOI: 10.1364/OE.21.026990

[7] Chen H, Zhang J, Zhao Y, et al. Demonstration of Cloud Resources Integrated Provisioning over Multi-Domain Software Defined Optical Networks. Cannes, France: ECOC2014; 2014. DOI: 10.1109/ECOC.2014.6964148

[8] Jing R, Zhang C, Zhao Y, et al. Experimental Demonstration of Hierarchical Control over Multi-Domain OTN Networks Based on Extended Openflow. Los Angeles, USA: OFC2015; 2015. DOI: 10.1364/OFC.2015.W4J.4

[9] Zhao Y, Zhang J, Ji Y, et al. Experimental demonstration of software defined data center optical networks with Tbps end-to-end tunability. Optical Fiber Technology. 2015;**25**:72–79. DOI: 10.1016/j.yofte.2015.08.005

[10] Open Networking Foundation. SDN Architecture Overview. White paper, version 1.0; 2013

[11] Jinno M, Takara H, Sone Y, et al. Multiflow optical transponder for efficient multilayer optical networking. IEEE Communications Magazine. 2012;**50**(5):56–65. DOI: 10.1109/MCOM.2012.6194383

[12] Geisler DJ, Fontaine NK, Scott RP, et al. Bandwidth scalable, coherent transmitter based on the parallel synthesis of multiple spectral slices using optical arbitrary waveform generation. Optics Express. 2011;**19**(9):8242–8253. DOI: 10.1364/OE.19.008242

[13] Jinno M, Takara H, Kozicki B, et al. Spectrum-efficient and scalable elastic optical path network: Architecture, benefits, and enabling technologies. IEEE Communications Magazine. 2009;**47**:66–73. DOI: 10.1109/MCOM.2009.5307468

[14] Wang Y, Cao X, Pan Y. A study of the routing and spectrum al-location in spectrum-sliced elastic optical path networks. In: Proceedings of INFOCOM; 2011. pp. 1503–1511. DOI: 10.1109/INF COM.2011.5934939

[15] Christodoulopoulos K, Tomkos I, Varvarigos E. Elastic bandwidth allocation in flexible OFDM-based optical networks. IEEE Journal of Lightwave Technology. 2011;**29**:1354–1366. DOI: 10.1109/JLT.2011.2155990

[16] Sone Y, Hirano A, Kadohata A, et al. Routing and spectrum assignment algorithm maximizes spectrum utilization in optical networks. In: Proceedings of ECOC; 2011. Mo.1.K.3. pp. 1–3. DOI: 10.1364/ECOC.2011.Mo.1.K.3

[17] Jinno M, Kozicki B, Takara H, et al. Distance-adaptive spectrum resource allocation in spectrum-sliced elastic optical path network. IEEE Communications Magazine. 2010;**48**:138–145. DOI: 10.1109/MCOM.2010.5534599

[18] Christodoulopoulos K, Tomkos I, Varvarigos E. Dynamic bandwidth allocation in flexible OFDM-based networks. Los Angeles, USA: OFC; 2011. DOI: 10.1364/OFC.2011.OTuI5

[19] Castro A, Velasco L, Ruiz M, et al. Dynamic routing and spectrum (re) allocation in future flexgrid optical networks. Computer Networks. 2012;**56**(12): 2869–2883. DOI: 10.1016/j.comnet.2012.05.001

[20] Zhang L, Lu W, Zhou X, et al. Dynamic RMSA in spectrum-sliced elastic optical networks for high-throughput service provisioning. In: 2013 International Conference on Computing, Networking and Communications (ICNC). IEEE; 2013. pp. 380–384. DOI: 10.1109/ICCNC.2013.6504113

[21] Shen G, Yang Q, You S, et al. Maximizing time-dependent spectrum sharing between neighbouring channels in CO-OFDM optical networks. In: 13th International Conference on Transparent Optical Networks (ICTON). Cartagena, Spain; 2011. DOI: 10.1109/ICTON.2011.5970769

[22] Klinkowski M, Ruiz M, Velasco L, et al. Elastic spectrum allocation for time-varying traffic in flexgrid optical networks. IEEE Journal on Selected Areas in Communications. 2013;**31**(1):26–38. DOI: 10.1109/JSAC.2013.130104

[23] Christodoulopoulos K, Tomkos I, Varvarigos E. Time-varying spectrum allocation policies and blocking analysis in flexible optical networks. IEEE Journal on Selected Areas in Communications. 2013;**31**(1):13–25. DOI: 10.1109/JSAC.2013.130103

[24] Zhu Y, Ammar MH. Algorithms for Assigning Substrate Network Resources to Virtual Network Components. Barcelona, Spain: INFOCOM; 2006. DOI: 10.1109/INFOCOM.2006.322

[25] Yu M, Yi Y, Rexford J, et al. Rethinking virtual network embedding: Substrate support for path splitting and migration. ACM SIGCOMM Computer Communication Review. 2008;**38**(2):17–29. DOI: 10.1145/1355734.1355737

[26] Szeto W, Iraqi Y, Boutaba R. A multi-commodity flow based approach to virtual network resource allocation. San Francisco, USA: GlobeCom; 2003. DOI: 10.1109/GLOCOM.2003.1258787

[27] Yang H, Zhao Y, Zhang J, et al. Cross Stratum Optimization of Application and Network Resource Based on Global Load Balancing Strategy in Dynamic Optical Networks. Los Angeles, USA: OFC; 2012. DOI: 10.1364/NFOEC.2012.JTh2A.38

[28] Yang H, Zhang J, Zhao Y, et al. CSO: cross stratum optimization for optical as a service. IEEE Communications Magazine, 2015, 53(8):130-139. DOI: 10.1109/MCOM.2015.7180520

[29] Zhao Y, Li Y, Wang X, et al. Experimental demonstration of bandwidth on demand (BoD) provisioning based on time scheduling in software-defined multi-domain optical networks. Optical Fiber Technology. 2016;**31**:147–155. DOI: 10.1016/j.yofte.2016.07.001

[30] Zhao Y, Li S, Song Y, et al. Virtual optical network provisioning with unified service logic processing model for software defined multi-domain optical network. Optical Engineering. 2015;**54**(12):126110. DOI: 10.1117/1.OE.54.12.126110

[31] Wang W, Zhao Y, He R, et al. Continuity aware spectrum allocation schemes for virtual optical network embedding in elastic optical networks. Optical Fiber Technology. 2016;**29**:28–33. DOI: 10.1016/j.yofte.2016.01.008

[32] Wang W, Lin Y, Zhao Y, et al. First demonstration of virtual transport network services with multi-layer protection schemes over flexi-grid optical networks. IEEE Communications Letter. 2016;**20**(2):260–263. DOI: 10.1109/LCOMM.2015.2509066

Chapter 2

Network Virtualization Over Elastic Optical Networks: A Survey of Allocation Algorithms

Danilo Bórquez-Paredes, Alejandra Beghelli and Ariel Leiva

Additional information is available at the end of the chapter

http://dx.doi.org/10.5772/intechopen.68340

Abstract

Network virtualization has emerged as a paradigm for cloud computing services by providing key functionalities such as abstraction of network resources kept hidden to the cloud service user, isolation of different cloud computing applications, flexibility in terms of resources granularity, and on-demand setup/teardown of service. In parallel, flex-grid (also known as elastic) optical networks have become an alternative to deal with the constant traffic growth. These advances have triggered research on network virtualization over flex-grid optical networks. Effort has been focused on the design of flexible and virtualized devices, on the definition of network architectures and on virtual network allocation algorithms. In this chapter, a survey on the virtual network allocation algorithms over flexible-grid networks is presented. Proposals are classified according to a taxonomy made of three main categories: performance metrics, operation conditions and the type of service offered to users. Based on such classification, this work also identifies open research areas as multi-objective optimization approaches, distributed architectures, meta-heuristics, reconfiguration and protection mechanisms for virtual networks over elastic optical networks.

Keywords: optical fibre networks, flexible-grid/elastic networks, network virtualization, resource allocation algorithms

1. Introduction

Cloud computing has emerged as a new network paradigm [1]. Built on the success of grid computing applications, cloud computing implements the idea of 'computing as a utility' in a more commercially-oriented vision. Thus, the customer pays per use of computing facilities under the conditions stated in a service level agreement (SLA), having dynamic scaling of resources and transparent access to network services, unaware of the location and hardware/software characteristics of the required resources [2]. Apart from high bandwidth, cloud computing applications require the following functionalities from the underlying physical network [1]:

- Abstraction: The technology/implementation specific details of the physical network resources are hidden to the users, due to the "computing as a utility" philosophy.
- Isolation: Different cloud computing applications should not interfere with each other in the access to common physical resources.
- Flexible resource granularity: The amount of resources (storage, processing power and bandwidth) required by different cloud computing applications might vary significantly.
- On-demand setup/tear down: For efficiency, network resources should be set-up/torn down with a highly dynamic, rapidly reconfigurable and programmable network environment, something not possible with the current status of Internet [3].
- Resiliency: Grid/cloud computing applications should continue running in spite of failures affecting the optical network.

Network virtualization, which extends the well-known concepts of server and storage virtualization to networks, is envisaged as a key enabling technology for cloud computing services. As such, the benefits of running cloud applications on top of virtual networks (as opposed to on top of virtual servers alone, as usually done [4, 5]), was evidenced by several preliminary studies on network virtualization for cloud computing. In Ref. [6], resource allocation of cloud-based data centres services was proposed by abstracting the service requests as virtual network requests. In Ref. [7], a network virtualization platform that acts as a mediator between the cloud user requirements and the physical resources was proposed. In Ref. [8], a new network architecture based on network virtualization was proposed for cloud computing applications where the geographic location of servers is relevant. In Ref. [9], a network operator perspective was given about the convenience of network virtualization as an enabler for cloud computing. Nowadays, the benefits of network virtualization for cloud services are well identified in terms of cost, agility, resilience and multi-tenancy [10–12].

The underlying network over which network virtualization takes place is of fundamental importance to guarantee a good service. Arguably, the two most important requirements regarding the underlying network are the bandwidth capacity and the variety in the bandwidth granularity of connections, to allow for a high number of cloud computing applications with different bandwidth requirements. Both requirements would be naturally provided by flexible-grid optical networks [13, 14]. By overcoming the rigid spectrum allocation of current fixed-grid wavelength-division multiplexing (WDM) networks, elastic optical networks would make better use of the band C by allocating each connection the bandwidth just required. Depending on the bit rate and the modulation format, a gain in bandwidth usage between 33 and 100% could be achieved by using flexible-grid networks instead of a fixed one operating with a spectral width of 50 GHz [15]. Finally, flex grid would allow a wide bandwidth granularity of connections: bit rates from 10 Gbps to 1 Tbps are envisaged [13].

Given the impact that network virtualization is expected to have on the ever-increasing cloud computing area and the potential for significant bandwidth increase and bandwidth granularity offered by flexible-grid optical networks, in this survey, we review the efforts on network virtualization over optical flexible-grid networks.

The remaining chapter is as follows: Section 2 reviews the fundamental concepts of network virtualization and flexible-grid optical networks; Section 3 discusses the main challenges in the area of network virtualization over flexible-grid optical networks; Section 4 presents a taxonomy of the proposals found in the literature to allocate virtual networks over a flexible-grid underlying transport network; and Section 5 concludes the chapter highlighting the open research lines in the area.

2. Fundamental concepts

2.1. Network virtualization

Network virtualization refers to the creation of different isolated virtual networks on top of a common physical substrate. The isolation feature means that the information transmitted through a particular virtual network cannot be retrieved or affected by other existing virtual networks and the operation of the different virtual networks cannot affect the operation of the physical substrate [16].

Among the main features of network virtualization environments, we found several of the requirements imposed by cloud computing applications, namely, coexistence of different virtual networks, isolation between coexisting virtual networks, programmability, dynamicity, flexibility and heterogeneity [17].

By implementing cloud applications on virtual networks (i.e. one virtual network for each different cloud computing application), several benefits can be identified:

- Resource allocation based on maximum load could be avoided, leading to a more cost-effective operation, as the virtual network associated to the cloud application would request just the resources needed for proper operation. Some virtual network environments have even considered the possibility of reconfiguring the virtual network during operation (e.g. exploiting the feature of on-line virtual server migration) to adapt to time-variant requirements from the applications [2, 18].
- Isolation between different cloud applications for access to common physical resources
- Resiliency against node/server failures, due to the server-migration feature of virtualization environments
- Implementation of proprietary non-standard protocols for specific cloud applications requirements

Network virtualization has been envisaged as a very useful tool in network research and industry. In research, the test of new routing algorithms, network protocols or network controllers can be done by establishing a virtual network, without interrupting the normal operation of a physical network or deploying a physical network for tests. Thus, the production network may become the testbed [19]. An early example of this type of use was PlanetLab [20–22], established in 2002 for distributed systems and network research. Other efforts have been GENI in USA [23], FEDERICA and OneLab2 in Europe [24, 25], Akari in Japan [26] and FIBRE in a joint effort between Brazil and Europe [27]. For a review of several precursor experimental initiatives, see Ref. [17]. In an industry, network virtualization can offer separate networks for different units in a company, differentiation of services based on bandwidth usage (e.g. voice and video) or a rapid and flexible creation of sub-networks for different projects [28, 29]. For example, in a data centre each client can have its own topology and control its traffic flows. Finally, different service providers can share the same network infrastructure being unaware of the others.

As a way of illustration, **Figure 1** shows a schematic of a network virtualization system. The lower part shows the physical substrate, made of five nodes and six bidirectional links. The available capacity of physical links, measured in capacity units (c.u.), is shown next to each link. The upper part shows two of the virtual networks (three-node rings with three bidirectional links each) that have been established on the physical network. The capacity unit required by the virtual links are shown near to each link. Dotted lines represent the mapping between virtual and physical links. Both virtual networks can have virtual links established over the same physical link and a virtual link can require more than one physical link to be established. For the sake of clarity, the mapping of the nodes is not shown but it can be deduced by identifying the physical nodes at the extreme of the physical links associated to the virtual links. The decision about whether establishing a new virtual network is possible or not and what virtual link/node is established in what physical link/node is made by a virtual network allocation algorithm.

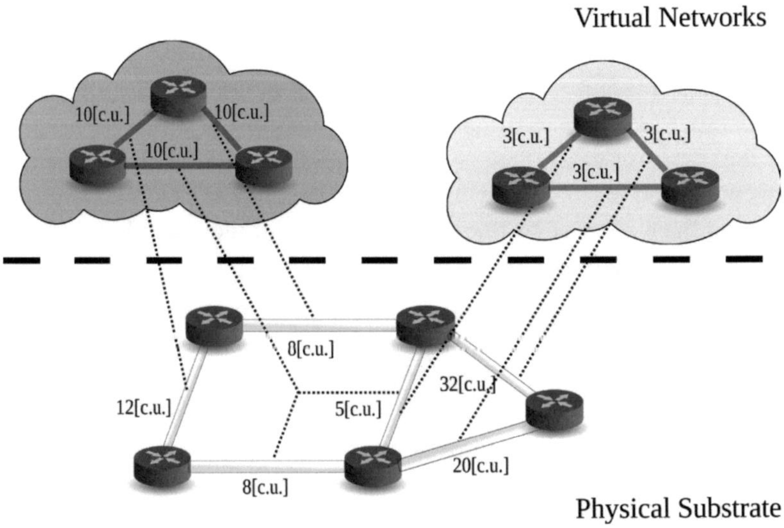

Figure 1. Two virtual networks established on the same physical substrate.

2.2. Mathematical modelling for network virtualization

The physical network is modelled by a directed graph $\mathcal{P} = (Np, Lp, R\,p^t, Cp)$, where Np and Lp are the sets of physical nodes and links, respectively; $R\,p^t$ is the set of resources of type t in the physical nodes (for example, storage and processing resources; $t \in \mathbb{N}$) and Cp the set of resources at the physical links (optical bandwidth).

Analogously, the i-th virtual network can be modelled by a directed graph $\mathcal{V}_i = (N\,v_i, L\,v_i, R\,v_i^t, C\,v_i)$, where $N\,v_i$ is the set of virtual nodes and $L\,v_i$ the set of virtual links; $R\,v_i^t$ is the set of resources of type t required by each virtual node of the virtual network $\mathcal{V}_i$ (e.g. storage and processing resources) and $C\,v_i$ is the set of resources required by the virtual links (optical bandwidth).

The information required to execute the resource allocation algorithm is as follows:

1. $\mathcal{A} = \{\mathcal{V}_i\}$: Set containing the identification of all virtual networks $\mathcal{V}_i$ already established over the physical network $\mathcal{P}$.
2. $\mathcal{S}n\,v_k$: Set of all virtual nodes already established on the physical node $k \in Np$.
3. $\mathcal{S}l\,v_m$: Set of virtual links already established on physical link $m \in Lp$.

Every time the resource allocation algorithm must process a new virtual network request, at least the following two constraints must be met to be able to accept such request:

$$r_k^t \geq \sum_{\forall\, nv \in \mathcal{S}nv_k} r_{nv}^t \;;\; \forall t \tag{1}$$

$$c_m \geq \sum_{\forall\, lv \in \mathcal{S}lv_m} c_{lv} \tag{2}$$

where r_k^t is the total number of resources of type t in physical node k, r_{nv}^t is the number of resources of type t allocated to virtual node nv, c_m is the total number of resources in physical link m and c_{lv} is the number of resources allocated to virtual link lv.

Eqs. (1) and (2) forbid that the number of resources allocated to the virtual nodes/links established in a particular physical node/link exceed the capacity of that node/link.

Additionally, depending on the type of physical network, extra constraints might appear on the allocation of resources to the virtual links. In the case of an optical network, fixed and flexible-grid networks impose different constraints. We review these two types of optical networks and their associated constraints in the following.

2.3. Fixed-grid optical network

In a circuit-switched optical network, each circuit is carried by an optical channel/carrier, based on the wavelength division multiplexing (WDM) technique. Currently, such optical channels operate in the range 1530–1565 nm, known as band C.

In a fixed-grid optical network, the optical carriers are determined by their central frequency and use a fixed amount of spectrum. According to the specification ITU-T G.694.1 [30], the selectable spectrum widths are 12.5 GHz, 25 GHz, 50 GHz and 100 GHz. Once a spectrum width is selected, all optical channels in a link are established with such spectral width. Depending on the selected spectral width, the central frequency used by the n-th optical channel is given by the following equation:

$$193.1 + n \times W \text{ THz} \tag{3}$$

where $W \in \{0.0125; 0.025; 0.05; 0.1\}$ denotes the spectral width selected and n is an integer number whose range depends on the spectral width as follows: $n \in [-123, 227]$ for $W = 0.0125$; $n \in [-61, 113]$ for $W = 0.025$; $n \in [-30, 56]$ for $W = 0.05$; $n \in [-15, 28]$ for $W = 0.1$.

Figure 2 shows an example of the spectral usage of a fixed-grid link where six optical channels have been established: two optical channels at 10 Gbps using the on-off keying (OOK) modulation format, three channels at 40 Gbps modulated with dual polarization-quadrature phase shift keying (DP-QPSK) and one channel at 100 Gbps, also modulated with DP-QPSK. The spectral width of each channel is equal to 50 GHz and the central frequencies are determined by Eq. (3). It is common practice to identify the channels by their equivalent wavelength as well. Thus, in **Figure 2**, the corresponding wavelength of each channel has been written between brackets under the central frequency.

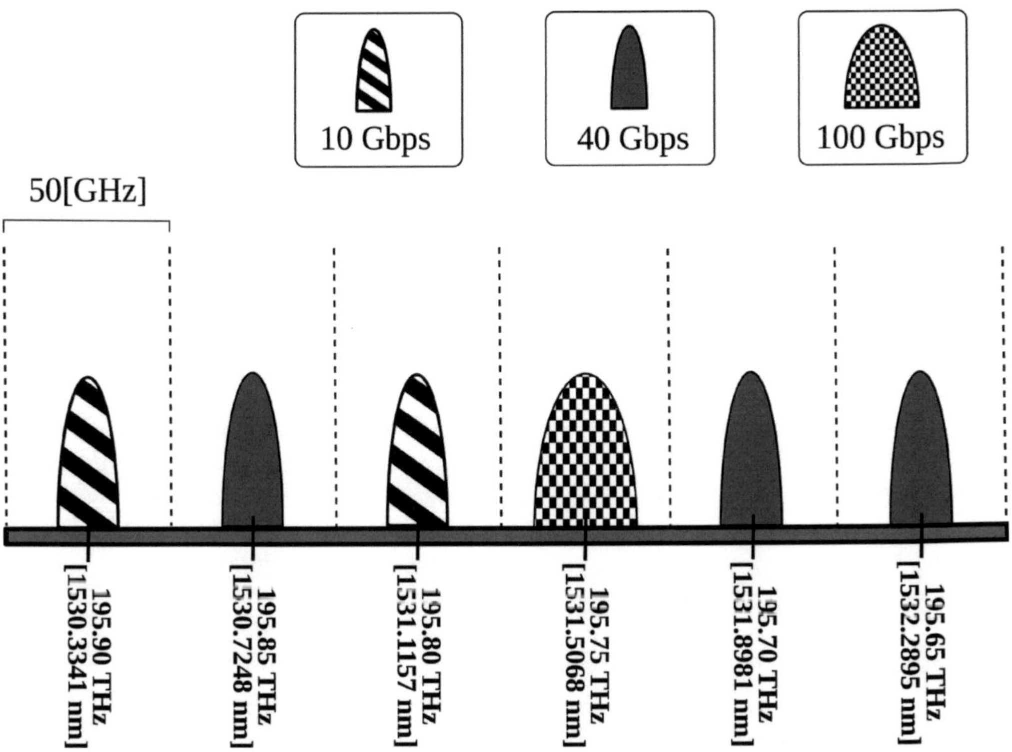

Figure 2. Frequency allocation to different transmission rate optical channels in a fixed-grid link.

In fixed-grid optical networks (in the absence of wavelength converters), the wavelength continuity constraint must be met. That is, the optical channel used by the virtual link must use the same central frequency and spectral width in all the physical links used. In networks operating with multiple transmission rates (as shown in **Figure 3**), additional constraints to deal with the signal degradation of higher bit rates channels mainly due to cross-phase modulation [31–33] may be required: for example, some channels should be left unused as guard bands or an optical reach (the maximum distance an optical signal can travel without exceeding a threshold on the bit error rate) be established.

The main drawback of fixed-grid optical networks is the inefficient spectrum usage [34], as observed in **Figure 2**, where channels are allocated more spectrum than effectively required: both a 10 Gbps OOK-modulated channel and a 40 Gbps channel modulated with DP-QPSK require a bandwidth equal to 25 GHz [34, 35], whereas a 100 Gbps channel modulated with DP-QPSK requires just 37.5 GHz [34]. To increase the spectrum usage, the flexible allocation of it has been proposed [14, 34]. This type of networks is known as flexible-grid or elastic optical networks.

2.4. Flexible-grid optical networks

In a flexible-grid optical network, the spectral width of a channel can be varied depending on the data transmission requirements [36]. Thus, the spectrum is divided in small units, typically of 12.5 GHz, known as frequency slot units (FSU) [34]. By using a different number of contiguous FSUs, different spectral widths can be achieved [37, 38] depending on the transmission requirements of the signal, such as the modulation format and the bit rate.

As a way of illustration, **Figure 3** shows the same six channels of **Figure 2**, now operating in a flexible-grid system. The numbers of 12.5 GHz FSUs required are 2, 2 and 3 for the 10, 40 and 100 Gbps channels, respectively. Thus, the flexible-grid allocation uses just 54.2% of the spectrum originally required (162.5 GHz instead of 300 GHz).

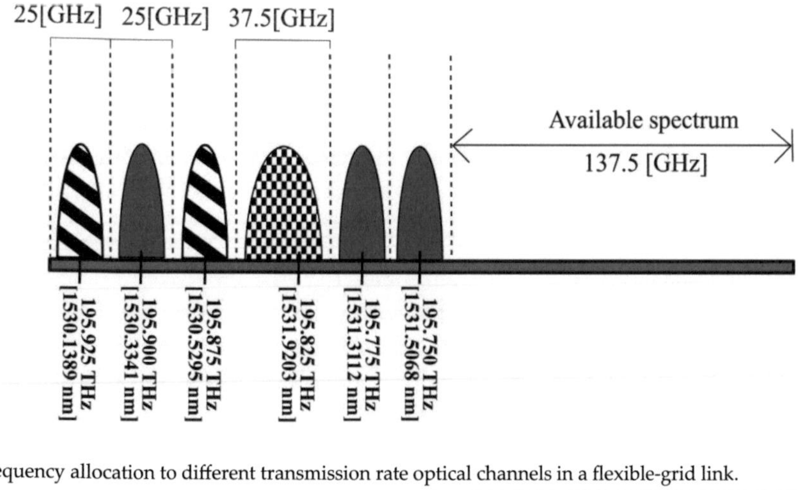

Figure 3. Frequency allocation to different transmission rate optical channels in a flexible-grid link.

Single-carrier and multi-carrier (super-channel) can be used to create an optical connection. In the latter, the overall bit rate is achieved through lower-rate sub-carriers. Examples of these systems are Co-WDM, Nyquist-WDM and time frequency packing [34, 39, 40]. In general, multi-carrier systems require a lower number of FSUs and exhibit a longer optical reach than single-carrier systems with the same total bit rate and modulation format [41, 42].

Regarding the modulation formats, there are bi-level and multi-level types. In a bi-level modulation format, as OOK and binary phase shift keying (BPSK) [42], the symbol rate equals the bit rate. In a multi-level modulation format, as QPSK and x-quadrature amplitude modulation (x-QAM) [41, 42], the symbol rate is lower than the bit rate of the bi-level type, leading to a lower requirement of FSUs. However, the optical reach of multi-level modulation formats is lower than that of bi-level [34, 36], highlighting a trade-off between number of FSUs and optical reach [34, 43, 44].

Once the number of FSUs required by a virtual link has been determined, the establishment of such link must meet at least two additional constraints: FSU continuity and FSU contiguity constraints. The FSU continuity constraint is analogous to the wavelength continuity constraint (exactly the same FSUs must be used in every physical link selected to establish a virtual link). The FSU contiguity constraint imposes that, if more than one FSU is required to establish a virtual link, then these FSU must be contiguous in the spectrum [45].

The sequence of physical links used to establish a virtual link meeting the FSU continuity and contiguity constraints is known as a spectrum path.

3. Research challenges in virvtual network allocation over flexible-grid optical networks

In the following, the main challenges in the research area of network virtualization over flexible-grid optical networks are discussed.

3.1. Performance metrics

A performance metric allows defining the quality of an algorithm to carry out its task. Thus, usually the (single) objective of an algorithm is the maximization or minimization of a performance metric. However, for a complex algorithm such as a virtual network allocation algorithm, there are several performance metrics that could be optimized.

Most published results have focused on minimizing the virtual network request rejection rate [46–56, 64]. The main advantage of using the performance metric is that it allows evaluating the ability of the algorithm to accommodate new virtual networks on the physical substrate. However, given that the blocking depends on many parameters (the physical and virtual network topologies, the capacity availability in physical nodes and links, the capacity requirements of virtual nodes and links [50–52]), to identify the best algorithm is necessary knowing exactly the network configuration where the algorithm will operate (something difficult to achieve in dynamic scenarios) or running extensive simulation experiments with different network configurations (a time-consuming task).

Instead of registering the blocking ratio, a computationally simpler metric consists on registering the number of virtual network establishment requests received when the first blocking (rejection) occurs [48, 49]. A good algorithm would aim at registering such event at the latest possible instant. If used in conjunction with the blocking ratio, the first blocking metric can give information about the instant when the network starts saturating (when the first blocking occurs) and the dynamic of the system once such saturation state is reached.

Maximizing the traffic carried by the physical network due to the established virtual networks has also been the objective of some algorithms [54, 57–59]. As with the blocking ratio, the value of this performance metric depends on the topologies of the physical and virtual networks as well as the capacity of physical nodes/links and capacity requirements of virtual nodes/links, which makes difficult drawing general conclusions about the quality of different algorithms. Additionally, the lack of information about the number of virtual networks rejected does not allow measuring the quality of the service offered to the users. Thus, it should be used in conjunction with the blocking ratio.

Guaranteeing a given level of availability (e.g. 0.99999) to a virtual network has not been addressed by the proposed virtual network allocation algorithms to date, although availability (the fraction of time that a service is in operative state) is one of the most important quality of service metrics in a service level agreement (SLA). However, some efforts have been carried out in guaranteeing operation under specific failure conditions [49, 53, 59, 61, 62].

All previous performance metrics somehow aim to evaluate the capacity of the algorithm to offer a good quality of service. However, the main challenge in evaluating the performance of complex algorithms is selecting a performance metric that can capture the quality of the service offered to the user as well as the cost in achieving such quality.

To offer physical resources to a virtual network, the service provider incurs expenditure and operational costs due to the acquisition and maintenance of transponders, regenerators, optical cables, optical amplifiers and ROADMs (reconfigurable optical add drop multiplexer) [60]. Thus, algorithms aiming at minimizing the cost have also been studied. This metric has been mostly used in static scenarios [46, 56, 61, 62], and it is useful for the network planning stage. In dynamic scenarios, it can be used to determine the cost per virtual network, the total cost of providing the network virtualization service during a period of time or the cost incurred to achieve a given performance in terms of blocking ratio or traffic carried.

To date, quality-of-service-related metrics and cost have been studied separately. The algorithm is designed to minimize/maximize one of them whilst the other one is just measured. Thus, a multi-objective optimization approach that evaluates quality (as blocking or availability) and the cost incurred to achieve the required quality would deliver more realistic information about the best algorithm alternative from a network operator perspective.

3.2. Network virtualization dynamics characterization

To date there are no commercial network virtualization systems over flexible-grid optical networks. In Ref. [63], an experimental system is reported, but traffic is artificially generated. Therefore there are no empirical statistics that help to model the structure (virtual topologies and their capacity requirements) and dynamic of such system. In terms of structure, it would be useful knowing how to model the virtual topologies and their capacity requirements. Such knowledge would facilitate the evaluation of allocation algorithms in terms of simulation, the only technique used so far to evaluate performance of dynamic systems.

In terms of structure, different works make different assumptions regarding the topologies of the virtual networks and their capacity requirements. **Table 1** summarises the main models used to characterize the virtual topologies. In it, the name of each physical and virtual topology is given along with its number of nodes ($|Np|$, $|Nv|$) and links ($|Lp|$, $|Lv|$). When a number lower than one is provided for $|Lv|$, it means that the probability interconnection between a node pair is given. The column 'Node/Link requirement' corresponds to the percentage of usage of the physical node and link by any virtual node and link, respectively. The symbol '-' implies that such information is not found in the chapter.

As most works (15 of 17) use a medium-sized physical network (NSFNet or DTNet) for evaluation, future works should consider at least one of these topologies as the physical substrate to facilitate comparison among different proposals. No pattern can be observed in terms of the virtual topologies, with most works using mesh topologies with different degrees of connectivity. Regarding resource requirements, all proposals require no more than 10% of the physical node/link resources. The rest uses percentages of a few units.

Regarding dynamism, the most used distribution to model the virtual network request inter-arrival time is the exponential [46, 50–56, 64]. The holding time is usually modelled by an exponential distribution [52, 55], a deterministic value [64] or infinite (to model incremental traffic) [48, 49].

3.3. Physical impairments

It is expected that flexible-grid optical networks can accommodate channels (used to implement virtual links) at rates from 10 Gbps to 1 Tbps. Such channels, in the same way as fixed-grid channels, will be affected by several physical impairments that degrade the quality of the signal transmission. Additionally to typical physical impairments, as attenuation, chromatic dispersion, four-wave mixing (FWM) and amplified spontaneous emission (ASE) noise [65], in elastic optical networks the non-linear effect of cross phase modulation (XPM) takes relevance because of the existence of channels with different modulation formats in the same link. Due to the XPM effect, channels using intensity-based modulation formats (e.g. OOK typically used in 10 Gbps channels) interfere negatively in the quality of the signal of phase-modulated channels (e.g. BPSK and QPSK, used for higher bit rate channels) [66].

Most previous works have not considered this situation, with some of them assuming an ideal physical substrate [50, 54] whereas others have resorted to simplified models. For instance, in Refs. [48, 49, 51, 53, 57, 61, 62], the degradation is summarized in the figure of the maximum optical reach of signals, in Refs. [46, 56, 58, 59], the use of guard bands to all channels is used to simulate an ideal substrate, whereas in Refs. [47, 52, 55, 64], guard bands (to all channels or selectively added to channels most affected by the XPM degradation) are added to the limitation of the optical reach.

Physical topology (|Np|, |Lp|)	Virtual topology (|Nv|, |Lv|)	Node/link requirement	Work
NSFNet (14, 21)	Mesh (2–6, 0.5	-/0.35–3.5%	[46]
	Mesh (-, E%)	-/-	[47]
	- (2–5, -)	0.32–2.5%/0.31–2.5%	[48]
	Mesh (5–7, -)	1%/0.62–0.93%	[52]
	Ring (5–7, 5–7)	10%/0.62–0.93%	
	- (-, -)	-/2–16%	[53]
	Mesh (2–5, 0.5)	0.29–0.86%/0.67–3.33%	[54]
	Ring (3–7, 3–7)	1%/0.63–15.94%	[55]
	Mesh (3–5, 3–10)	1%/0.625–3.125%	[56]
	Mesh (4–7, 4–14)	1%/1.25–5%	[58]
	Ring (2–4,2–4)	-/-	[59]
	Mesh (5,7)	2–4%/-	[62]
6-node (6, 8–10)	Mesh (2–3, 0.5)	2–6%/2–6%	[51]
	Ring (3, 3)	10%/1.25–5%	[58]
	- (4–5, 3.6–3.8)	-/-	[61]
	Mesh (3–4,0.5)	1–4%/-	[64]
DTNet (14, 23)	Mesh (3–4. 0.5)	0.5–5%/0.5–5%	[51]
	Mesh (3–10, 0.5)	1–4%/-	[58]
	Ring (2–4,2–4)	-/-	[59]
	Mesh (2–5, 1–15)	-/0.85–3.12%	[61]
	Mesh (3–4, 0.5)	0.5–5%/0.5–5%	[64]
US network (24,)	- (2–5, -)	0.06–2%/0.31–2.5%	[50]
Random (50, 141)	Mesh (2–10, 0.5)	0.5–10%/0.5–10%	[51, 52]
CORONET (75, 99)	- (-,11.4)	-/-	[54]
ARPANET (20, 32)	Mesh (3–7, 0.5)	0.2–1.2%/0.5–5%	[58]

Table 1. Characteristics of virtual network requests used in the literature.

3.4. Resource allocation to virtual networks

The selection of the physical nodes and links to be allocated to a virtual network is a $\mathcal{NP}$-Hard problem [67]. Thus, most proposals solving this problem over flexible-grid optical networks have resorted to heuristics [46–59, 61, 62, 64] and a few of them have proposed integer linear models [51, 58, 59, 61, 64], but mostly in the context of a static scenario where the random nature of the virtual network requests is not a problem.

Much work is still needed in identifying the features of good performing heuristics to allocate virtual networks as well as evaluating the performance of meta-heuristics.

3.5. Spectrum fragmentation

Under dynamic operation, as a result of the resource release from virtual network that depart from the network, voids in the spectrum are generated. A void is a set of contiguous available FSUs between portions of allocated FSUs (or between a portion of allocated FSUs and the beginning/end of the band), as shown in **Figure 4**.

Due to the FSU contiguity constraint, the existence of these voids is problematic, as they fragment the spectrum. As a result, a virtual link could not be implemented due to the lack of enough contiguous FSUs, leading to a higher blocking ratio. For example, in the situation depicted in **Figure 4**, although three FSUs are available, a virtual link requiring three FSUs could not be established because of the contiguity constraint.

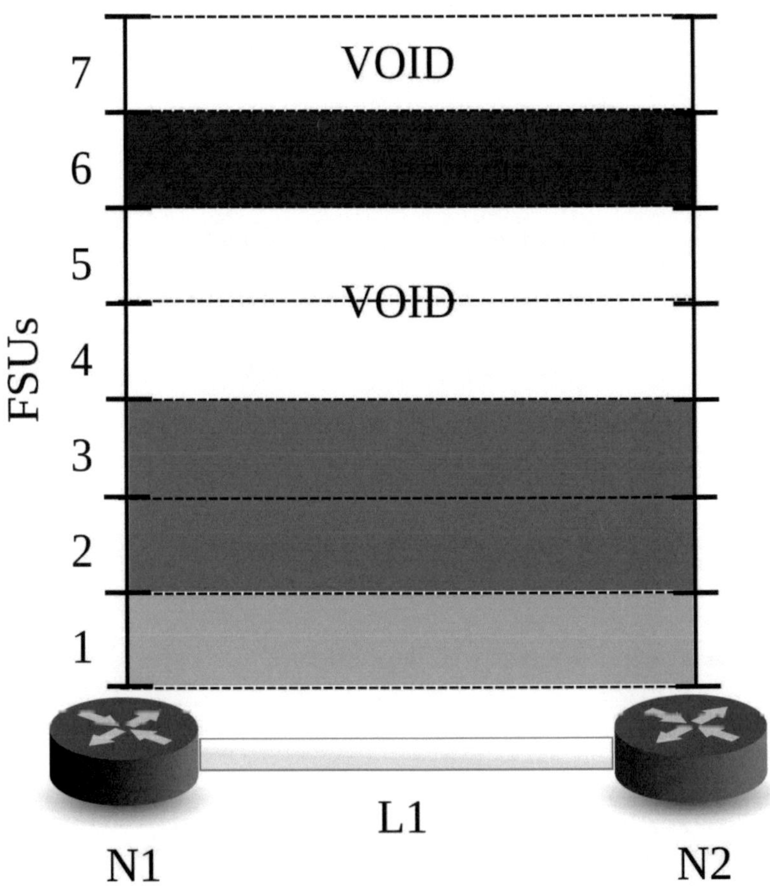

Figure 4. Spectrum fragmentation.

To decrease the spectrum fragmentation, the re-allocation of FSUs to the different channels in a link has been proposed in the area of flexible-grid networks by Refs. [68–72]. In Ref. [73], the impact of avoiding fragmentation on the blocking ratio can be seen.

In Ref. [54], a technique of spectrum defragmentation in the area of virtual networks over flexible-grid optical networks is reported, showing that the blocking ratio decreases with respect to an algorithm without defragmentation. However, defragmentation is costly as computation time and additional resources must be used to apply it. This highlights a trade-off between the blocking ratio decrease and the frequency of defragmentation. Further research on the interplay of allocation algorithms and defragmentation techniques is required.

4. Taxonomy

Figure 5 shows a comprehensive classification of the resource allocation algorithms in the area of network virtualization over flexible-grid optical networks. The taxonomy includes current proposals, but it is generic enough as to include algorithms not studied yet.

In the taxonomy, each possible algorithm is defined by three main dimensions: its performance metric, its operation conditions and the type of service offered to the user. In the following each of these dimensions are described as well as the different choices available in each one of them.

4.1. Performance metric

The most commonly used performance metric in the literature is the blocking ratio [46–56, 64]. Although the use of the same metric would facilitate comparison, due to the different assumptions made on the physical and virtual topologies, a direct comparison is not always possible.

The variant of blocking, first blocking, has been used in Refs. [48, 49]. Remaining metrics used in reported works are the traffic carried by the physical networks [54, 57–59] and cost-related metrics [46, 56, 61, 62].

Although published works do not explicitly mention the performance metric of availability, few works make assumptions on the operation conditions of the network that allow guaranteeing 100% availability. In Refs. [53, 59, 61, 62], only single link failures are assumed. Thus, the allocation of two link-disjoint spectrum paths to implement each virtual link is enough to ensure the operation of every virtual network. In Ref. [49], single link/node failures are assumed and, then by allocating two node/link-disjoint spectrum paths to each virtual link, a 100% availability is provided. Note that if the system violates the assumptions on the type of failure that can occur (e.g. a double link failure occurs in a system designed to tolerate single link failure), 100% availability cannot be guaranteed anymore.

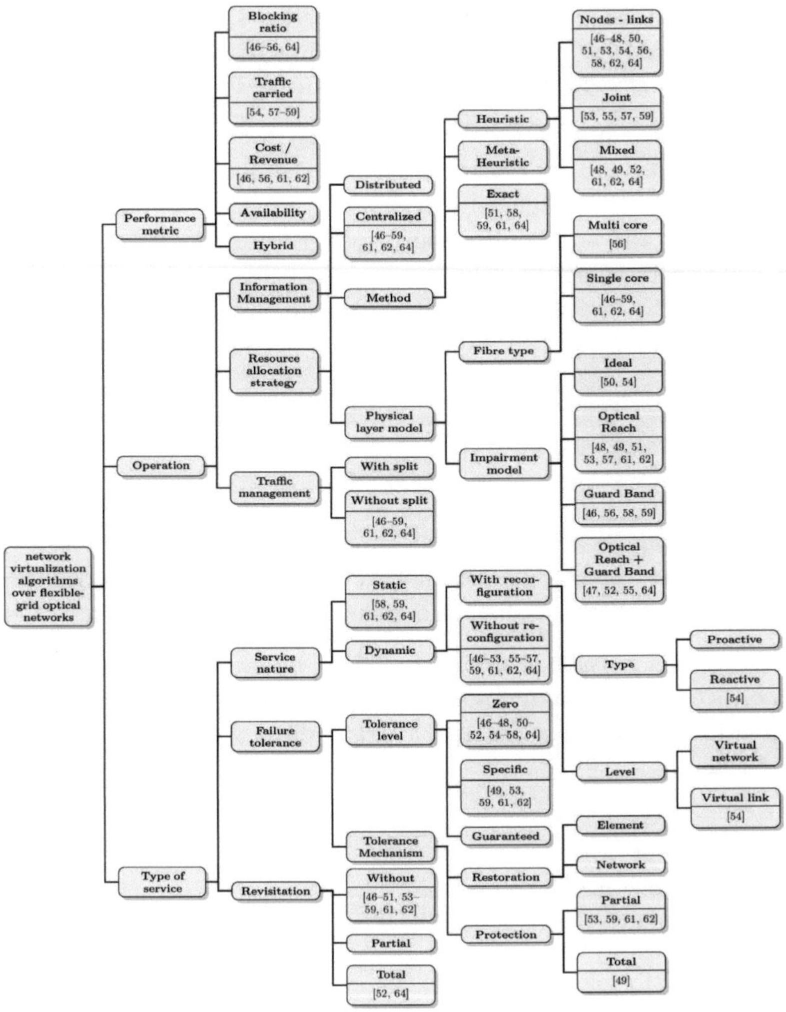

Figure 5. Taxonomy of network virtualization algorithms over optical flexible-grid networks.

Guaranteeing availability under any type of failure has not been researched in the area of network virtualization over flexible-grid networks neither the combination of quality and cost performance metrics.

4.2. Operation

4.2.1. Information management

To date, all proposals implicitly assume centralized information management [46–59, 61, 62, 64]. That is, a central entity has global knowledge of the network status and the resource allocation algorithm is executed every time a new virtual network request is generated. In fact, the first proposals for architecture with a virtual network controller are based on a centralized scheme, as the one proposed in Ref. [74]. Centralized systems are suitable when the time between successive requests is long enough for the central controller to execute the resource allocation algorithm.

In Ref. [75], different distributed virtual network allocation approaches are discussed in the context of packet networks. Results show that a distributed operation reduces the delay in mapping a virtual network and the number of messages required to be exchange to coordinate the allocation. In Ref. [76], the impact of a distributed virtual network reconfiguration approach on the interruption time of the service is studied in the context of fixed-grid networks. Although the distributed operation has advantages in terms of resilience against failures, lower computation times and network congestion due to message exchange, it has increased complexity in terms of control plane network (more controllers), synchronization of messages and a potential decreased performance due to the obsolescence of information. These aspects are yet to be studied in network virtualization systems over flexible-grid networks.

4.2.2. Resource allocation strategy

The virtual network allocation strategy must consider two aspects: the method used to solve the problem of embedding the virtual network on the physical network and the model used to characterize the constraints of the physical substrate.

There are three general methods to solve the problem of the virtual network embedding:

a. Exact methods: These are the techniques that find the global optimal solution to a problem. However, they are computationally complex and thus, they are usually applied only to small instances of the problems with slow dynamics. In real dynamic systems, where a solution must be found in short-time scales, this type of method is not feasible. However, in simulation environments, an integer linear programming (ILP) model can be solved for each virtual network request to be used as a benchmark, as done in Ref. [51]. In the area of virtual network allocation over flexible-grid networks, most ILP models have been used to solve the problem in a static scenario (virtual networks permanently established, not allocated on-demand). Works in Refs. [51, 58, 59, 61, 64] apply ILP to allocate a set of predefined virtual networks on a small physical network (six nodes) with the objective of minimizing cost.

b. Meta-heuristics: These are the generic algorithms capable to adapt to different problems by adjusting their parameters and configurations. Usually, they find very good quality solutions, but cannot guarantee the optimum solution as exact methods do. Work in Refs. [77, 78] proposed the use of genetic algorithms and ant colony to solve the problem of virtual network embedding in conventional networks, respectively. No works have been reported on flexible-grid networks as the physical substrate, neither in static nor in dynamic scenarios.

c. Heuristics: These are ad-hoc algorithms that do not guarantee a global optimum solution, designed for a specific problem. However, they are computationally simpler than the previous techniques. Most works in the area of network virtualization over flexible-grid networks resort to heuristics [46–59, 61, 62, 64], mainly focused on dynamic scenarios.

Normally, heuristics designed to solve complex problems, divide the original problem in sub-problems easier to solve separately. This approach is applied in this area as well. The original problem of mapping a virtual network is divided in node mapping (allocation of a physical node to a virtual node) and link mapping (allocation of a spectrum path to a virtual link). Most proposals map nodes first to then establish the virtual links connecting them [46–48, 50, 51, 53, 54, 56, 58, 62, 64].

To map the nodes and links, the heuristic must define the order in which the virtual and physical nodes/links are processed. To do so, a ranking is elaborated for each set of physical/virtual nodes/links and the first element in the ranking of virtual nodes/links is attempted to be mapped in the first element of the ranking of the physical nodes/links. The most common criterion to build the physical node ranking is the amount of available resources [48, 49, 58]. A function of the computing capacity and the nodal degree [50], a function of the number of sub-carriers of each transponder in the physical node and the slice capability of the physical node [46] and the node index [64] have also been used. Criteria to rank the virtual nodes are the amount of resources required [48, 58], the nodal degree [50] or the node index. The case where the virtual nodes must be established in specific physical nodes (defined in the virtual network establishment request), as in Ref. [47], is a particular case of a node/link mapping, as all virtual nodes are established in the specified physical nodes (if enough resources are available) before establishing the virtual links.

Physical links can be ranked in terms of their distance [48, 50, 58], cost [64] or number of available FSUs. Finally, virtual links are ranked in terms of their FSU requirements [47, 53, 58, 64].

Given that the solution found by solving the node/link mapping sub-problems sequentially is expected to be of lower quality than solving the original problem, an attempt to solve both problems jointly was proposed in Refs. [53, 55, 57, 59]. In these works, a sub-set of all possible mapping patterns for the nodes of a virtual network are evaluated and the one using the lowest slot layer (slot layer of a mapping pattern is the highest FSU used) [57], lowest cost [53] or best Hamming-inspired distance [55] is selected.

Finally, the approach of alternating the allocation of virtual nodes and links (mixed) has also been studied in Refs. [48, 49, 52, 61, 62, 64]. For example, in Ref. [61], the virtual nodes at the ends of each virtual link are mapped to then map the virtual link, showing results close to the ILP approach in a static scenario.

Apart from the FSU continuity and contiguity constraints, the solution methods can use one of several models to characterize additional constraints of the physical substrate. To date, the following models have been used:

a. Ideal, where no signal degradation is assumed [50, 54].

b. Optical-reach-based, this is the simplest model where the maximum distance covered by a spectrum path is determined solely by the modulation format and the bit rate, as in Refs. [48, 49, 51, 53, 57, 61, 62].

c. Guard-band-based, where a given number of FSUs might be left unused between channels of different bit rate, as in Refs. [46, 56, 58, 59].

d. Optical reach and guard band, where the optical reach is determined by the modulation format and the bit rate. Since the optical reach can decrease due to effect of neighbouring signals, by adding (selectively or not) guard bands between channels [47, 52, 55, 64] such detrimental effect can be mitigated.

4.2.3. *Traffic management*

In the context of packet networks, the split of traffic of a virtual link into several paths in the physical substrate has been proposed as a way of increasing the probability of accepting a virtual network establishment request [79]. In a flexible-grid optical network where a virtual link requiring M contiguous FSUs must be established but no path has more than $x < M$ contiguous FSUs, such situation could be solved by establishing the virtual link along several spectrum paths in such a way that the total number of FSUs used along all the paths equal M. Such mechanism could be enabled by recently introduced sliceable or multi-flow transponders [80, 81]. This approach has not been explored in the area of network virtualization over flexible-grid networks.

4.3. Type of service

4.3.1. *Service nature*

The service provider can offer a static or dynamic service. In the former case, the virtual network demands are known *a priori* and they are established permanently, whether they are used to transmit information or not [58, 59, 61, 62, 64]. In the latter case, virtual networks are established and released on demand.

In a dynamic service, spectrum experiences fragmentation. As a result, even when there is an enough number of FSUs to accommodate a new virtual network, these FSUs might not meet the contiguity constraint, leading to the rejection of requests. To decrease spectrum fragmentation, some dynamic systems reconfigure the established connections. Several works have evaluated the impact of reconfiguration on point-to-point connections on flexible-grid optical networks [68–72]. As expected, reconfiguration decreases blocking [54] at the expense of higher complexity of the control plane.

There are two types of reconfiguration techniques: proactive or reactive [82]. The former re-allocate resources before a blocking condition occurs, either in a synchronous or asynchronous way. In Refs. [69–71], pro-active reconfiguration algorithms are presented for point-to-point connections over flexible-grid optical networks. Reconfiguration may take place every time a given number of virtual networks request has been received. No proactive systems have been reported in network virtualization over flexible-grid networks. Reactive reconfiguration techniques re-allocate resources only when a new request cannot be accepted. In Ref. [54], a reactive reconfiguration method to re-allocate virtual networks over fixed-grid networks is presented, getting lower rejection rates than not reconfiguring at low-medium loads.

Reconfiguration can be applied at two different levels: re-allocation of complete virtual networks or re-allocation of a sub-set of virtual links/nodes, as in Ref. [54] in flexible-grid networks or [83] in fixed-grid networks. None of these cases has been studied in network virtualization systems over flexible-grid optical networks.

4.3.2. Fault tolerance

A network virtualization service can offer different levels of fault tolerance: zero, specific or guaranteed. Most works reported to date have studied systems without fault tolerance at all [46–48, 50–52, 54–58, 64]. In that case, the occurrence of any type of failure interrupts the operation of the virtual networks operating over the physical component affected by the failure. A specific survivability system is capable of continuing operation in spite of the occurrence of specific types of failures. Normally, these systems are designed to survive the most common failure events (e.g. a cable cut) and remain unprepared for unlikely events (as a node failure). In the area of network virtualization over flexible-grid networks, the algorithm proposed in Refs. [53, 59, 61, 62] can survive only to single link failures, whereas Ref. [49] can survive single link or node failure. Finally, a guaranteed survivability system ensures that limits on downtime are not exceed, no matter what the type of failure, as done in Refs. [84, 85] in a context different from network virtualization. If such condition is violated, the service provider is enforced to pay an economic compensation to the user. Such approach has not been explored in the area of network virtualization over flexible-grid networks.

Fault tolerance mechanisms can also be classified as proactive (protected systems) or reactive (restored systems). Protected systems allocate backup resources when the primary resources for the virtual network are allocated [49, 53, 59, 61, 62]. Therefore, upon failure occurrence, the time to recover from failure is shorter than reactive systems. Protected systems can allocate a complete backup virtual network (total protection) [49] or backup to some components (partial protection, e.g. only virtual links have backup resources) [53, 59, 61, 62]. Protected systems can also be classified as dedicated or shared. In the former, backup resources are dedicated to the corresponding primary resource. In the latter, a backup resource is shared among several primary resources. No research has been reported on the area of shared protection for virtual networks over flexible-grid networks. Restored systems allocate resources to the virtual networks affected by a failure only once the failure has occurred; as a result, the recovery time is longer, but a lower amount of backup resources are required. Restoration can be carried out for complete virtual networks or only for the part of them affected by the failure. Restoration on virtual networks over flexible-grid networks has not been researched yet.

4.3.3. Revisitation

Revisitation allows the establishment of two virtual nodes from the same virtual network in the same physical node [16]. Revisitation has been proposed in the context of overlay networks [86] as a way of emulating larger networks on small testbeds. In virtual network systems over flexible-grid networks, revisitation has been used in Ref. [64] and the impact of it on blocking was studied in Ref. [52] showing a decrease of blocking ratio of two orders of magnitude with respect to the same algorithm without revisitation.

http://dx.doi.org/10.5772/intechopen.68340

Revisitation has been little researched in the literature, probably because a real application for it has not been found yet. For example, for research on new Internet protocols, delay and bandwidth utilization are two key metrics that could not be measured if two virtual nodes are hosted in the same physical node. For cloud replication services would not be useful either, as the replicas must be allocated to geographically different sites. However, it is mentioned as one of the four key architectural principles of network virtualization in Ref. [16], where it would be useful to help the service providers to manage highly complex tasks and facilitate virtual networks management.

In **Table 2**, a summary of the virtual network resource allocation proposed to date is presented. For each algorithm, all the dimensions presented in the taxonomy of **Figure 5** are specified.

	Performance metric	Operation		Type of service			
		Information management	Resource allocation strategy	Traffic management	Service nature	Failure tolerance	Revisitation
[46]	Blocking ratio cost/revenue	Centralized	Nodes-links Guard band	Without split	Dynamic without reconfiguration	Zero	Without
[47]	Blocking ratio	Centralized	Nodes-links Optical reach + Guard band	Without split	Dynamic without reconfiguration	Zero	Without
[48]	Blocking ratio	Centralized	Nodes-links Mixed Optical reach	Without split	Dynamic without reconfiguration	Zero	Without
[49]	Blocking ratio	Centralized	Nodes-links Optical reach	Without split	Dynamic without reconfiguration	Specific, total protection	Without
[50]	Blocking ratio	Centralized	Nodes-links Ideal	Without split	Dynamic without reconfiguration	Zero	Without
[51]	Blocking ratio	Centralized	Exact Nodes-links Optical reach	Without split	Dynamic without reconfiguration	Zero	Without
[52]	Blocking ratio	Centralized	Mixed Optical reach + Selective guard band	Without split	Dynamic without reconfiguration	Zero	Total
[53]	Blocking ratio	Centralized	Nodes-links Joint Optical reach	Without split	Dynamic without reconfiguration	Specific, partial protection	Without
[54]	Blocking ratio traffic carried	Centralized	Nodes-links Ideal	Without split	Dynamic with reconfiguration in virtual links	Zero	Without

	Performance metric	Operation		Type of service			
		Information management	Resource allocation strategy	Traffic management	Service nature	Failure tolerance	Revisitation
[55]	Blocking ratio	Centralized	Mixed Optical reach + Selective guard band	Without split	Dynamic without reconfiguration	Zero	Without
[56]	Blocking ratio cost/revenue	Centralized	Nodes-links Guard band	Without split	Dynamic without reconfiguration	Zero	Without
[57]	Traffic carried	Centralized	Joint Optical Reach	Without split	Dynamic without reconfiguration	Zero	Without
[58]	Traffic carried	Centralized	Exact Nodes-links Guard band	Without split	static	Zero	Without
[59]	Traffic carried	Centralized	Exact Joint Guard band	Without split	Dynamic without reconfiguration	Specific	Without
[61]	Cost/revenue	Centralized	Exact Mixed Optical reach	Without split	Static	Specific, partial protection	Without
[62]	Cost/revenue	Centralized	Nodes-links Mixed Optical reach	Without split	Dynamic without reconfiguration	Specific	Without
[64]	Blocking ratio	Centralized	Exact Nodes-links Mixed Optical reach +Guard band	Without split	Static Dynamic without reconfiguration	Zero	Total

Table 2. Summary of the characteristics of the algorithms reviewed.

5. Conclusions

Network virtualization has emerged as an enabling technology for cloud computing services. Such services would push even further the limits on bandwidth utilization, where flexible-grid optical networks will be the key to increase the network capacity of actually deployed optical networks. As a result, a new area of research focused on network virtualization over flexible-grid networks has emerged.

On such area, the research efforts focus on three main lines: design of flexible and virtualized devices, definition of network architectures and virtual network allocation algorithms.

In this chapter, a survey on the virtual network allocation algorithms over flexible-grid networks has been presented along with a classification of all possible proposals of algorithms by means of taxonomy. Such classification allowed the identification of several aspects that must be further investigated in the area:

- Multi-objective optimization approaches that allow to select resource allocation algorithms with a good compromise between quality and cost.
- The design and evaluation of distributed virtual network allocation algorithms.
- The application of meta-heuristics (as genetic algorithms, ant colony, etc.) to solve the virtual network allocation problem over flexible-grid networks.
- The study of the impact of traffic split on the performance of virtual network allocation algorithms.
- The effect and complexity of reconfiguration on the performance of network virtualization systems.
- The design and evaluation of shared protection mechanisms.
- The design and evaluation of shared protection and restored fault tolerance mechanisms.

Acknowledgements

Financial support from CONICYT through the Fondecyt Project 1130388 and 11140386 and the PhD Scholarship,CONICYT-PCHA/Doctorado Nacional/2015-21150525 is gratefully acknowledged.

Author details

Danilo Bórquez-Paredes[1*], Alejandra Beghelli[1] and Ariel Leiva[2]

*Address all correspondence to: danilo.borquez@uai.cl

1 Science and Engineering Faculty, Universidad Adolfo Ibañez, Chile

2 Electrical Engineering School, Pontificia Universidad Católica de Valparaíso, Chile

References

[1] Develder C, De Leenheer M, Dhoedt B, Pickavet M, Colle D, De Turck F, Demeester P. Optical networks for grid and cloud computing applications. Proceedings of the IEEE. 2012;**100**(5):1149–1167. DOI: 10.1109/JPROC.2011.2179629

[2] Gharbaoui M, Martini B, Castoldi P. Anycast-based optimizations for inter-data-center interconnections [Invited]. IEEE/OSA Journal of Optical Communications and Networking. 2012;**4**(11):B168–B178. DOI: 10.1364/JOCN.4.00B168

[3] GIGAOM. Why the Cloud Needs New Forms of Networking [Internet]. 2011. Available from: https://gigaom.com/2011/06/23/structure-network-virtualization-openflow/[Accessed: 28 December 2016]

[4] Alhazmi K, Abu Sharkh M, Ban D, Shami A. A map of the clouds: Virtual network mapping in cloud computing data centers. In: IEEE 27th Canadian Conference on Electrical and Computer Engineering; 4-7 May 2014; Toronto, Canada. IEEE; 2014. pp. 1–6. DOI: 10.1109/CCECE.2014.6901053

[5] Xiao L, Jiang W, Chen F, Guo L, Hu Y. A survey of cloud computing data virtualization service. Applied Mechanics and Materials. 2013;**441**:1016–1019. DOI: 10.4028/www.scientific.net/AMM.441.1016

[6] Sun G., Yu H., Anand V., Li L, Di H. Optimal provisioning for virtual network request in cloud-based data centers. Photonic Network Communications. 2012;**24**(2):118—131. DOI: 10.1007/s11107-012-0372-0

[7] Baroncelli F., Martini B., Castoldi P. Network virtualization for cloud computing. Annals of telecommunications—annales des télécommunications. 2010;**65**(11-12):713–721. DOI: 10.1007/s12243-010-0194-y

[8] Hao F, Lakshman T, Mukherjee S, Song H. Enhancing dynamic cloud-based services using network virtualization. SIGCOMM Computer Communication Review 2010;**40**(1):67–74. DOI: 10.1145/1672308.1672322

[9] Dang Tran F. Network Virtualization as an Enabler for Cloud Computing: A Telco Perspective [Internet]. 2009. Available from: http://www.itcspecialistseminar.com/presentations/itcss09_DangTran_pres.pdf [Accessed: 24 February 2017]

[10] I. Morita. Network control and virtualization for cloud and mobile services from carrier's point of view. In: Optical Fiber Communication Conference; 22–26 Mar 2015; Los Angeles, California, USA. OSA; 2015. p. W1G.3. DOI: 10.1364/OFC.2015.W1G.3

[11] Harter I, Schupke D, Hoffmann M, Carle G. Network virtualization for disaster resilience of cloud services. IEEE Communications Magazine. 2014;**52**(12):88–95. DOI: 10.1109/MCOM.2014.6979957

[12] Alshaer H. An overview of network virtualization and cloud network as a service. International Journal of Network Management. 2015;**25**(1):1–30. DOI: 10.1002/nem.1882

[13] Tomkos I, Palkopoulou E, Angelou M. A survey of recent developments on flexible/elastic optical networking. In: 14th International Conference on Transparent Optical Networks; 2-5 July 2012; England: Coventry; 2012. pp. 1–6. DOI: 10.1109/ICTON.2012.6254409

[14] Jinno M, Takara H, Kozicki B, Tsukishima Y, Yoshimatsu T, Kobayashi T, Miyamoto Y., Yonenaga K, Takada A, Ishida O, Matsuoka S. Demonstration of novel spectrum-efficient elastic optical path network with per&whyphen;channel variable capacity of 40 Gb/s to over 400 Gb/s. In: 34th European Conference on Optical Communication; 21-25 September 2008; Brussels, Belgium. IEEE; 2008. pp. 1–2. DOI: 10.1109/ECOC.2008.4729581

[15] Mayoral A, Gonzalez de Dios O, Lopez V, Fernandez-Palacios J. Migration steps towards flexi-grid networks. In: Future Network and Mobile Summit; 3-5 July 2013; Lisboa, Portugal. IEEE; 2013. pp. 1–9. DOI: 10.1364/JOCN.6.000988

[16] Chowdhury N, Boutaba R. A survey of network virtualization. Computer Networks. 2010;**54**(5):862–876. DOI: 10.1016/j.comnet.2009.10.017

[17] Chowdhury N, Boutaba R. Network virtualization: State of the art and research challenges. IEEE Communications Magazine. 2009;**47**(7):20–26. DOI: 10.1109/MCOM.2009.5183468

[18] Houidi I, Louati W, Zeghlache D, Papadimitriou P, Mathy L. Adaptive virtual network provisioning. In: 2nd ACM SIGCOMM Workshop on Virtualized Infrastructure Systems and Architectures; 30 August-03 September 2010; New Delhi, India. ACM; 2010. pp. 41–48. DOI: 10.1145/1851399.1851407

[19] Sherwood R, Gibb G, Yap K, Appenzeller G, Casado M, McKeown N, Parulkar G. Can the production network be the testbed? In: 9th USENIX Conference on Operating Systems Design and Implementation; 4-6 October 2010; Vancouver, Canada. USENIX Association; 2010. pp. 1–6.

[20] Bavier A, Bowman M, Chun B, Culler D, Karlin S, Muir S, Peterson L, Roscoe T, Spalink T, Wawrzoniak M. Operating system support for planetary-scale network services. In: 1st Conference on Symposium on Networked Systems Design and Implementation; 29-31 March 2004; San Francisco, California, USA. pp. 19–19. DOI: 10.1145/1113361.1113362

[21] PlanetLab. An Open Platform for Developing, Deploying and Accessing Planetary-Scale Services [Internet]. 2007. Available from: http://www.planet-lab.org/ [Accessed: 28 December 2016]

[22] Anderson T, Peterson L, Shenker S, Turner J. Overcoming the Internet impasse through virtualization. The Computer Journal. 2005;**38**(4):34–41. DOI: 10.1109/MC.2005.136

[23] GENI. Global Environment for Network Innovations [Internet]. 2006. Available from: http://www.geni.net [Accessed: 28 December 2016]

[24] FEDERICA. Federated E-Infrastructure Dedicated to European Researchers Innovating in Computing Network Architechtures [Internet]. 2008. Available from: http://www.fp7-federica.eu/ [Accessed: 28 December 2016]

[25] OneLab2. An Open Federated Laboratory Supporting Network Research for the Future Internet [Closed project] [Internet]. 2008. Available from: http://cordis.europa.eu/project/rcn/87273_en.html [Accessed: 28 December 2016]

[26] NICT. "AKARI" Architecture Design Project for New Generation Network [Internet]. 2006. Available from: https://web.archive.org/web/20130210083738/http://akari-project.nict.go.jp/eng/index2.htm [Accessed: 28 December 2016]

[27] Future Internet Brazilian Environment for Experimentation [Internet]. 2011. Available from: http://fibre.org.br/ [Accessed: 28 December 2016]

[28] CISCO. Easy Virtual Network—Simplifying Layer 3 Network Virtualization [Internet]. 2013. Available from: http://www.cisco.com/c/en/us/products/collateral/ios-nx-os-software/layer-3-vpns-l3vpn/whitepaper_c11-638769.html [Accessed: 28 December 2016]

[29] VMWARE. NSX for vSphere Network Virtualization Design Guide ver 3.0 [Internet]. 2014 [Updated: 2015]. Available from: https://communities.vmware.com/docs/DOC-27683 [Accessed: 28 December 2016]

[30] ITU-T. Spectral grids for WDM applications: DWDM frequency grid [Internet]. 2002 [Updated: 2012]. Available from: https://www.itu.int/rec/T-REC-G.694.1/en [Accessed: 24 February 2017]

[31] Sambo N, Secondini M, Cugini F, Bottari G, Iovanna P, Cavaliere F, Castoldi P. Modeling and distributed provisioning in 10-40-100-Gb/s multirate wavelength switched optical networks. Journal of Lightwave Technology. 2011;**29**(9):1248–1257. DOI: 10.1109/JLT.2011.2122245

[32] Chowdhury P, Tornatore M, Nag A, Ip E, Wang T, Mukherjee B. On the design of energy-efficient mixed-line-rate (MLR) optical networks. Journal of Lightwave Technology. 2012;**30**(1):130–139. DOI: 10.1109/JLT.2011.2177441

[33] Peng S, Nejabati R, Simeonidou D. Impairment-aware optical network virtualization in single-line-rate and mixed-line-rate WDM networks. IEEE/OSA Journal of Optical Communications and Networking. 2013;**5**(4):283–293. DOI: 10.1364/JOCN.5.000283

[34] Gerstel, M. Jinno, A. Lord, S. Yoo. Elastic optical networking: A new dawn for the optical layer? IEEE Communications Magazine. 2012;**50**(2):s12–s20. DOI: 10.1109/MCOM.2012.6146481

[35] Amaya N, Irfan M, Zervas G, Banias K, Garrich M, Henning I, Simeonidou D, Zhou YR, Lord A, Smith K, Rancano VJF, Liu S, Petropoulos P, Richardson DJ. Gridless optical networking field trial: Flexible spectrum switching, defragmentation and transport of 10G/40G/100G/555G over 620-km field fiber. In: 37th European Conference and Exhibition on Optical Communication; 18-22 September 2011; Geneva, Switzerland. pp. 1–3. DOI: 10.1364/OE.19.00B277

[36] Jinno M, Takara H, Kozicki B, Tsukishima Y, Sone Y, Matsuoka S. Spectrum-efficient and scalable elastic optical path network: architecture, benefits, and enabling technologies. IEEE Communications Magazine. 2009;**47**(11):66–73. DOI: 10.1109/MCOM.2009.5307468

[37] Christodoulopoulos K, Tomkos I, Varvarigos E. Dynamic bandwidth allocation in flexible OFDM-based networks. In: Optical Fiber Communication Conference; 6-10 March 2011; Los Angeles, California, USA. OSA; 2011. p. OTuI5. DOI: 10.1364/OFC.2011.OTuI5

[38] Patel A, Ji P, Wang T, Jue J. Optical-layer traffic grooming in flexible grid WDM networks. In: IEEE Global Communications Conference; 5-9 December 2011; Houston, Texas, USA. IEEE; 2011. pp. 1–6. DOI: 10.1109/GLOCOM.2011.6133618

[39] Sambo N, Meloni G, Paolucci F., Cugini F, Secondini M, Fresi F, Poti L, Castoldi P. Programmable transponder, code and differentiated filter configuration in elastic optical networks. Journal of Lightwave Technology. 2014;**32**(11):2079–2086. DOI: 10.1109/JLT.2014.2319859

[40] Cerutti I, Martinelli F, Sambo N, Cugini F, Castoldi P. Trading regeneration and spectrum utilization in code-rate adaptive flexi-grid networks. Journal of Lightwave Technology. 2014;**32**(23):4496–4503. DOI: 10.1109/JLT.2014.2359179

[41] Jinno M, Kozicki B, Takara H, Watanabe A, Sone Y, Tanaka T, Hirano A. Distance-adaptive spectrum resource allocation in spectrum-sliced elastic optical path network [Topics in Optical Communications]. IEEE Communications Magazine. 2010;**48**(8):138–145. DOI: 10.1109/MCOM.2010.5534599

[42] Lach E, Idler W. Modulation formats for 100G and beyond. Optical Fiber Technology. 2011;**17**(5):377–386. DOI: 10.1016/j.yofte.2011.07.012

[43] Eira A, Santos J, Pedro J, Pires J. Design of survivable flexible-grid DWDM networks with joint minimization of transponder cost and spectrum usage. In: 38th European Conference on Optical Communications; 16-20 September 2012; Amsterdam, Netherlands. IEEE; 2012. pp. 1–3. DOI: 10.1364/ECEOC.2012.P5.16

[44] Gringeri S, Basch E, Xia T. Technical considerations for supporting data rates beyond 100 Gb/s. IEEE Communications Magazine. 2012;**50**(2):s21–s30. DOI: 10.1109/MCOM.2012.6146482

[45] Rosa A, Cavdar C, Carvalho S, Costa J, Wosinska L. Spectrum allocation policy modeling for elastic optical networks. In: 9th International Conference on High Capacity Optical Networks and Enabling Technologies; 12-14 December 2012; Istanbul, Turkey. IEEE; 2012. pp. 242–246. DOI: 10.1109/HONET.2012.6421472

[46] J. Zhang, B. Mukherjee, J. Zhang, Y. Zhao. Dynamic virtual network embedding scheme based on network element slicing for elastic optical networks. In: 39th European Conference on Optical Communication; 22-26 September 2013; IET; 2013. London, England. pp. 1–3. DOI: 10.1049/cp.2013.1323

[47] Patel A, Ji P, Huang Y, Wang T. Distance-adaptive virtual network embedding in software-defined optical networks. In: OptoElectronics and Communications Conference held Jointly with the 18th International Conference on Photonics in Switching; 30 June-4 July 2013; Kyoto, Japan. IEEE; 2013. pp. 1–2.

[48] Patel A, Ye Z, Ji P. Cloud service embedding in software-defined flexible grid optical transport networks. In: Optical Fiber Communication Conference; 9-14 March 2014; San Francisco, California, USA. IEEE; 2014.p. M2H.2. DOI: 10.1364/OFC.2014.M2H.2

[49] Ye Z, Patel A, Ji P, Qiao C. Survivable virtual infrastructure mapping over transport software-defined networks (T-SDN). In: Optical Fiber Communication Conference; 9-14 March 2014; San Francisco, California, USA. IEEE; 2014. p. M3H.1. DOI: 10.1364/OFC.2014.M3H.1

[50] Gong L, Zhao W, Wen Y, Zhu Z. Dynamic transparent virtual network embedding over elastic optical infrastructures. In: IEEE International Conference on Communications; 9-13 June 2013; Budapest, Hungary. IEEE; 2013. pp. 3466–3470. DOI: 10.1109/ICC.2013.6655086

[51] Gong L, Zhu Z. Virtual optical network embedding (VONE) over elastic optical networks. Journal of Lightwave Technology. 2014;**32**(3):450–460. DOI: 10.1109/JLT.2013.2294389

[52] Barra E, Salinas R, Mora F, Tarifeno M, Beghelli A, Sambo N, Castoldi P. Virtual network provisioning over multi-line rate networks with fixed or flexible grid. In: 16th International Conference on Transparent Optical Networks; 6-10 July 2014; Graz, Austria. IEEE; 2014. pp. 1–4. DOI: 10.1109/ICTON.2014.6876709

[53] Khandaker F, Xie W, Jue J, Wang X, Zhang Q., She Q, Cankaya H, Palacharla P, Sekiya M. Survivable Virtual Optical Network mapping in spectrum and modulation format convertible flexible grid optical networks. In: Optical Fiber Communications Conference; 22-26 March 2015; Los Angeles, California, USA. pp. Th3J.1. DOI: 10.1364/OFC.2015.Th3J.1

[54] Shakya S, Pradhan N, Cao X, Ye Z, Qiao C. Virtual network embedding and reconfiguration in elastic optical networks. In: IEEE Global Communications Conference; 8-12 December 2014; Austin, Texas, USA. IEEE; 2014. pp. 2160–2165. DOI: 10.1109/GLOCOM.2014.7037128

[55] Gonzalez A, Barra E, Beghelli A, Leiva A. A sub-graph mapping-based algorithm for virtual network allocation over flexible grid networks. In: 17th International Conference on Transparent Optical Networks; 5-9 July 2015; Budapest, Hungary. IEEE; 2015. pp. 1–4. DOI: 10.1109/ICTON.2015.7193484

[56] Zhu R, Zhao Y, Zhang J, Yang H, Tan Y, Zhang J, Jue J, Wang N. Multi-dimensional resource virtualization in spectral and spatial domains for inter-datacenter optical networks. In: Optical Fiber Communication Conference; 20-22 March 2016; Anaheim, California, USA. IEEE; 2016. p. W2A.53. DOI: 10.1364/OFC.2016.W2A.53

[57] Wang X, Zhang Q, Kim I, Palacharla P, Sekiya M. Flexible virtual network provisioning over distance-adaptive flex-grid optical networks. In: Optical Fiber Communication Conference; 9-14 Mar 2014; San Francisco, California, USA. OSA; 2014 p. W3H.3. DOI: 10.1364/OFC.2014.W3H.3

[58] Zhang S, Shi L, Vadrevu C, Mukherjee B. Network virtualization over WDM and flexible-grid optical networks. Optical Switching and Networking. 2013;**10**(4):291–300. DOI: 10.1016/j.osn.2013.03.005

[59] Assis K, Peng S, Almeida R, Waldman H, Hammad A, Santos A, Simeonidou D. Network virtualization over elastic optical networks with different protection schemes. IEEE/OSA Journal of Optical Communications and Networking. 2016;**8**(4):272–281. DOI: 10.1364/JOCN.8.000272

[60] Zhou X, Nelson L, Magill P. Rate-adaptable optics for next generation long-haul transport networks. IEEE Communications Magazine. 2013;**51**(3):41–49. DOI: 10.1109/MCOM.2013.6476864

[61] Xie W, Jue J, Zhang Q, Wang X, She Q, Palacharla P, Sekiya M. Survivable virtual optical network mapping in flexible-grid optical networks. In: International Conference on Computing, Networking and Communications; 3-6 February 2014; Honolulu, Hawai. IEEE; 2014. pp. 221–225. DOI: 10.1109/ICCNC.2014.6785335

[62] Chen B. Power-aware virtual optical network provisioning in flexible bandwidth optical networks [Invited]. Photonic Network Communications. 2016;**32**(2):300–309. DOI: 10.1007/s11107-016-0609-4

[63] Wang W, Lin Y, Zhao Y, Zhang G, Zhang J, Han J, Chen H, Hou B, Ji Y, Zong L. First demonstration of virtual transport network services with multi-layer protection schemes over flexigrid optical networks. IEEE Communications Letters. 2016;**20**(2):260–263. DOI: 10.1109/LCOMM.2015.2509066

[64] Zhao J, Subramaniam S, Brandt-Pearce M. Virtual topology mapping in elastic optical networks. In: IEEE International Conference on Communications; 9-13 June 2013; Budapest, Hungary.IEEE; 2013. pp. 3904–3908. DOI: 10.1109/ICC.2013.6655167

[65] Morea A., Chong A., Rival O. Impact of transparent network constraints on capacity gain of elastic channel spacing. In: Optical Fiber Communication Conference; 6-10 March 2011; Los Angeles, California, USA. IEEE; 2011. pp. 1–3. DOI: 10.1364/NFOEC.2011.JWA062

[66] Sambo N, Castoldi P, Cugini F, Bottari G, Iovanna P. Toward high-rate and flexible optical networks. IEEE Communications Magazine. 2012;**50**(5):66–72. DOI: 10.1109/MCOM.2012.6194384

[67] Chowdhury N, Rahman M, Boutaba R. Virtual network embedding with coordinated node and link mapping. In: IEEE International Conference on Computer Communications; 19-25 April 2009; Rio de Janeiro, Brazil. IEEE; 2009. pp. 783–791. DOI: 10.1109/INFCOM.2009.5061987

[68] Patel A, Ji P, Jue J, Wang T. Defragmentation of transparent Flexible optical WDM (FWDM) networks. In: Optical Fiber Communication Conference; 6-10 March 2011; Los Angeles, California, USA. pp. 1–3. DOI: 10.1364/OFC.2011.OTuI8

[69] Chen C, Chen X, Zhang M, Ma S, Shao Y., Li S, Suleiman M, Zhu Z. Demonstrations of efficient online spectrum defragmentation in software-defined elastic optical networks. Journal of Lightwave Technology. 2014;**32**(24):4099–4109. DOI: 10.1109/JLT.2014.2364515

[70] Khodashenas P, Comellas J, Perelló J, Spadaro S. Correlation between traffic granularity and defragmentation periodicity in elastic optical networks. Transactions on Emerging Telecommunications Technologies. 2014;**26**(7):1011–1018. DOI: 10.1002/ett.2795

[71] Zhang M, Shi W, Gong L, Lu W, Zhu Z. Bandwidth defragmentation in dynamic elastic optical networks with minimum traffic disruptions. In: IEEE International Conference on Communications; 9-13 June 2013; Budapest, Hungary. IEEE; 2013. pp. 3894–3898. DOI: 10.1109/ICC.2013.6655165

[72] Cugini F, Paolucci F, Meloni G, Berrettini G, Secondini M, Fresi F, Sambo N, Poti L, Castoldi P. Push-pull defragmentation without traffic disruption in flexible grid optical networks. Journal of Lightwave Technology. 2013;**31**(1):125–133. DOI: 10.1109/JLT.2012.2225600

[73] Borquez D, Beghelli A, Leiva A. Deadlock-Avoiding vs. greedy spectrum allocation algorithms in dynamic flexible optical networks. In: 17th International Conference on

Transparent Optical Networks; 5-9 July 2015; Budapest, Hungary. pp. 1–4. DOI: 10.1109/ICTON.2015.7193486

[74] Donadio P, Fioccola G, Canonico R, Ventre G. A PCE-based architecture for the management of virtualized infrastructures. In: IEEE 3rd International Conference on Cloud Networking; 8-10 October 2014; Luxembourg. IEEE; 2014. pp. 223–228. DOI: 10.1109/CloudNet.2014.6968996

[75] Belbekkouche A, Hasan M, Karmouch A. Resource Discovery and Allocation in Network Virtualization. IEEE Communications Surveys Tutorials. 2012;**14**(4):1114–1128. DOI: 10.1109/SURV.2011.122811.00060

[76] Marquezan C, Nobre J, Granville L, Nunzi G, Dudkowski D, Brunner M. Distributed reallocation scheme for virtual network resources. In: IEEE International Conference on Communications; 14-18 June 2009; Dresden, Germany. IEEE; 2009. pp. 1–5. DOI: 10.1109/ICC.2009.5198934

[77] Sun G, Anand V, Yu H, Liao D, Li L. Optimal provisioning for elastic service oriented virtual network request in cloud computing. In: IEEE Global Communications Conference; 3-7 December 2012; Anaheim, California, USA. pp. 2517–2522. DOI: 10.1109/GLOCOM.2012.6503495

[78] Fajjari I, Aitsaadi N, Pujolle G, Zimmermann H. VNE-AC: Virtual network embedding algorithm based on ant colony metaheuristic. In: IEEE International Conference on Communications; 5-9 June 2011; Kyoto, Japan. IEEE; 2011. pp. 322–340. DOI: 10.1504/IJCNDS.2011.039538

[79] Zinner T, Tran-Gia P, Tutschku K, Nakao A. Performance evaluation of packet reordering on concurrent multipath transmissions for transport virtualisation. Journal of Communication Networks and Distributed Systems. 2011;**6**(3):322–340. DOI: 10.1504/IJCNDS.2011.039538

[80] Jinno M, Takara H, Sone Y, Yonenaga K, Hirano A. Multiflow optical transponder for efficient multilayer optical networking. IEEE Communications Magazine. 2012;**50**(5):56–65. DOI: 10.1109/MCOM.2012.6194383

[81] Sambo N, D'Errico A, Porzi C, Vercesi V, Imran M, Cugini F, Bogoni A, Poti L, Castoldi P. Sliceable transponder architecture including multiwavelength source. IEEE/OSA Journal of Optical Communications and Networking. 2014;**6**(7):590–600. DOI: 10.1109/JOCN.2014.6850200

[82] Ahmed J, Solano F, Monti P, Wosinska L. Traffic re-optimization strategies for dynamically provisioned WDM networks. In: 15th International Conference on Optical Network Design and Modeling; 8-10 February 2011; Bologna, Italy. IEEE; 2011. pp. 1–6.

[83] Bai H, Gu F, Liang K, Rahnamay-Naeini M, Khan S, Hayat M, Ghani N. Virtual network reconfiguration in optical cloud substrates. In: Optical Fiber Communications Conference; 9-14 March 2014; San Francisco, California, USA. IEEE; 2014. pp. 1–6. DOI: 10.1364/OFC.2014.M3H.7

[84] Mello D, Pelegrini J, Ribeiro R, Schupke D, Waldman H. Dynamic provisioning of shared-backup path protected connections with guaranteed availability requirements. In: 2nd International Conference on Broadband Networks; 7 October 2005; Boston, Massachusetts, USA. IEEE; 2005. pp. 1320–1327. DOI: 10.1109/ICBN.2005.1589761

[85] Tarifeño M, Beghelli A, Moreno E. Availability-driven optimal design of shared path protection in WDM networks. Networks. 2016; 68(3):224–237. DOI: 10.1002/net.21695

[86] Touch J. Dynamic Internet overlay deployment and management using the X-Bone. In: International Conference on Network Protocols; 14-17 November 2000; Osaka, Japan. IEEE; 2000. pp. 59–68. DOI: 10.1109/ICNP.2000.896292

Chapter 3

Multi-Period Attack-Aware Optical Network Planning under Demand Uncertainty

Konstantinos Manousakis, Panayiotis Kolios and Georgios Ellinas

Additional information is available at the end of the chapter

http://dx.doi.org/10.5772/intechopen.68491

Abstract

In this chapter, novel attack-aware routing and wavelength assignment (Aa-RWA) algorithms for multiperiod network planning are proposed. The considered physical layer attacks addressed in this chapter are high-power jamming attacks. These attacks are modeled as interactions among lightpaths as a result of intra-channel and/or inter-channel crosstalk. The proposed Aa-RWA algorithm first solves the problem for given traffic demands, and subsequently, the algorithm is enhanced in order to deal with demands under uncertainties. The demand uncertainty is considered in order to provide a solution for several periods, where the knowledge of demands for future periods can only be estimated. The objective of the Aa-RWA algorithm is to minimize the impact of possible physical layer attacks and at the same time minimize the investment cost (in terms of switching equipment deployed) during the network planning phase.

Keywords: physical layer attacks, routing and wavelength assignment, optical networks, multi-period planning, demand uncertainty

1. Introduction

In wavelength division multiplexed (WDM) optical networks, wavelength routing is used for establishing communication between source-destination pairs. In these networks, data are transmitted over all-optical WDM channels called lightpaths. A connection is established by utilizing a lightpath, which is determined by choosing a path between the source and the destination and allocating a wavelength on all the links of the path. The selection of the path and wavelength is an important optimization problem and is known as the routing and wavelength assignment (RWA) problem [1].

In WDM optical networks, transparent optical cross-connects (OXCs) are used in order to provide efficient space and wavelength switching functions [2]. An OXC takes as input signals at multiple wavelengths and some of these wavelengths can be dropped locally, while others pass through by switching them to the appropriate output ports. For the implementation of OXCs, wavelength selective switch (WSS) technology is used for the deployment of cost-effective and dynamic wavelength-switched networks [3].

In transparent optical networks, where data signals remain in the optical domain until they reach their destinations, connections are vulnerable to physical layer attacks. An attack is defined as an intentional action against the ideal and secure functioning of the network. One type of attack in optical networks is high-power jamming which can affect the signal through in-band jamming that is the result of intra-channel crosstalk or out-of-band jamming that is the result of inter-channel crosstalk and nonlinearities [4]. This type of attack propagates through the transparent network affecting several connections, and as a consequence, the localization of this kind of attack is a difficult problem. Due to the high bit rates of optical networks and the interaction of the connections, a jamming attack can potentially cause a huge amount of information loss. Therefore, the limitation of attack propagation is a crucial consideration in optical network planning. An overview of security challenges in communication networks can be found in Ref. [5].

Physical layer attacks in optical networks have been studied by several researchers [6–10]. In these works, the concept of attack-aware routing and wavelength assignment (Aa-RWA) is analyzed. Specifically, in Ref. [6], authors proposed an integer linear program (ILP) formulation and a tabu search heuristic algorithm for the routing sub-problem in optical networks in order to minimize the effect of out-of-band jamming and the gain competition caused in optical fibers and optical amplifiers, respectively. In Ref. [7], authors proposed ILP formulation and heuristic algorithms for the wavelength assignment sub-problem in optical networks in order to minimize the in-band jamming attack caused in optical nodes. In Ref. [8], authors proposed ILP and heuristic algorithms based on simulated annealing techniques in order to minimize the in-band and out-of-band jamming attacks. Moreover, in Ref. [9, 10], authors proposed a greedy randomized adaptive search procedure (GRASP) heuristic and an ILP formulation, respectively, for the placement of power equalizers in order to limit the jamming attack propagation in transparent optical networks.

Another important aspect in network planning that usually is not taken into account is the uncertainty of the connection requests. In most cases, the demands are considered to be known before network planning; however, in some cases, network planning must be performed for a period of time where the demand requests can only be forecasted with uncertainty. One approach to deal with demand uncertainty is by overprovisioning, essentially allocating many resources that can satisfy any traffic demand. However, this approach requires a high cost investment (capital expenditure—capex) from the network operators [11]. More sophisticated approaches to deal with demand uncertainty are necessary in order to achieve a cost-effective network investment strategy [12].

Stochastic programming (SP) [13] and robust optimization (RO) [14] are the main alternative techniques to deal with uncertain data both in a single period and in a multi-period decision making process. In SP, the probability distribution functions of the underlying stochastic parameters must be known. On the other hand, RO addresses the uncertain nature of the problem without making specific assumptions on probability distributions. The uncertain parameters are assumed to belong to a deterministic uncertainty set. RO adopts an approach that addresses uncertainty by guaranteeing the feasibility and optimality of the solution against all instances of the parameters within the uncertainty set.

In Ref. [15], authors apply robust optimization in order to incorporate the uncertainty of demands into the network upgrade problem. Under the robust network upgrade model, the network planning can be performed by tuning the trade-off between network cost and robustness level. Further, in Ref. [16], authors propose multi-period network planning approaches based on SP, where the demands are forecasted over periods of time and the network investments are performed based on these forecasts.

In this chapter, novel Aa-RWA algorithms are proposed to address the problem of multi-period network planning under demand uncertainty with the objective to minimize the impact of possible physical layer attacks and at the same time to minimize the network infrastructure investment cost. Physical layer attacks are modeled as interactions among connections through in-band and out-of-band channel crosstalk. Moreover, the investment cost is taken into account in this formulation via the number of WSSs required in order to minimize the impact of a possible physical layer attack.

The simulation results show that when the distribution of demands for all the time periods is taken into account in advance, better results can be obtained in terms of the number of WSSs required to be placed in the network nodes so as to minimize the impact of a jamming attack, compared to the case where the distribution is known only for the period under consideration.

The chapter is organized as follows. Section 2 describes the network architecture, while Section 3 describes the planning approaches for demand uncertainty. In Section 4, the physical layer attacks in optical networks are presented, and in Section 5, the problem of attack-aware RWA with given traffic demands is solved. This is followed in Section 6 by the attack-aware RWA under demand uncertainties. Performance results are presented in Section 7, while Section 8 presents some concluding remarks.

2. Network and node architecture

An optical network topology is represented by a connected graph $G = (V, E)$, where V denotes the set of optical cross-connects (nodes) and E denotes the set of (point-to-point) single-fiber links (edges). Each fiber link is able to support a common set $C= \{1,2,\ldots,W\}$ of, W, distinct wavelengths. Source-destination pairs are equipped with transmitter-receiver pairs, also known as transponders (TSP), in order to transmit/receive data. Optical nodes currently deployed in optical networks are based on two architectures. The first architecture utilizes a broadcast-and-select (BS) configuration and the second a route-and-select (RS) configuration. Both of these optical node architectures consist of two stages and can remotely configure all transit traffic and only differ in the implementation of their first stage. The building components of these node architectures are the WSSs. A WSS can steer each optical channel present on its input port toward one of its output ports according to the desired routing choice.

BS-based nodes (**Figure 1**) include a splitter first stage (1 × N) that implicitly provides a broadcast capability toward all outputs. In a BS-based architecture, the WSS functionality (second stage) resembles a multiplexer (it switches each individual wavelength to a certain output). Although this is a simple and popular architecture, the loss introduced by the power splitters limits its scalability and can only be utilized in network nodes with small degrees.

RS architecture nodes (**Figure 2**) on the other hand have a WSS first stage (1 × N) that provides on-demand routing to the required output. The basic advantage of the RS-based architecture with respect to the BS-based architecture is that the through loss is not dependent on the degree of the node. However, it requires additional WSSs at the input stage, which makes it more costly to be implemented.

Both implementations have a WSS second stage (N × 1) that provides the selection of the wavelengths at the output fibers, allowing full switching flexibility (any wavelength from any incoming fiber can pass through or any wavelength from the add/drop terminals can be added/dropped).

In order to deal with the losses introduced by the power splitters of the BS-based architecture and the high cost of the RS-based architecture, a hybrid architecture can also be used (**Figure 3**). This architecture contains either splitters (1 × N) or WSSs (1 × N) at the input ports as can be seen in **Figure 3**. In essence, hybrid nodes are constructed by replacing splitters with WSSs at the input stage of the BS-based nodes.

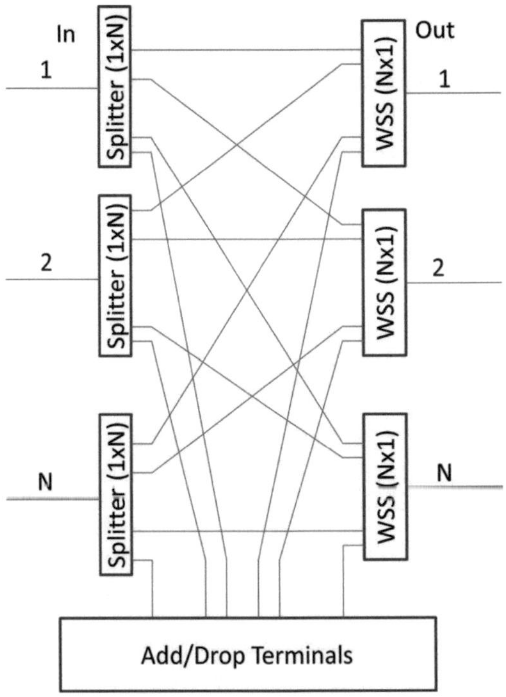

Figure 1. Broadcast-and-select-based node architecture.

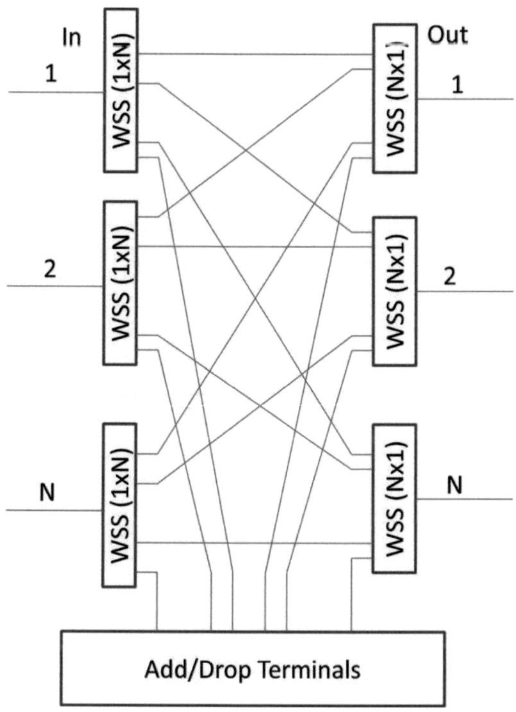

Figure 2. Route and-select-based node architecture.

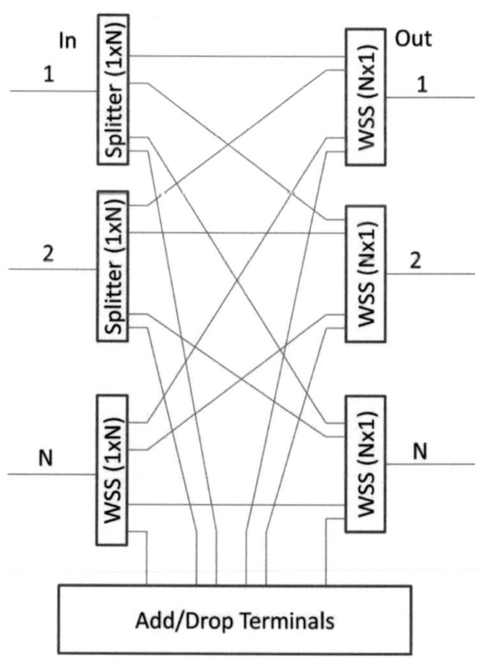

Figure 3. Hybrid node architecture.

Depending on the network traffic, it is envisioned that a fraction of the network nodes will be BS-based, other nodes will be RS-based and the rest will be hybrid nodes. The objective of the proposed algorithms of this chapter is to use hybrid nodes in order to minimize the lightpath interactions and at the same time to minimize the network cost. This means that WSSs are placed only in some of the input ports and specifically only at the locations that are necessary in order to allow only the necessary wavelengths to pass through the WSS and avoid all crosstalk interactions. Thus, by using hybrid nodes and not RS-based nodes, we can minimize the network cost while at the same time eliminating crosstalk interactions and consequently protecting the network against jamming attacks.

3. Planning approaches for demand uncertainty

In order to provide cost-efficient network solutions, it is necessary to plan optical networks over a long-time horizon. When dealing with optical networks, where the cost to build the network is high and the investment that takes place should last for a long time, sophisticated planning decisions must take place to ensure that the network infrastructure will not require any major upgrades over a predetermined amount of time. The problem becomes more involved in the case of future traffic demand forecasts that include uncertainty, as network planning decisions must be taken without the exact knowledge of future traffic demands. In this case, these decisions will be based on estimations. In the remaining of this chapter, the proposed multi-period network planning approaches with uncertain traffic demands are discussed. The planning approaches assume that for the first period, the demands follow a known distribution and for the periods that follow the demands are increased based on a multiplicative factor.

The multi-period network planning problem in this chapter will be investigated for two different period-planning types as detailed below.

3.1. Incremental network planning

This approach considers the demands of the next period and optimizes the investment cost in each period. Therefore, the solution is calculated sequentially for each period. The solution can be optimal for each period but not jointly for all the periods under consideration. Once the solution is provided for one period, then this solution affects the solution of the periods that follow. This is due to the fact that the solution of one period is assumed to be fixed and the solutions of the periods that follow are now based upon the previously found solutions.

3.2. Multi-period network planning

This approach considers the demands of all periods and optimizes the investment cost from the beginning of the planning period, that is the multi-period approach minimizes the network cost over all periods at once. Therefore, the demand distribution for every time period is necessary. This approach can calculate an optimal overall solution and provide decisions for the investment strategy of network operators.

4. Physical layer attacks

In general, the physical layer attacks in transparent optical networks can be grouped in two main categories: eavesdropping and service disruption.

In eavesdropping, the purpose of an attacker is to passively analyze the traffic in the network after gaining access to the information through an unauthorized observation method. To gain mid-span access to the fiber, the eavesdropper has to cut through and strip away the cable's outer jacket to access the individual fibers in its center.

Service disruption can be performed through high-power jamming attacks and can be classified into three sub-categories based on the effects it inflicts on the signal:

i. in-band jamming which is the result of intra-channel crosstalk,

ii. out-of-band jamming that is the result of inter-channel crosstalk and nonlinearities, and

iii. gain competition in optical amplifiers, where a high-power jamming signal can increase its own power, thus resulting in reduction in the gain of the rest of the co-propagating channels on the same fiber.

These types of attacks propagate through the transparent network affecting several connections, and as a consequence, the localization of an attack is a difficult problem. Due to the high bit rates of optical networks and the interaction of the connections, a jamming attack can cause a huge amount of information loss. Therefore, the limitation of attack propagation is a crucial consideration in designing transparent WDM optical networks.

The focus of this study is to deal with service disruption and especially with in-band and out-of-band jamming attacks.

4.1. In-band jamming attack

High-power in-band jamming attack is an attack that can be performed through the intra-channel crosstalk effect. Intra-channel crosstalk is the effect of power leakage between lightpaths crossing the same switch and using the same wavelength due to non-ideal isolation of the inputs/output ports of the switching fabric. Intra-channel crosstalk cannot be filtered out, since the interfering signal is on the same wavelength as the one affected. Thus, a high-power jamming signal can cause significant leakage inside the switches between lightpaths that are on the same wavelength as the attacking signal.

Figure 4 illustrates an example of a high-power jamming attack in node n_1 of the network through lightpath (p_1, w_i). In this figure, the attacker uses the lightpath (p_1, w_i) in order to attack the network. The attacking signal initially affects lightpath (p_0, w_i), through intra-channel

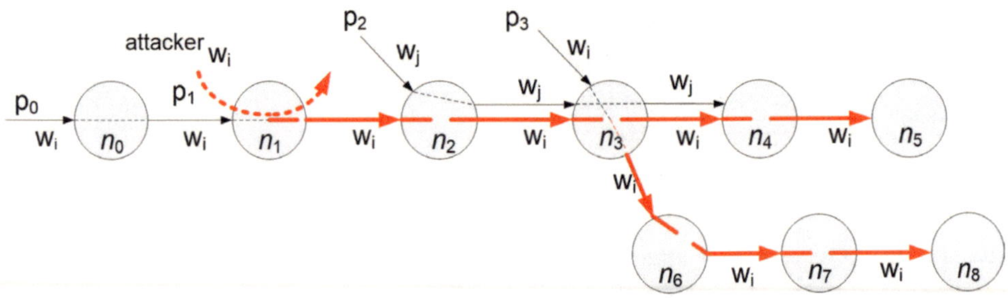

Figure 4. High-power in-band jamming attack propagation.

crosstalk because this lightpath uses the same wavelength and is crossing the same node as the attacking lightpath. In turn, lightpath (p_0, w_i) becomes an attacker too called "secondary attacker". Thus, lightpath (p_0, w_i) spreads the attack further to lightpath (p_3, w_i).

4.2. Out-of-band jamming attack

High power out-of-band jamming attack is an attack that can be performed through the inter-channel crosstalk effect. Inter-channel crosstalk results due to the power leakage between adjacent channels.

Figure 5 illustrates the high-power out-of-band signal propagation through the inter-channel crosstalk effect. In this case, lightpath (p_1, w_{i+1}) is used by an attacker in order to attack the network. Lightpath (p_1, w_{i+1}) then affects lightpath (p_0, w_i) as the two lightpaths co-propagate along the same fiber utilizing adjacent wavelengths. Then, the affected lightpath (p_0, w_i) becomes a "secondary attacker" and affects lightpath (p_3, w_{i-1}).

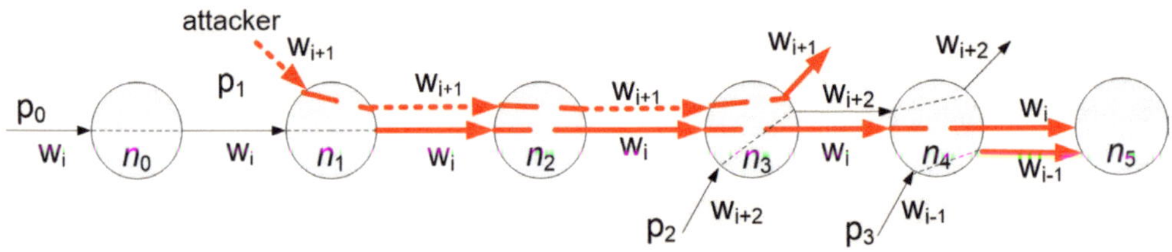

Figure 5. High-power out-of-band jamming attack propagation.

5. Attack-aware routing wavelength assignment

In this section, a heuristic algorithm is presented for the Aa-RWA with given demands in order to minimize the propagation of physical layer attacks. The algorithm aims at minimizing the interactions among lightpaths in order to avoid the propagation of high-power jamming attacks, in terms of affected lightpaths through intra- and inter-channel crosstalk. As discussed above, with these types of attacks, an affected lightpath can also affect other lightpaths, thus spreading the attack to other parts of the network. The goal of the Aa-RWA techniques is then to minimize as much as possible the spread of any attack that can occur in the network.

The proposed heuristic approach solves the problem by sequentially serving one-by-one the connections and consists of two phases. In the first phase, k candidate paths are calculated for each requested connection. In the second phase, the algorithm establishes the connections sequentially with the objective to minimize the number of in-band and out-of-band lightpath interactions.

5.1. Finding candidate paths

In the first phase, k candidate paths are identified for serving each requested connection. These paths are selected by employing a k-shortest path algorithm. The k-shortest path algorithm pre-calculates for each source-destination pair (s, d) a set of k candidate paths P_{sd} as follows: first, the shortest path is calculated using Dijkstra's algorithm, and then, the cost of the links which belong to the shortest path is doubled and Dijkstra's algorithm is executed again. This procedure is repeated until k paths are found. After a subset P_{sd} of candidate paths for each source-destination pair (s, d) is computed, the total set of computed paths is given as input to the next phase of the algorithm.

5.2. Attack-aware RWA

This section describes the heuristic algorithm for establishing the connections, one-by-one, in some particular order with the objective to minimize the lightpath interactions through the crosstalk effect.

5.2.1. Definitions

Each link l of the network is characterized by a Boolean wavelength availability vector $BWAV_l(i)$, $1 \le i \le W$, whose i^{th} element is equal to 0 if the ith wavelength of link l is utilized by a connection and is equal to 1, otherwise. W is the number of wavelengths that each fiber is able to support.

Each path p is characterized by a Boolean wavelength availability vector $BWAV_p(i)$, $1 \le i \le W$. The $BWAV_p$ consisting of links $l \in p$ is defined as the Boolean AND operation to the $BWAV_l$ of these links in each of the wavelengths of the $BWAV_l$ vectors.

$$BWAV_p = AND_{l \in p}(BWAV_l) \quad (1)$$

Thus, the element $BWAV_p(w)$ is equal to 1 if wavelength w is available over path p. The above equation enforces the wavelength continuity constraint among the links comprising a path. Each element $BWAV_p(i)$ represents a lightpath (p, w) between source-destination pairs (s, d).

5.2.2. Algorithm description

The aim of the heuristic algorithm is to establish Λ_{sd} lightpaths for (s, d) under the current utilization state of the network, given in the form of the wavelength availability vectors $BWAV_l$, for all l and the established lightpaths up to that point. The objective of the Aa-RWA heuristic algorithm is to minimize the number of lightpaths that interact with other lightpaths through intra- and inter- channel crosstalk and thus to minimize the propagation of high-power jamming signal attacks.

The wavelength utilization $BWAV_p$ of the candidate pre-calculated paths for the source-destination pair (s, d) is computed based on the $BWAV_l$ of the links. For each demand, the lightpath (p, w), from the set of candidate lightpaths with the smallest number of in-band and out-of-band channel interactions with the already established lightpaths, is chosen. To evaluate this, the wavelength availability vectors $BWAV_l$ are used to identify the interactions of established lightpaths. Then, the lightpath with the minimum sum of in-band and out-of-band channel interactions is established.

After establishing the lightpath (p, w), the corresponding $BWAV_l$ is updated. The algorithm at each step establishes a requested connection Λ_{sd}. If there are no available wavelengths, then the connection is blocked. Subsequently, the algorithm establishes lightpaths for all the connection requests in sequential order. The output of the algorithm is a set of established lightpaths in terms of paths and wavelengths. For each lightpath, the algorithm also returns two scalars that represent the number of inter-channel and the intra-channel interactions of this lightpath with the other established lightpaths.

6. Attack-aware routing and wavelength assignment under demand uncertainty for multi-period planning

As emphasized above, multi-period network planning is crucial in avoiding overprovisioning WSSs within hybrid nodes. As such, the aforementioned Aa-RWA algorithm is extended in this section to consider the demand forecasts of future time periods and in doing so to ensure that the WSS placement considers the changing network characteristics. In line with the most popular period-planning types available in the literature, the Aa-RWA algorithm is applied for both the incremental network planning case as well as the multi-period planning approach. In the former case, the Aa-RWA algorithm is applied in each step, the WSS placement for that step is decided, and the subsequent period considers the presence of those WSSs in the network when running the Aa-RWA algorithm for the next time period. In the multi-period approach on the other hand, the in-band and out-of-band interactions in each node are calculated for all time periods by the Aa-RWA algorithm and then statistical measures are used to assess the level of interaction and the extent to which a WSS is needed at a specific node.

In either case, the level of in-band and out-of-band interactions (and the subsequent decision on WSS placement) is strongly governed by the demand uncertainties and the assumptions made on growth year after year. The growth factor is assumed to be the mean value around a normally distributed random variable of the actual traffic growth between source destination pairs and thus Monte Carlo simulations are conducted to investigate the overall performance under independent trials. Details of the network setup and the exact values considered are detailed in Section 7.

6.1. Incremental Aa-RWA network planning

In incremental Aa-RWA network planning, there is knowledge for the demand distribution for only one period at a time (the period under consideration). For this reason, decisions are taken only for the current period. The flowchart of the proposed algorithm is given in **Figure 6**. The algorithm takes as input N independent sets of demands. For each one of the N sets, the algorithm solves the problem according to the deterministic Aa-RWA algorithm as presented in Section 5 and produces N outputs with metrics related to in-band and out-of-band interactions. These metrics associate two values for each input port of every network node. Specifically, these values count the number of lightpaths that interact though in-band and out-of-band crosstalk in the specific input port. Based on these values, the algorithm specifies the ports where WSSs should be placed. The assumption in this work is that in every period a maximum number of m WSSs can be placed due to budget constraints. The input ports where the WSSs are placed are chosen according to the maximum mean values of the in-band and out-of-band interactions. Subsequently, the output of each period contains the established lightpaths, and the next period takes as input the already established lightpaths and the placement of the WSSs from the previous period. The same procedure is followed for every period during the entire time horizon under consideration.

6.2. Multi-period Aa-RWA network planning

In multi-period Aa-RWA network planning, there is a priori knowledge for the demand distribution for all the time periods under consideration. Therefore, decisions are taken based on the traffic estimate for all time periods. The flowchart of the proposed algorithm is given in **Figure 7**. The algorithm takes as input N independent sets of demands for every one of the T periods (increasing over time based on a multiplicative factor as previously mentioned).

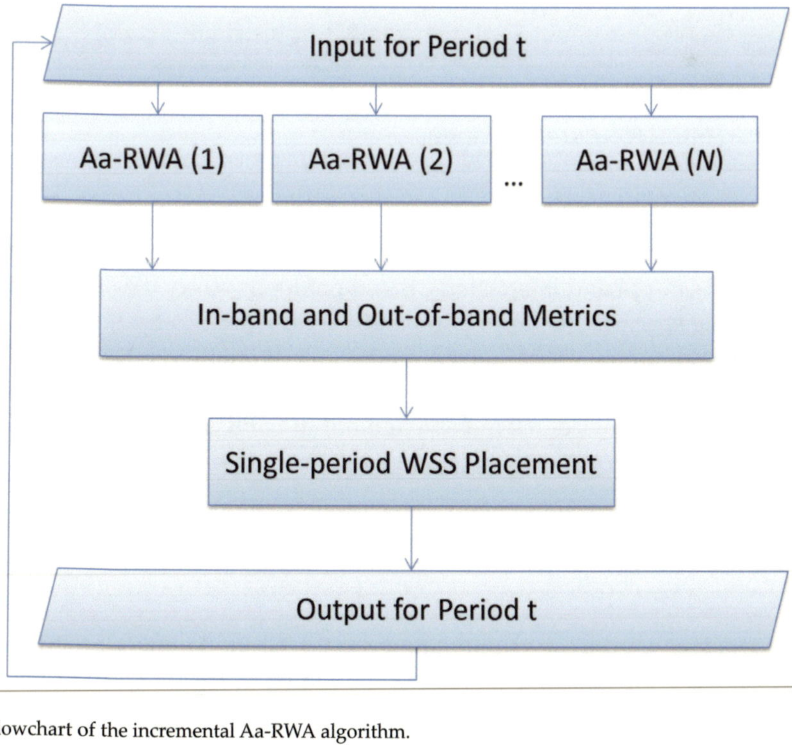

Figure 6. A flowchart of the incremental Aa-RWA algorithm.

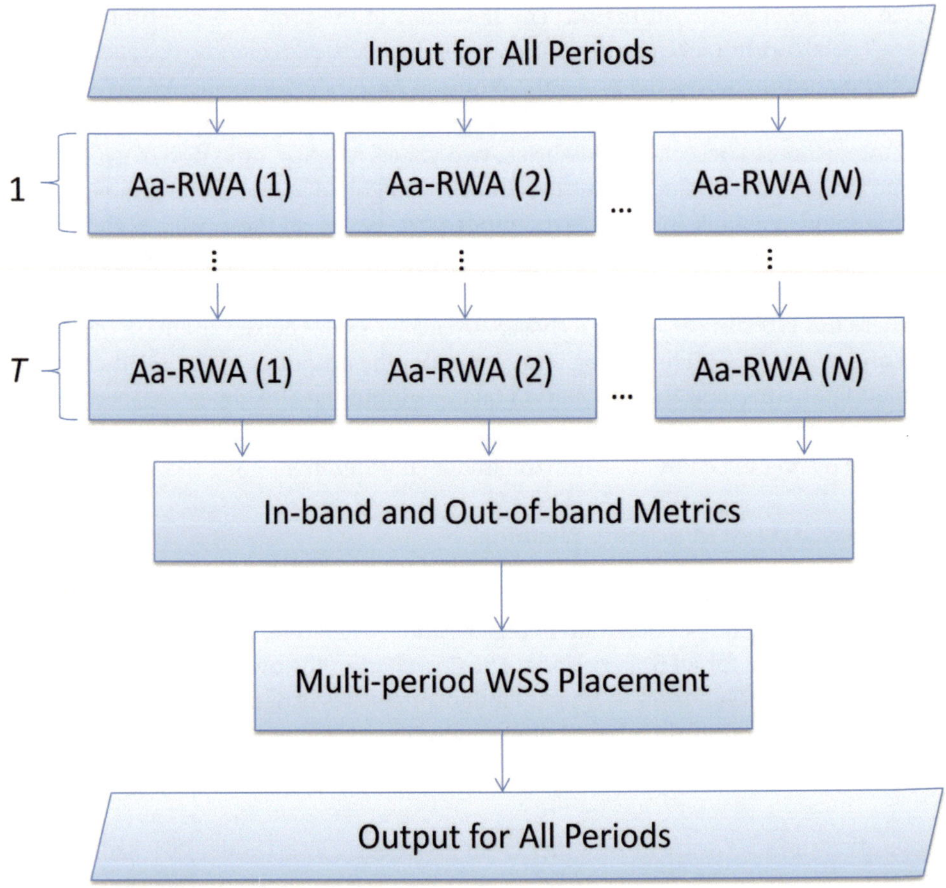

Figure 7. A flowchart of the multi-period Aa-RWA algorithm.

For each one of the N sets and for each time period, the algorithm solves the problem according to the deterministic Aa-RWA algorithm and produces $N*T$ outputs with metrics related to the in-band and out-of-band interactions. Based on these values, the "multi-period WSSs placement" module specifies the input ports and the time periods for the placement of the WSS. Again, the assumption is that in every period, a maximum number of m WSSs can be placed due to budget constraints. In this case, the placement of the WSSs is performed based on the maximum mean values of the in-band and out-of-band interactions over all instances and all periods.

7. Performance results

The network topology used in our simulations was the Geant-2 network topology [17] that has 34 nodes and 54 bidirectional links (108 fibers; shown in **Figure 8**). Each fiber is able to

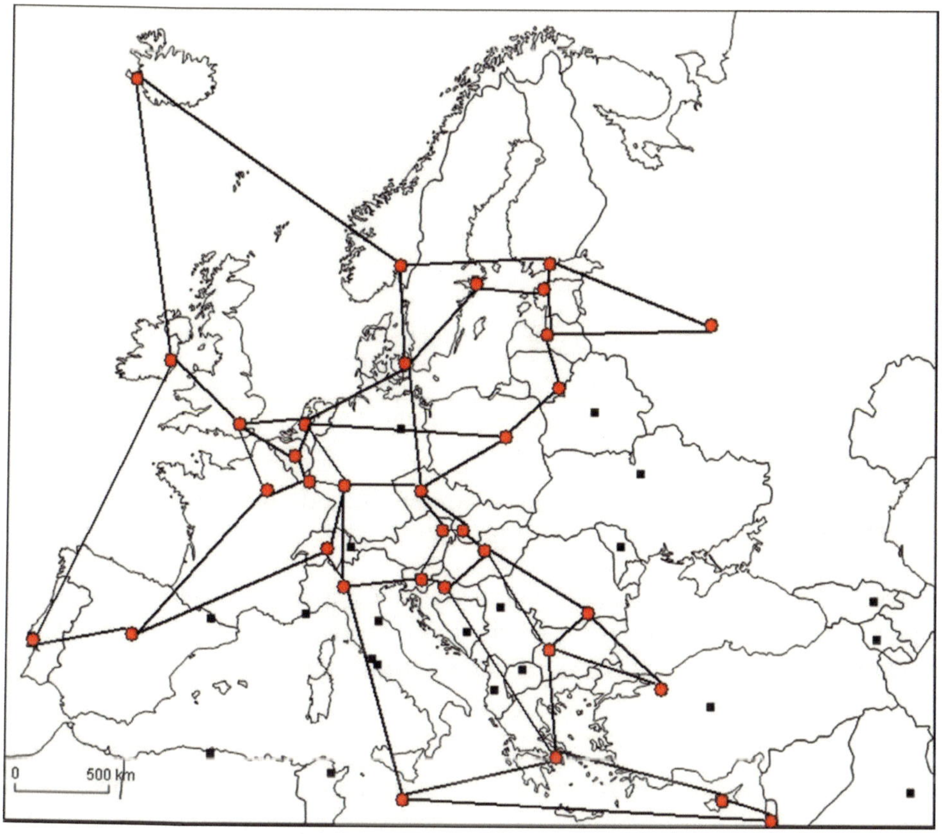

Figure 8. Geant-2 network topology: 34 nodes, 54 links.

support 80 wavelengths. The capacity of each wavelength was assumed equal to 10 Gbps. Initially, 50 different traffic matrices were produced with uniform distribution between source destination pairs and mean value equal to 1.35 Tbs of total requested capacity. Both algorithms (multi-period Aa-RWA and incremental Aa-RWA) were studied for five periods. The growth factor for each period was assumed to be equal to 1.5. The demand increase for each period applies for the source destination pairs that have a non-zero value at the initial traffic matrix. The algorithms for each source destination pair computed $k = 3$ alternative candidate paths.

In **Figure 9**, results for the multi-period Aa-RWA algorithm are depicted. Specifically, in **Figures 9(a)**, **(b)**, the mean values for inter-channel and intra-channel crosstalk for a horizon of five periods are presented, respectively. The mean values are the result of the 50 different traffic matrices. The inter-channel and intra-channel crosstalk per link (input port of a node) are the number of the interactions at this port. In **Figure 9**, the central mark of each box is the median, and the edges of the box are the 25th and 75th percentiles, the whiskers extend to the most extreme data points that are not considered outliers, and outliers are plotted individually.

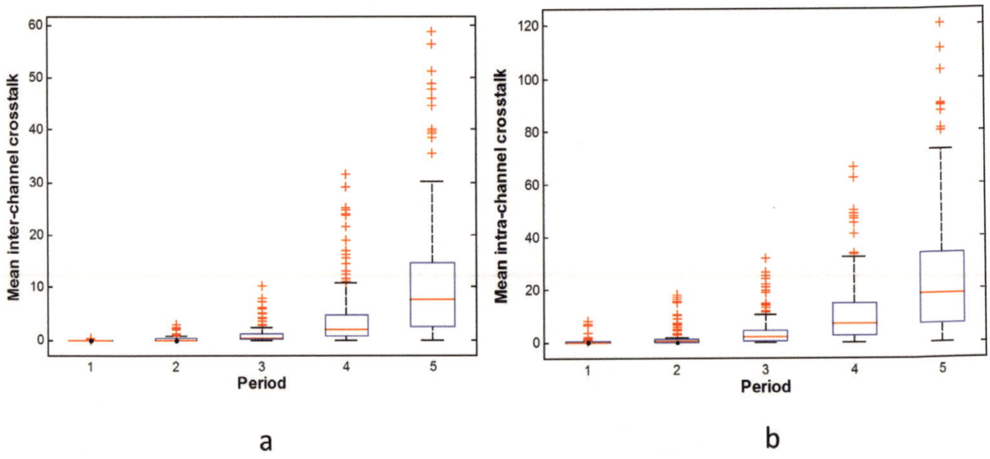

Figure 9. Mean values of multi-period Aa-RWA algorithm for (a) inter-channel crosstalk and (b) intra-channel crosstalk for a horizon of five periods.

Both inter- and intra-channel crosstalk increase exponentially with increasing traffic demands. However, as shown in **Figures 9(a)**, **(b)**, specific links experience significantly higher crosstalk than others. Therefore, the required WSSs can be placed only at the input ports of the nodes that experience high crosstalk.

Incremental Aa-RWA algorithm follows the same trend as the multi-period Aa-RWA algorithm (as illustrated in **Figure 10**). Note that the trend would be completely different in the case where an attack-unaware RWA algorithm was used. In that case, all the periods would experience high values of crosstalk as can be found from the results of [8]. These results are not presented here, since the scope of this chapter is to plan an optical network in order to deal with physical layer attacks and therefore an attack-unaware RWA algorithm is out of the scope of this study.

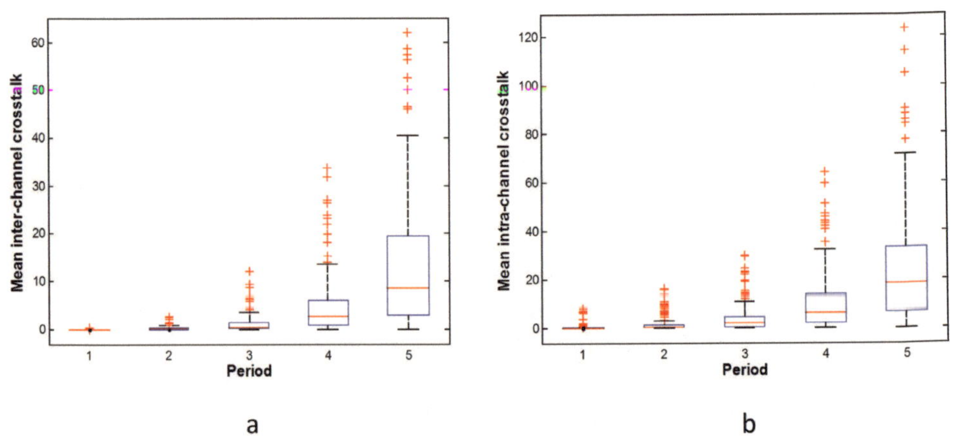

Figure 10. Mean values of incremental Aa-RWA algorithm for (a) inter-channel crosstalk and (b) intra-channel crosstalk for a horizon of five periods.

In **Figure 11**, the mean value of inter- and intra-channel crosstalk that the links experience during time period 5 is presented for the multi-period Aa-RWA algorithm. The results are presented in the form of histograms, where each column represents the number of links that have crosstalk between the ranges that are depicted in the x-axis of the histograms. From **Figure 11**, it is clear that a very small number of links have very high crosstalk, while the majority of links experience only a small crosstalk effect. This result offers a good indication that an addition of a small number of WSSs at the specific nodes where high crosstalk is experienced will significantly improve the performance of the network, thus minimizing the effect of a jamming attack. Note that the larger the number of links that appear in the leftmost bar, the smaller the crosstalk effect at the input ports of these nodes. Therefore, the best algorithms will be those where their histograms are more left shifted.

In **Figure 12**, the same histograms are presented for the case of the incremental Aa-RWA algorithm. Compared to the previous results of the multi-period case, the crosstalk effect of the incremental updating results to slightly increased inter-channel crosstalk and comparable intra-channel crosstalk. Nevertheless, the same crosstalk trends are observed here as well, where a small number of links experience significant crosstalk, while the rest of the links experience significantly lower crosstalk.

In **Figure 13**, the total number of required WSSs in order to minimize the impact of crosstalk effect per period is presented for the two proposed algorithms. For each period, the algorithms decide to place a WSS at the input port of a link when the mean values of the inter- and intra-channel crosstalk are above a certain threshold. Based on these decisions, the multi-period Aa-RWA algorithm requires less number of WSSs per period as compared to the incremental Aa-RWA algorithm. This is due to the fact the routing and wavelength assignment of the multi-period algorithm takes into account the future traffic demands, and the decisions are more appropriate. On the other hand, the incremental algorithm may decide to place a WSS in one period, and in future periods, there will be demands that would not be able to be established over already placed WSSs due to insufficient number of wavelengths. Thus, there would be not enough choices for efficient routing and wavelength assignment.

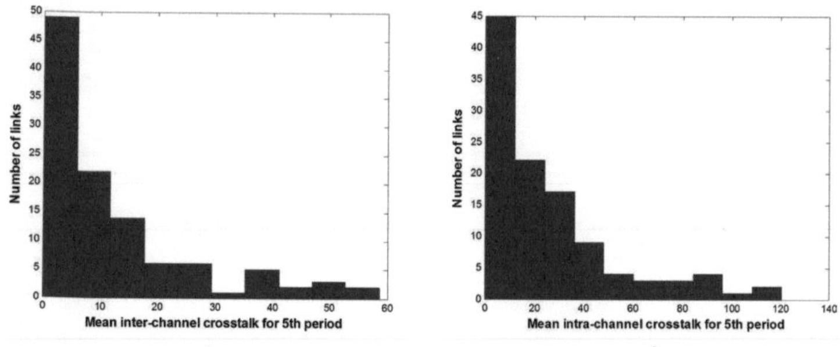

Figure 11. Histogram for link (input ports of nodes) distribution related to (a) inter-channel and (b) intra-channel crosstalk interactions for multi-period Aa-RWA algorithm for the fifth period.

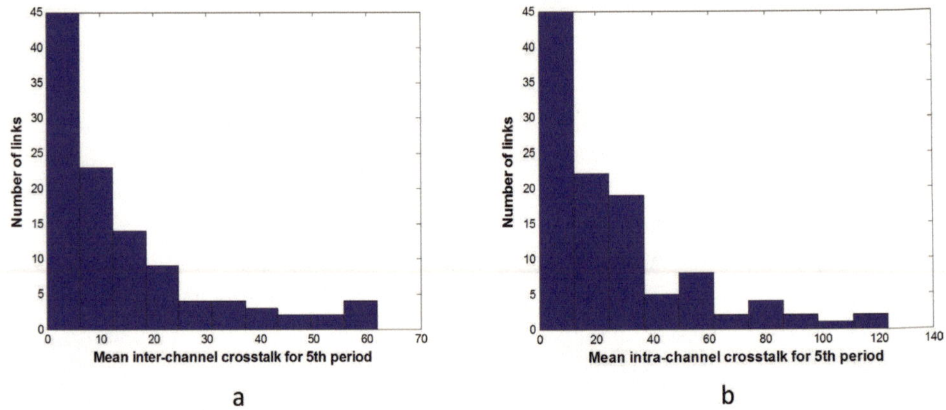

Figure 12. Histogram for link (input ports of nodes) distribution related to (a) inter-channel and (b) intra-channel crosstalk interactions for incremental Aa-RWA algorithm for the fifth period.

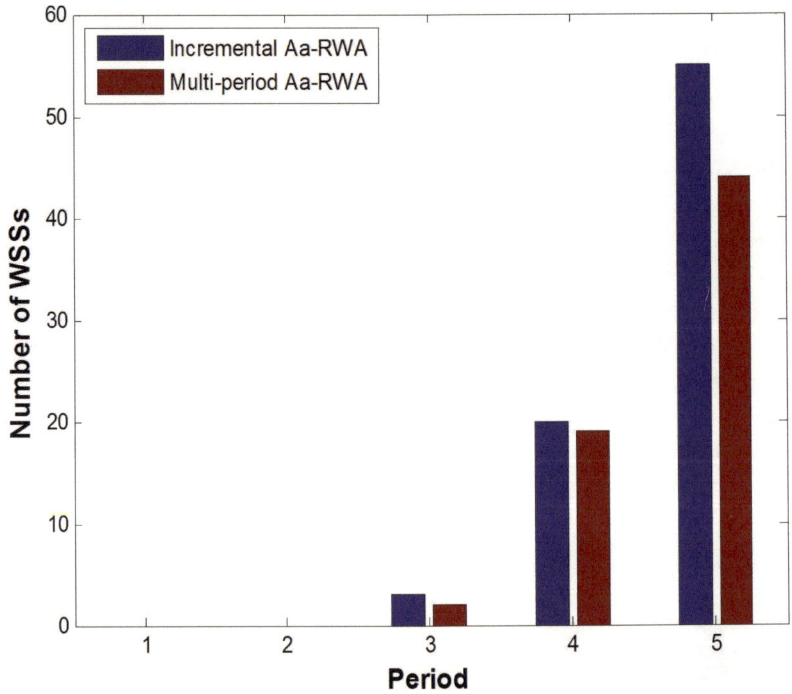

Figure 13. Number of required WSSs per period for the incremental Aa-RWA and the multi-period Aa-RWA algorithms.

8. Conclusions

This chapter proposed new attack-aware RWA algorithms for the multi-period planning of optical networks under demand uncertainty. These algorithms decide on the placement of wavelength selective switches at the input ports of network nodes and the period that the placement should be performed. The decisions are taken based on the distribution of the demands with the objective to minimize the impact of physical layer attacks over all periods. The algorithm that takes into account jointly all the time periods has a better performance than the algorithm that takes into account the periods in a sequential manner, resulting in a smaller number of required WSSs to be placed in the network so as to minimize the effect of a jamming attack.

Acknowledgements

This chapter is based upon work from COST Action CA15127 ("Resilient communication services protecting end user applications from disaster-based failures: RECODIS") supported by COST (European Cooperation in Science and Technology). This work has received funding from the People Program (Marie Curie Actions) of the European Union's Seventh Framework Program (FP7/2007–2013) under REA Grant Agreement n° 630853.

Author details

Konstantinos Manousakis*, Panayiotis Kolios and Georgios Ellinas

*Address all correspondence to: manousakis.konstantinos@ucy.ac.cy

Department of Electrical and Computer Engineering, KIOS Research Center for Intelligence Systems and Networks, University of Cyprus, Cyprus

References

[1] Ramaswami R, Sivarajan K. Routing and wavelength assignment in all-optical networks. IEEE/ACM Transactions on Networking. 1995;**3**(5):489–500. DOI: 10.1109/90.469957

[2] Ramaswami R, Sivarajan K, Sasaki G Optical Networks: A Practical Perspective. 3nd ed. Morgan Kaufmann; 2010. ISBN: 978-0-12-374092-2 Burlington, MA 01803, USA

[3] Kaman V, Helkey R, Bowers J. Multi-degree ROADM's with agile add-drop access. In: Proceedings of the IEEE Photonics in Switching; San Francisco, CA. 2007. DOI: 10.1109/PS.2007.4300729

[4] Mas C, Tomkos I, Tonguz OK. Failure location algorithm for transparent optical networks. IEEE Journal on Selected Areas in Communication. 2005;**23**(8):1508–1519. DOI: 10.1109/JSAC.2005.852182

[5] Furdek M, et al. An overview of security challenges in communication networks. In: Proceedings of the 8th IEEE International Workshop on Resilient Networks Design and Modeling (RNDM); Halmstad, Sweden. 2016. pp. 43–50. DOI: 10.1109/RNDM.2016.7608266

[6] Skorin-Kapov N, et al. A new approach to optical networks security: Attack-aware routing and wavelength assignment. IEEE/ACM Transactions on Networking. 2010;**18**(3): 750–760. DOI: 10.1109/TNET.2009.2031555

[7] Skorin-Kapov N, Furdek M, Pardo R, Mariño P. Wavelength assignment for reducing in-band crosstalk attack propagation in optical networks: ILP formulations and heuristic algorithms. European Journal of Operational Research. 2012;**222**(3):418–429. DOI: 10.1016/j.ejor.2012.05.022

[8] Manousakis K, Ellinas G. Attack-aware planning of transparent optical networks. Optical Switching and Networking. 2016;**19**(2):97–109. DOI: 10.1016/j.osn.2015.03.005

[9] Jirattigalachote A, Skorin-Kapov N, Furdek M, Chen J, Monti P, Wosinska L. Sparse power equalization placement for limiting jamming attack propagation in transparent optical networks. IEEE/OSA Journal of Optical Communications and Networking. 2011;**8**(4):249–258. DOI: 10.1016/j.osn.2011.06.008

[10] Manousakis K, Ellinas G. Equalizer placement and wavelength selective switch architecture for optical network security. In: Proceedings of the IEEE Symposium on Computers and Communication; Cyprus. 2015. DOI: 10.1109/ISCC.2015.7405631

[11] Verbrugge S, Colle D, Pickavet M, Demeester P. Common planning practices for network dimensioning under traffic uncertainty. In: Proceedings of the 4th IEEE International Workshop on the Design of Reliable Communication Networks (DRCN); October 2003; Banff, Alberta, Canada. pp. 317–324. DOI: 10.1109/DRCN.2003.1275372

[12] Leung D, Grover WD. Capacity planning of survivable mesh-based transport networks under demand uncertainty. Photonic Network Communications. 2005;**10**(2):123–140. DOI: 10.1007/s11107-005-2479-z

[13] Birge JR, Louveaux F. Introduction to Stochastic Programming. New York: Springer-Verlag; 2011. ISBN: 978-1-4614-0236-7

[14] Bertsimas D, Sim M. The price of robustness. Operations Research. 2004;**52**(1):35–53. DOI: 10.1287/opre.1030.0065

[15] Aparicio-Pardo R, Pavon-Marino P, Mukherjee B. Robust upgrade in optical networks under traffic uncertainty. In: Proceedings of the 16th IEEE International Conference on Optical Network Design and Modelling (ONDM); Colchester. 2012. pp. 1–6. DOI: 10.1109/ ONDM.2012.6210204

[16] Kronberger C, Schondienst T, Schupke DA. Impact and handling of demand uncertainty in multiperiod planned networks. In: Proceedings of the IEEE International Conference on Communications (ICC); Kyoto. 2011. pp. 1–6. DOI: 10.1109/icc.2011.5962890

[17] GÉANT2 Network. Available from: http://www.geant2.net/

Chapter 4

The Future Electrical Multiplexing Technique for High Speed Optical Fibre

Rahmat Talib and Mohammad Faiz Liew Abdullah

Additional information is available at the end of the chapter

http://dx.doi.org/10.5772/intechopen.68407

Abstract

Advancement in transmission technology based on fiber optic such as multiplexing technique is an attractive research area for future development of high capacity and high speed optical communication system. Typical electrical based multiplexing such as electrical time division multiplexing (ETDM) and duty cycle division multiplexing (DCDM) have difficulty to fulfil the requirements of modern fiber optic communication with practical solution. Multi slot amplitude coding (MSAC) is the latest multiplexing technique that has been proposed as an alternative to ETDM and DCDM. The results show that the spectral width is reduced by around 25%, not less than 55% improvement of chromatic dispersion (CD) tolerance, 0.6 dB better receiver sensitivity, and 1.5 dB better optical signal to noise ratio (OSNR) compared to DCDM for 30 Gbit/s transmission capacity. The spectral width for 3 × 10 Gbit/s, 4 × 10 Gbit/s and 5 × 10 Gbit/s MSAC is 60 GHz, which indicates improvement of spectral efficiency. This advantage is not possible to be achieved through ETDM technique. In addition, 10 GHz clock signal can be extracted from the MSAC signal which is important for recovery circuit at receiver since it is similar to symbol rate.

Keywords: optical fibre communication, multiplexing, optical modulation

1. Introduction

The current trend of internet use for social applications, such as facebook and twitter, is a strong indication of the substantial demand for information and communication technology. Therefore, behind the success story of these popular social applications is the capability of telecommunication infrastructure to handle huge amount of information transfer worldwide.

Optical communications have been widely used for this important task. Advancement in high capacity and high speed optical communication system has always become an important topic of discussion among the communications community. One of the important techniques in optical communication system is to realize high capacity data transportation through multiplexing technique. The technology is moving forward, where innovative alternative to the existing multiplexing method is indispensable to accomplish the future needs in optical communication implementation.

Generally, multiplexing is required to share the huge bandwidth of well-known fibre optic medium with many users, hence provides more cost efficient for practical implementation for high capacity data transformation. Typical implementation of multiplexing can be done in electrical or optical domain. Electrical-based multiplexing is very important technique due to the capability of electronic technology to switch the data with high speed, efficiency and reliability [1].

In this chapter, we report the investigation of system performance for recent electrical-based multiplexing technique known as multi-slot amplitude coding (MSAC) for high speed optical communication link. This work is very important in order to justify the benefit of MSAC as an innovative multiplexing compared to other electrical-based multiplexing such as electrical time division multiplexing (ETDM) and duty cycle division multiplexing (DCDM).

2. Electrical time division multiplexing (ETDM)

ETDM is a technique that adopting electronic circuit to execute multiplexing process in electrical domain. In EDTM concept, the bit from multiple input tributaries is arranged as a single output tributary by allocating all the bits with smaller time slot as compared to the time slot of input channel. As a result, the bit rate or speed of output tributary is higher than the input bit rate so that all the bits at the input can be transferred correctly to the output. The output tributary bit rate, R_T of ETDM is

$$R_T = NR \tag{1}$$

Where N is the number of input tributaries and R is the input tributary bit rate. The bit duration T_b of the input tributary is given by

$$T_b = \frac{1}{R} \tag{2}$$

The basic concept of EDTM implementation based on three input tributaries depicted in **Figure 1**. In this figure, ETDM operates with three input tributaries so the output bit rate of EDTM is $3R$ bits/s and the output bit duration becomes $T_b/3$. The output of ETDM can be used to modulate the light source for data transmission. Due to simple and efficient operation of ETDM technology, it is extensively adopted for commercial application for synchronous optical network (SONET) , synchronous digital hierarchy (SDH) and optical transport network (OTN). Currently, high speed electronic-based ETDM has been reported to support more than 40 Gbit/s serial data using advance material and state-of-the-art technology [2].

http://dx.doi.org/10.5772/intechopen.68407

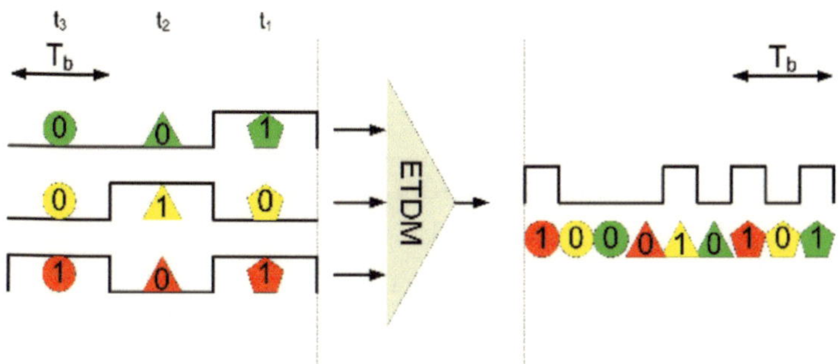

Figure 1. ETDM concept.

3. Duty cycle division multiplexing (DCDM)

Multiplexing process also can be realized using DCDM concept, and it has been proposed as an alternative technique to ETDM [3]. In order to implement DCDM concept, basic components can be used which are return-to-zero (RZ) convertor and electrical adder. Since, RZ convertor has the ability to adjust the duty cycle (DC) parameter, hence various DCs of binary signal can be obtained. In DCDM, each input tributary is applied to RZ convertor with the different predefined DC parameter. The outputs for all RZ convertors are combined by an electrical adder to generate the DCDM signal. This technique requires a synchronize bit and identical amplitude of non-return-to-zero (NRZ) format for all input tributary. **Figure 2** shows an example of DCDM signal generation for three tributaries. **Figure 2**a–c shows the 8 bits sequence (all possible combination) of input tributary 1, 2 and 3, respectively. **Figure 2d–f** shows the RZ format with 25% DC of tributary 1, 50% DC of tributary 2 and 75% DC of tributary 2, respectively, after NRZ to RZ conversion. Combination of signals in **Figure 2d–f** using electrical adder generates DCDM signal as shown in **Figure 2g**. Based on this multiplexing,

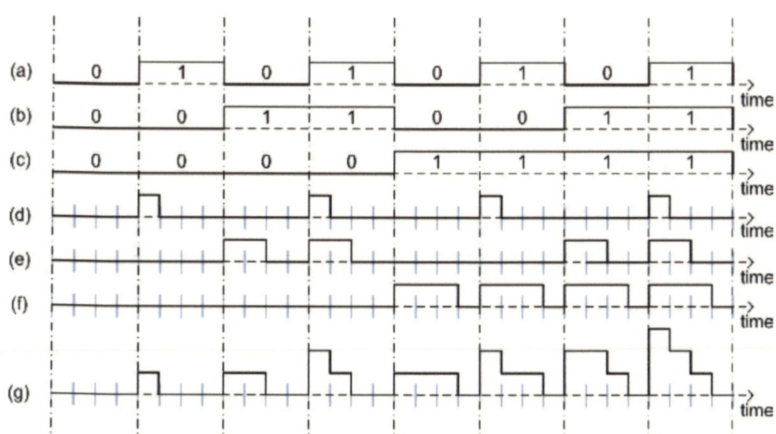

Figure 2. An example of DCDM signal.

a unique signal waveform or symbol can be obtained. There is a transition at the beginning of the DCDM symbol because RZ conversion turns the signal from high to low for bit 1 s.

4. Multi-slot amplitude coding (MSAC)

MSAC technique is a latest multiplexing concept to enhance the utilization of signal level for generating waveform or symbol compared to DCDM and ETDM [4]. In this technique, the multiplexer converts all possible combination bits of each tributary as MSAC symbol based on predefine translation rule. MSAC symbols can be obtained based on two parameters, which are the number of slots, *S* and number of signal levels, *M* as depicted in **Figure 3**. Assuming an equal duration of slot, the slot duration of MSAC symbol, *Td* is given by

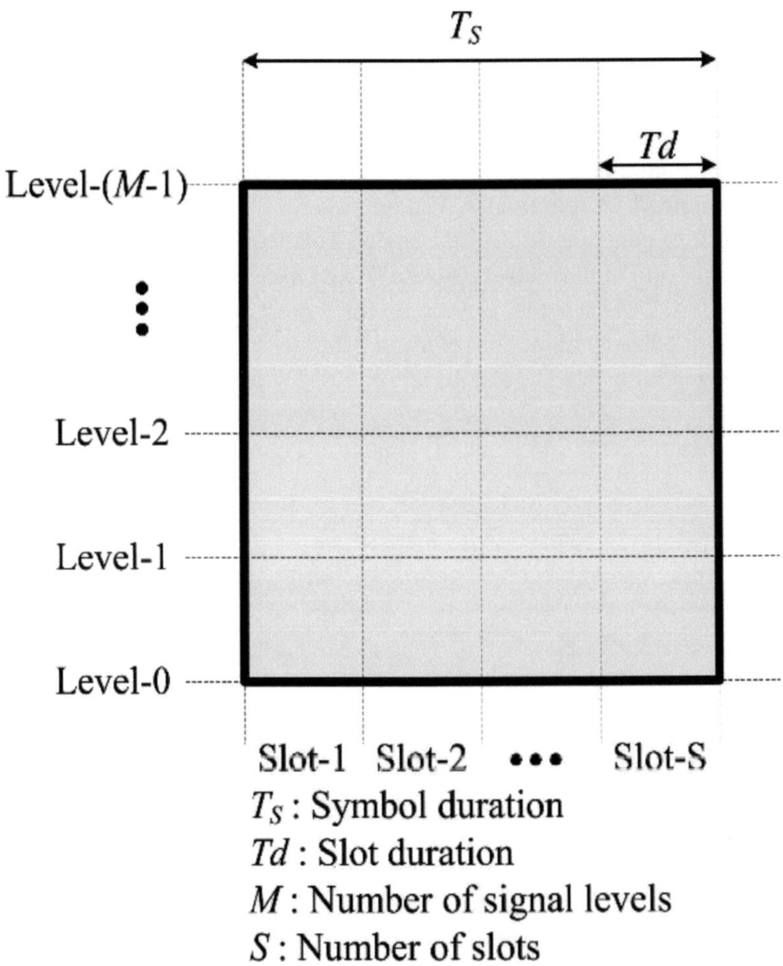

Figure 3. General format of symbol for MSAC.

$$Td = \frac{T_S}{S} \tag{3}$$

where T_S is the symbol duration. T_S is similar to the bit duration of the input tributary, T_b. Therefore,

$$T_S = T_b = \frac{1}{R} \tag{4}$$

where R is the input tributary bit rate.

For equal signal level spacing, the maximum amplitude of MSAC, A is

$$A = (M-1)\Delta \tag{5}$$

where Δ is amplitude spacing. **Figure 4** displays an example of MSAC symbol for three users. There are nine symbols ($x_1(t)$ to $x_9(t)$) for $S = 3$ and $M = 3$. Note that MSAC symbol allocates the first slot in symbol format as zero level.

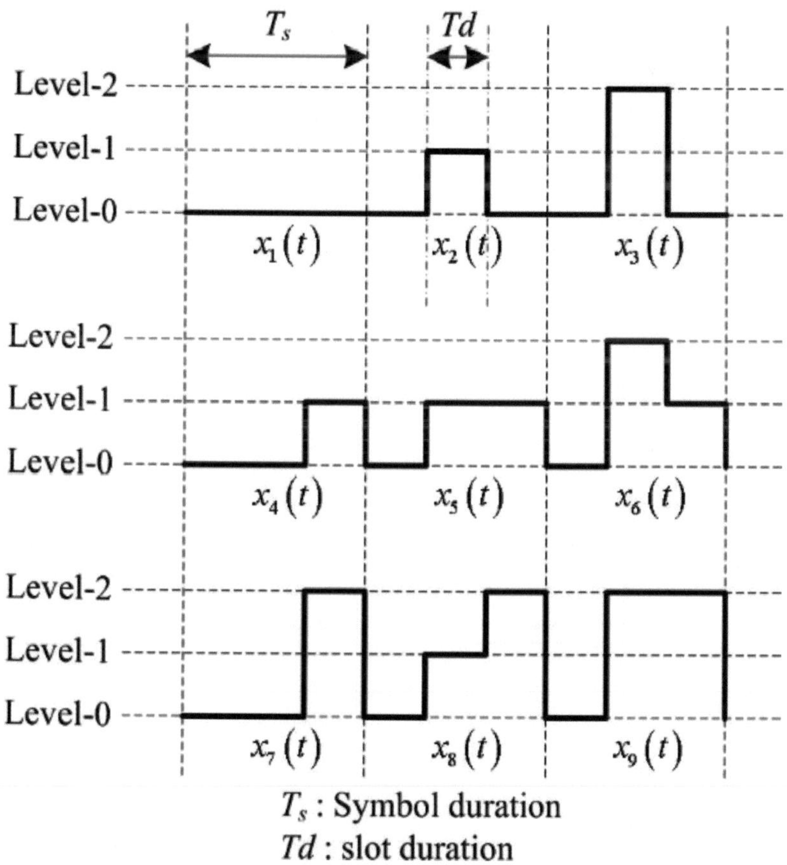

Figure 4. Example of symbol waveform for $M = 3$ and $S = 3$.

5. Optical communication system simulation using MSAC

Figure 5 shows the simulation setup for *N* tributaries MSAC in optical communication system. This setup consists of transmitter section, transmission section and receiver section. The component for the transmitter section is *N* pulse pattern generators, MSAC multiplexer, external modulator and continuous wave laser. The number of pulse pattern generator will depend on number of tributary. Each tributary consists of a pulse pattern generator for generating pseudo random binary signal (PRBS). Note that *Tr*1, *Tr*2 and *TrN* represent tributary 1, tributary 2 and tributary *N*, respectively. Each pulse pattern generator has a common clock signal in order to obtain synchronize binary data stream as input signal to the MSAC multiplexer. MSAC multiplexer model implements a conversion process based on the rule. The signal from MSAC multiplexer then modulates the light from a continuous wave laser (CW LD) at 1550 nm, using an external modulator. The optical power of CW LD is fixed at 0 dBm. The external modulator is based on an amplitude modulator (AM) model. Transmission section consists of an optical attenuator and optical fibre. The modulated optical signal is fed into an optical attenuator. In optical attenuator, the signal input electrical field for both polarizations is attenuated. This optical attenuator is used to control the amount of launch optical power from the AM. For this simulation, optical fibre is based on a single mode fibre model, and it is placed after the optical attenuator. The propagation of modulated optical signal in single mode fibre model is based on the Schrödinger equation [5]. The receiver section consists of an optical amplifier, PIN photodiode, electrical low-pass filter and clock and data recovery. An optical amplifier that acts as a pre-amplifier is placed before PIN photodiode in order to boost the signal. Note that the optical amplifier also introduces ASE noise. The received optical signal is then converted to an electrical signal using a PIN photodiode. In this simulation, the signal is corrupted with typical noises in optical system such as shot noise and thermal noise. The PIN photodiode output signal is filtered with a Gaussian low-pass filter (LPF). In order to

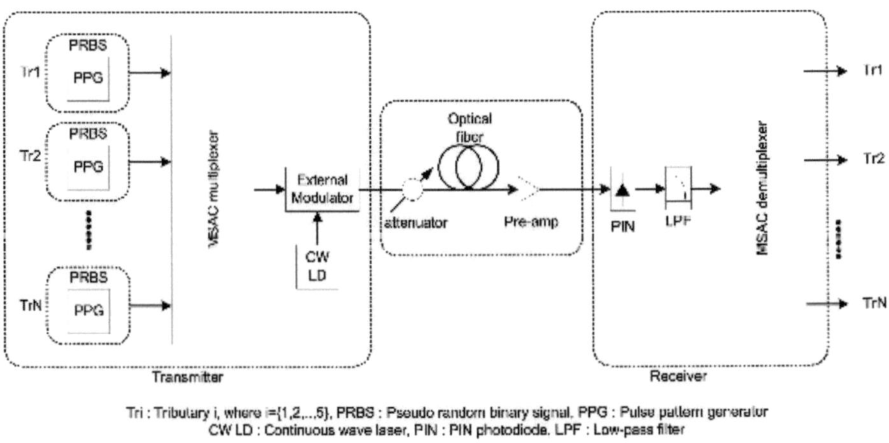

Figure 5. MSAC system setup.

optimize the system performance, the cut-off frequency of filter is set at 0.75 *BW*, where BW is the first null bandwidth of baseband MSAC signal. The filtered electrical signal is fed into the clock and data recovery module to regenerate each tributary data stream. In clock and data recovery module, data recovery process is implemented using MATLAB programming based on the recovery rules of MSAC demultiplexer.

6. Results and discussion

6.1. Optical signal spectrum, spectral width and clock recovery frequency

Figure 6a–c illustrates the optical spectrum of MSAC 3 × 10, 4 × 10 and 5 × 10 Gbit/s, respectively. This spectrum has been observed at the transmitter side (after external modulator) using an optical spectrum analyser.

From this figure, simulated null-to-null spectral width of MSAC 3 × 10, 4 × 10 and 5 × 10 Gbit/s is around 60 GHz. This result shows that spectral width of MSAC remains the same even though *N* is increasing from 3 to 5. This is because they have similar number of slots in a symbol. In this case, number of slots is three, thus slot period is 1/(3 × 10 Gbit/s). This slot determines the width of main-lobe of optical spectrum. Therefore, increasing aggregate capacity from 30 to 50 Gbit/s using this technique will not affect the required optical bandwidth. Moreover, the modulation speed remains unchanged because of the fixed slot interval.

Another important characteristic of the optical signal spectrum is to visualize the clock information in which the frequency is similar to the symbol rate. Note that impulse or spike in optical signal spectrum means high clock frequency in the signal. As shown in **Figure 6a**, there are seven impulses in the null-to-null spectral width, where f_0 is impulse at optical carrier (1550 nm). Note that the spectral is symmetrical at f_0. Besides that the impulses appear at 10 GHz (f_1) and 20 GHz (f_2) away from f_0 on the right side of the main-lobe. The impulse at 10 GHz (f_1) can be used to recover the clock frequency by a clock recovery circuit in receiver. Note that these impulse frequencies are similar for 4 × 10 (**Figure 6b**) and 5 × 10 Gbit/s (**Figure 6c**).

6.2. Spectral efficiency

In optical communication system based on wavelength division multiplexing (WDM) technology, more than one WDM channel can be propagated in optical fibre. Each WDM channel is separated with channel spacing such as 25, 50, 100 or 200 GHz. Note that spectral efficiency in WDM optical system can be calculated based on formula in [6]. Although the simulation is based on 1550 nm or single channel wavelength, MSAC technique can be implemented over WDM system as well. In order to operate with WDM system, WDM channel spacing must be greater than the spectral width of MSAC signal in order to avoid serious interference between adjacent channels. Since the spectral width of MSAC signal is around 60 GHz, therefore 100 or 200 GHz WDM channel spacing can be used. Assuming that WDM channel spacing is 100 GHz, these are corresponding to 0.3, 0.4 and 0.5 bit/s/Hz of spectral efficiency for 3 × 10, 4 × 10 and 5 × 10 Gbit/s, respectively.

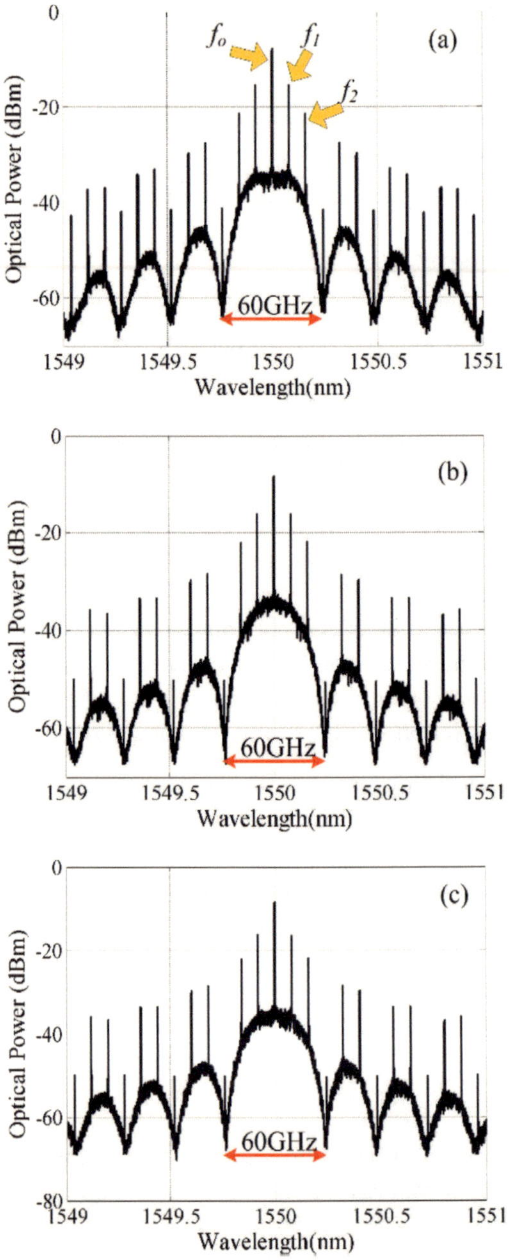

Figure 6. Optical spectrum of MSAC (a) 3 × 10, (b) 4 × 10 and (c) 5 × 10 Gbit/s.

6.3. Received optical power

Bit error rate (BER) estimation in this simulation is based on probability of error method. BER versus received optical power of MSAC at 3 × 10, 4 × 10 s and 5 × 10 Gbit/s has been plotted as in **Figure 7**. Note that this simulation is based on back-to-back setup (optical fibre is not included). As a result, the performance of MSAC system is limited by noises from optical amplifier and PIN

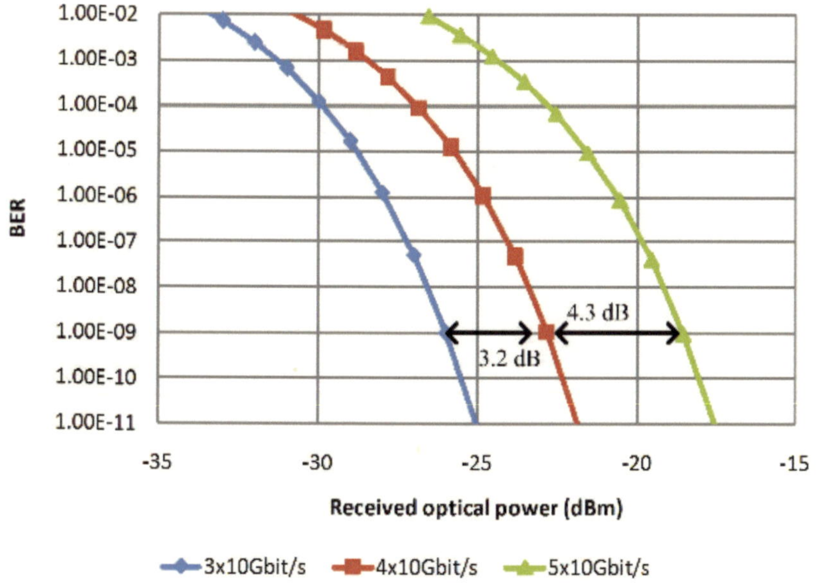

Figure 7. BER versus received optical power of 3 × 10, 4 × 10 and 5 × 10 Gbit/s MSAC.

photodiode. Inter symbol interference is minimized by setting the cut-off frequency of low-pass filter at 0.75 *BW*. From this figure, the required received optical power or receiver sensitivity of MSAC 3 × 10, 4 × 10 and 5 × 10 Gbit/s at BER of 10^{-9} is −26.0, −22.8 and −18.5 dBm, respectively.

This simulation results show that 3 × 10 Gbit/s has the lowest received optical power, whereas 5 × 10 Gbit/s has the highest received optical power. Power penalty of around 3.2 and 7.5 dB is observed for 4 × 10 and 5 × 10 Gbit/s at BER of 10^{-9}, respectively, compared to 3 × 10 Gbit/s. The reason for this penalty is due to the number of signal levels increased. MSAC 3 × 10 Gbit/s uses three signal levels; therefore, it has the lowest received optical power, whereas MSAC 5 × 10 Gbit/s uses six signal levels, thus it has the highest received optical power.

6.4. Optical signal to noise ratio (OSNR)

Figure 8 shows the BER versus OSNR of MSAC at 3 × 10, 4 × 10 and 5 × 10 Gbit/s. From this figure, the required OSNR of MSAC 3 × 10, 4 × 10 and 5 × 10 Gbit/s at BER of 10^{-9} is 25.5, 28.8 and 33.2 dB, respectively. Power penalty due to adding tributary is around 3.3 and 7.7 dB for 4 × 10 and 5 × 10 Gbit/s, respectively. As expected, MSAC system requires more OSNR in order to maintain the BER performance when tributaries increase because the number of levels increases. Note that the variation between power penalty for OSNR and received optical power is small. This indicates that received optical power and OSNR are important parameters for reliable communication.

6.5. Chromatic dispersion tolerance

Chromatic dispersion (CD) is one of the optical fibre impairment in high speed optical communication system. The CD effect in silica fibre will degrade the BER performance due to

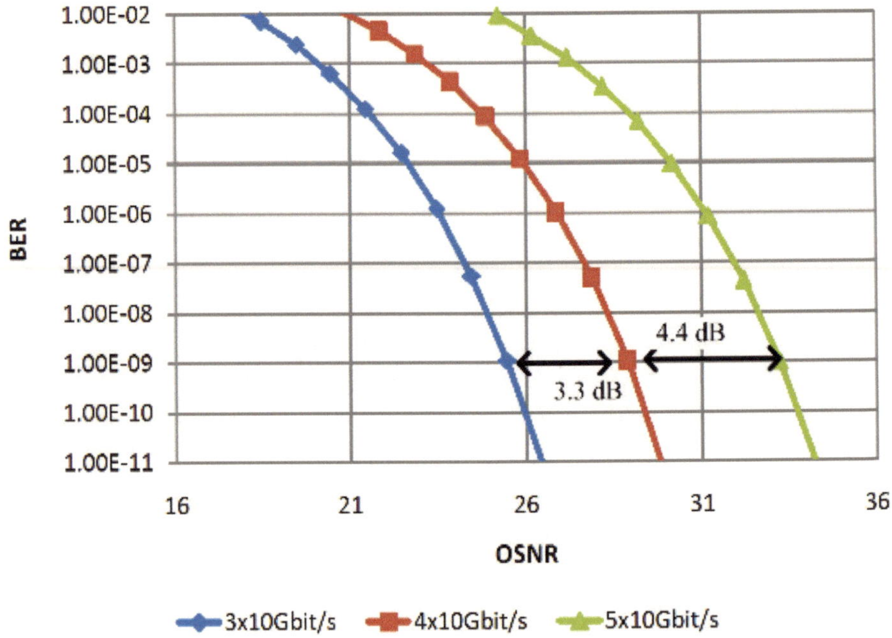

Figure 8. BER versus OSNR of 3 × 10, 4 × 10 and 5 × 10 Gbit/s MSAC.

evolution of signal shape. In order to achieve high performance quality, the accumulated CD must not exceed the allowable CD tolerance. CD tolerance is determined by estimating the amount of CD (positive and negative) at the target BER of 10^{-9}.

In order to determine the chromatic dispersion tolerance, a standard single mode fibre model is included in this simulation setup. The dispersion parameter of optical fibre model is varied from negative dispersion to positive dispersion. Other fibre impairments effect such as attenuation, self-phase modulation (SPM) and non-linear fibre effect are ignored so that the system performance is determined by fibre dispersion only.

Figure 9 depicts the BER versus dispersion of MSAC 3 × 10, 4 × 10 and 5 × 10 Gbit/s. Chromatic dispersion tolerance of MSAC 3 × 10, 4 × 10 and 5× 10 Gbit/s at BER of 10^{-9} is ±164.5, ±149.5 and ±69 ps/nm, respectively. This comparison shows that CD tolerance decreases when the number of signal levels increased. Note that optical power at the highest level for higher number of signal levels of MSAC is high compared to MSAC with small number of signal levels for BER of 10^{-9}. In terms of dispersion mechanism, pulses at higher signal level experience highest energy loss compared to signal at lower level, thus reduces the eye opening and induces a power penalty. Therefore, at similar average power, MSAC with higher number of signal levels losses its CD robustness capability.

6.6. The effect of signal level spacing

In previous work, the separation of signal level or level spacing in MSAC symbol format is equally spaced. The level spacing between level i and level i-1 is Δ. This means that level spacing

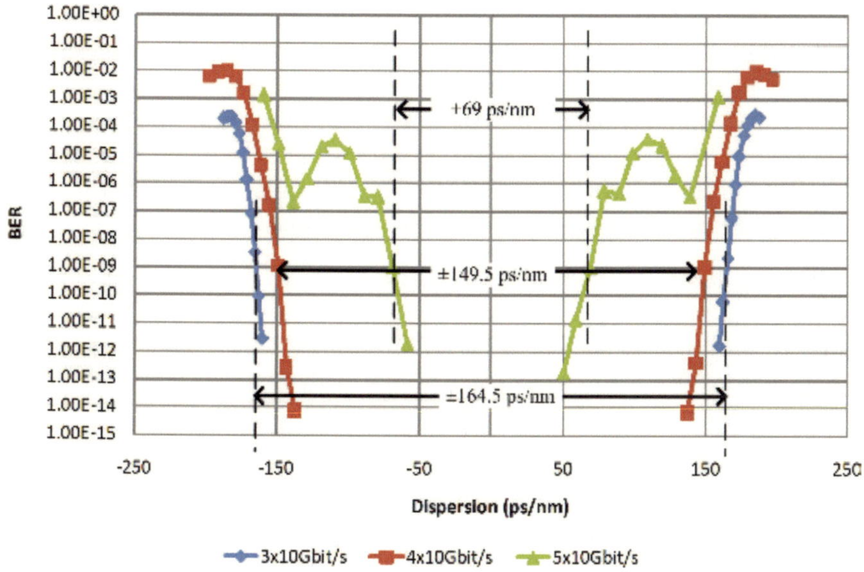

Figure 9. BER versus dispersion of 3 × 10, 4 × 10 and 5 × 10 Gbit/s MSAC.

ratio is 0, 0.5 and 1 for level 0, level 1 and level 2, respectively. This setting is, therefore, known as equal level spacing (ELS). **Figure 10** shows the BER of MSAC against normalized signal spacing of level 1 between 0.1 and 0.6. The optimum level spacing (OLS) is observed at normalized signal spacing of level 1 of 0.31. As expected, higher level spacing for upper level compared to lower level spacing in order to achieve equal probability of error region for each signal level.

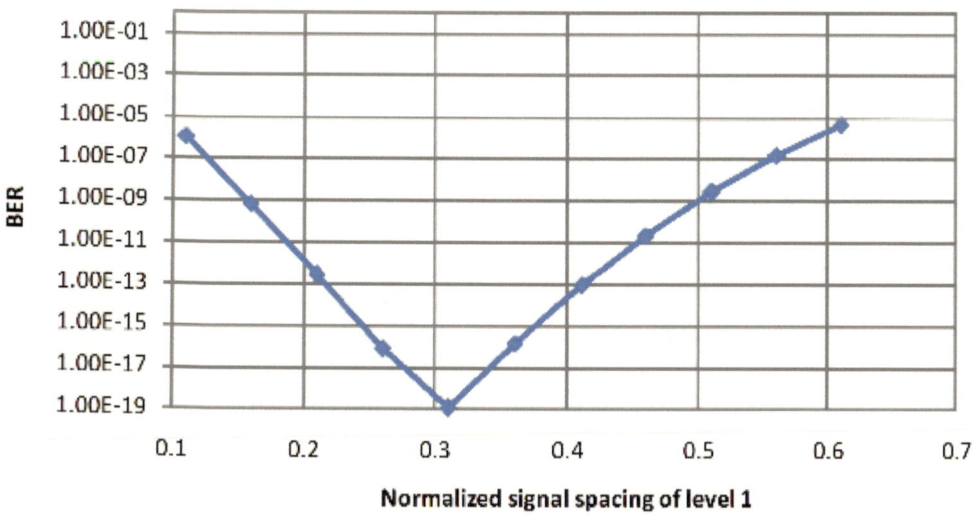

Figure 10. BER of MSAC against normalized signal spacing of level 1 between 0.1 and 0.6.

Figures 11 and **12** show the comparison between equal level spacing (ELS) and optimize level spacing (OLS) of back to back (b2b) BER performance of 3 × 10 Gbit/s MSAC system based on received optical power and OSNR, respectively. From the figure, the received optical powers of −26 and −29.5 dBm are obtained at BER of 10^{-9} for ELS and OLS, respectively. This is an improvement of 3.5 dB when OLS MSAC is adopted at receiver. In term of OSNR, ELS requires 25.5 dB, in contrast OLS requires 21.8 dB. This clearly shows that there is OSNR improvement when OLS method is applied, with improvement around 3.7 dB.

The performance in terms of dispersion tolerance for ELS and OLS method is plotted as shown in **Figure 13**. Based on that figure, ELS and OLS methods are capable of tolerating positive and negative chromatic dispersion of 329 and 311 ps/nm at BER of 10^{-9}. The reduction of 18 ps/nm is observed for OLS method compared to ELS method.

The effect of the signal level is also investigated for the MSAC system with setup of 4 × 10 Gbit/s. In general, the approach to determine the optimum level is similar for previous setup (3 × 10 Gbit/s). Since 4 × 10 Gbit/s MSAC has four signal levels, both signal levels 1 and 2 are adjusted while signal levels 0 and 3 are fixed. It is found that the optimum level spacing is achieved when the normalized signal level 1 and signal level 2 are 0.183 and 0.51, respectively.

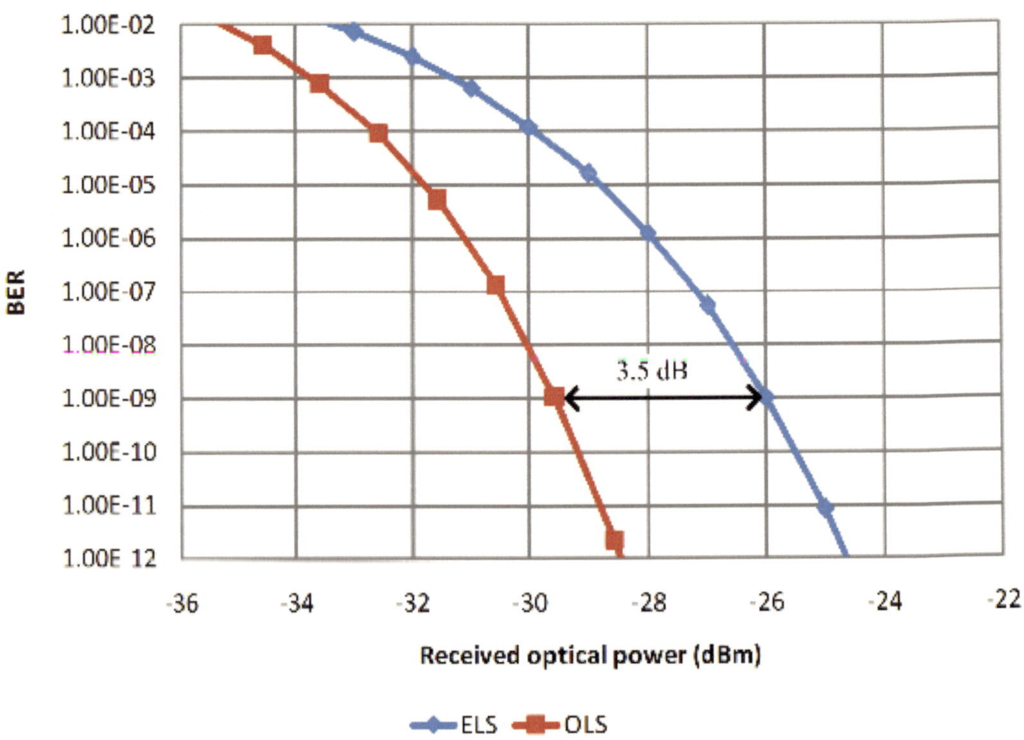

Figure 11. Received optical power comparison between ELS and OLS of b2b 3 × 10 _Gbit/s MSAC system.

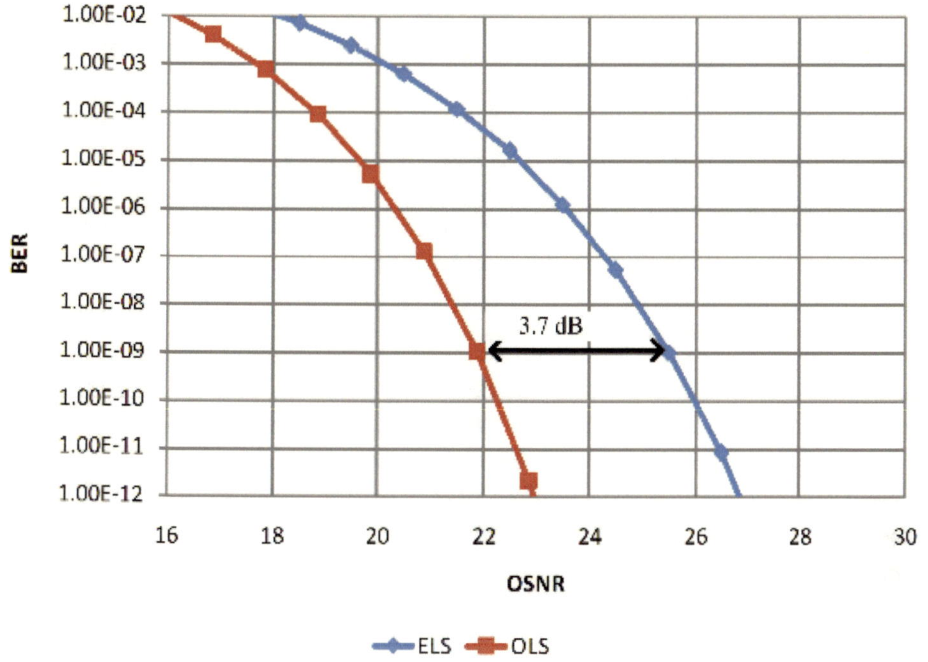

Figure 12. OSNR comparison between ELS and OLS of b2b 3 × 10 Gbit/s MSAC system.

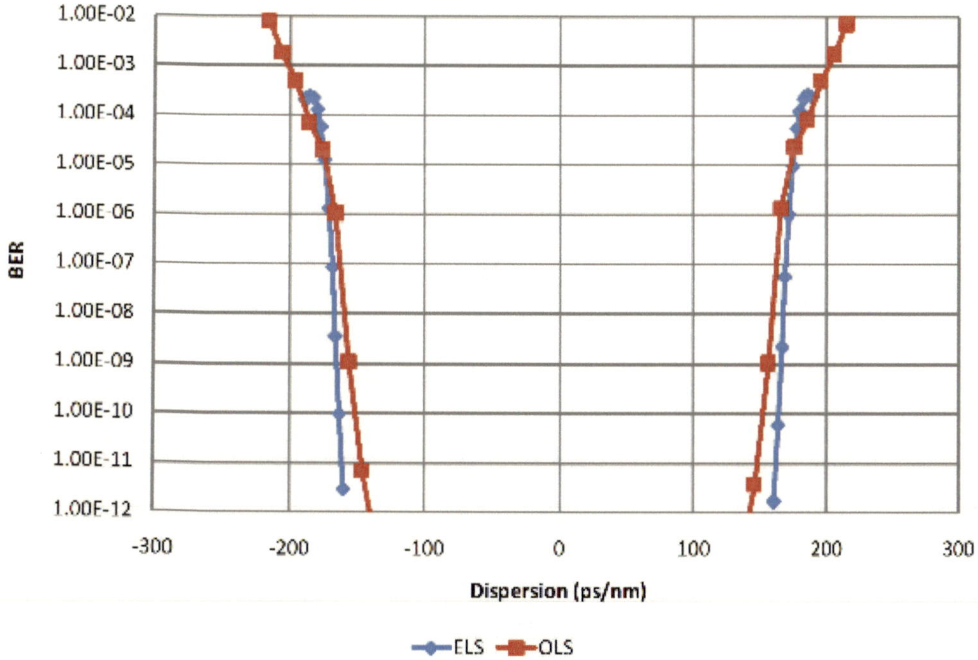

Figure 13. Dispersion tolerance comparison between ELS and OLS of b2b 3 × 10 Gbit/s MSAC system.

Figure 14 shows the comparison between ELS and OLS in terms of the received optical power for b2b 4 × 10 Gbit/s MSAC system. The received optical powers of ELS and OLS MSAC are −22.8 and −27.1 dBm, respectively, at BER of 10^{-9}. The improvement of receiver sensitivity around 4.3 dB is observed. The comparison between ELS and OLS in terms of OSNR is shown in **Figure 15**. Based on this graph, OSNR is 28.8 dB for ELS, whereas 24.5 dB for OLS. This result shows that the OSNR improvement has been achieved with similar amount for receiver sensitivity improvement. **Figure 16** shows the comparison between ELS and OLS in terms of CD tolerance. The CD tolerance is ±149.5 and ±73.1 ps/nm for ELS and OLS, respectively.

6.7. Performance comparison

The performance comparison between other electrical-based multiplexing is made according to transmission capacities, which are 30 and 40 Gbit/s. For this comparison, other versions of DCDM such as DCDM-amplitude distribution controller (DCDM-ADC), DCDM with dual drive Mach Zehnder modulator (DD-MZM), new multiplexed pattern DCDM (NMP-DCDM) and absolute polar DCDM (AP-DCDM) are included. **Table 1** shows the performance comparison between MSAC with various types of DCDM. It is very clear that MSAC with OLS has better performance compared to DCDM-ADC, DCDM with DD-MZM and NMP-DCDM. Note that level spacing for DCDM-ADC was optimized by installing a controller, whereas DCDM with DD-MZM was optimized using a dual drive Mach Zehnder modulator. MSAC with OLS offers better performance in terms of CD tolerance compared to AP-DCDM; however, the receiver sensitivity and spectral width are almost similar. Another advantage of

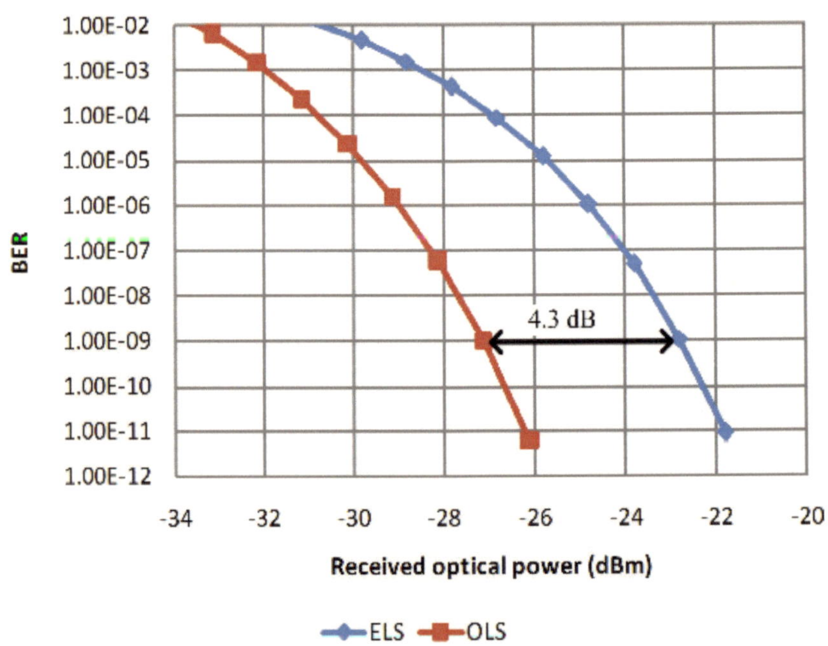

Figure 14. Received optical power comparison between ELS and OLS of b2b 4 × 10 Gbit/s MSAC system.

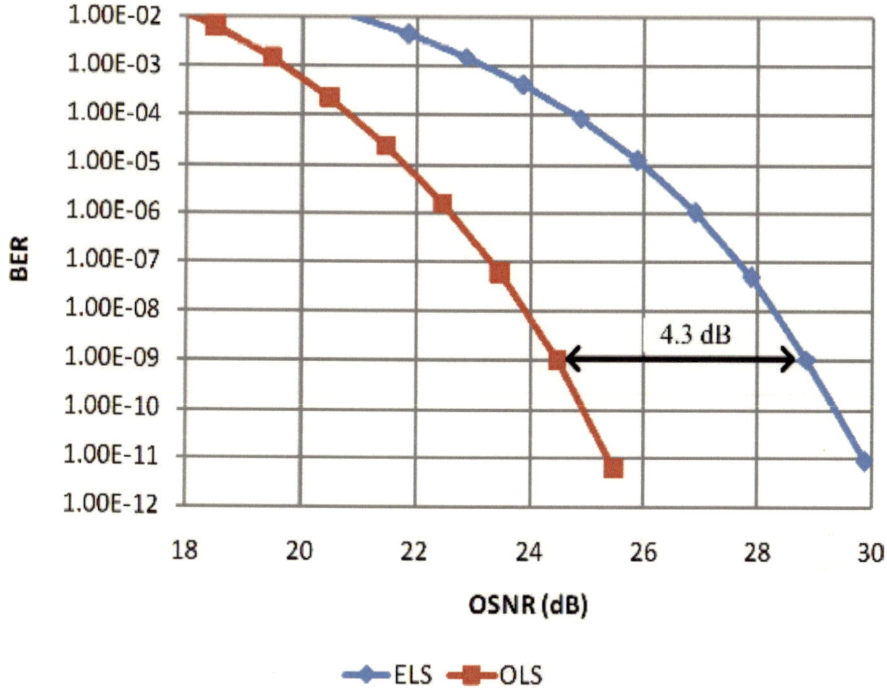

Figure 15. OSNR comparison between ELS and OLS of b2b 4 × 10 Gbit/s MSAC system.

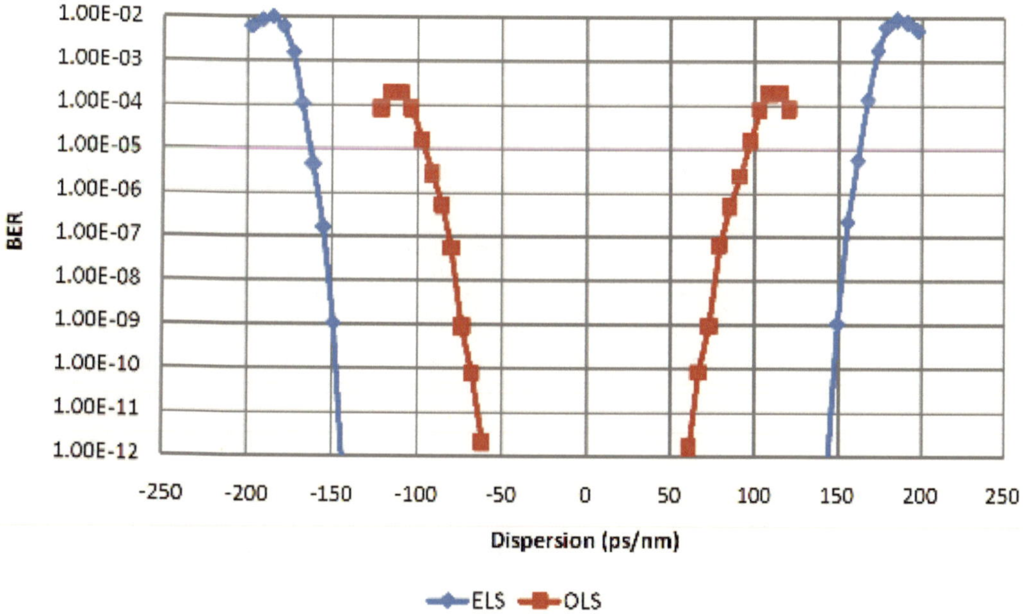

Figure 16. Dispersion tolerance comparison between ELS and OLS of b2b 4 × 10 Gbit/s MSAC system.

Technique	Guard slot	RS BER @ 10^{-9}	OSNR BER @ 10^{-9}	Spectral width (GHz)	CD tolerance (ps/nm)	
DCDM-ADC	Yes	−28.9	23.35	80	±93.5	[7]
DCDM with DD-MZM	Yes	−28.85	23.3	80	−59 to +100	[8]
AP-DCDM	No	−29	22.82	60	±109	[9]
NMP-DCDM	Yes	−28.4	23.5	–	–	[10]
MSAC with OLS	Yes	−29.5	21.8	60	±155	

Abbreviations: RS, receiver sensitivity; DCDM-ADC, DCDM amplitude distribution controller; DD-MZM, dual drive Mach Zehnder modulator; AP-DCDM, absolute polar DCDM; NMP-DCDM, new multiplexed pattern DCDM; NA, not available.

Table 1. Performance comparison between various DCDM version and MSAC for 3 × 10 Gbit/s system setup.

MSAC compared to AP-DCDM is high frequency component at symbol rate. This feature is obtained because MSAC symbols are equipped with guard slot, like DCDM or RZ, therefore, increases the transition density in the modulated signal. As a result, a complex circuit is not required for recovering clock frequency in receiver. This comparison indicates that MSAC provides significant advantages against DCDM and AP-DCDM for multiplexing high speed data stream in electrical domain.

Table 2 shows the performance comparison between DCDM-ADC, AP-DCDM and MSAC for optical transmission system with 40 Gbit/s setup. In general, the performance for all parameters for MSAC with OLS is better than DCDM-ADC. This finding is consistence with the result in **Table 1**. The performance of MSAC is also better than AP-DCDM version with guard slot. The receiver sensitivity for MSAC is inferior when comparing with the optimized level spacing AP-DCDM using DD-MZM, but it has better performance in terms of the spectral width and CD tolerance. Moreover, upgrading MSAC from 3 (3 × 10 Gbit/s) to 4 tributaries (4 × 10 Gbit/s) is achieved without extra spectral width. For ETDM, spectral width of 160 GHz

Technique	Guard slot	RS BER @ 10^{-9}	OSNR BER @ 10^{-9}	Spectral width (GHz)	CD tolerance (ps/nm)	
DCDM-ADC	Yes	−26	26.38	100	±58	[7]
AP-DCDM	Yes	−26.8	25.8	100	±62	[11]
AP-DCDM with DD-MZM	No	−31	NA	80	±68	[12]
MSAC with OLS	Yes	−27.1	24.5	60	±73.1	

NA, not available.

Table 2. Performance comparison between DCDM, AP-DCDM and MSAC for 4 × 10 Gbit/s setup.

is required for RZ format and 80 GHz for NRZ format for 40 Gbit/s system [13]. This means that ETDM is not bandwidth efficient compared to MSAC.

7. Conclusion

The evaluation of MSAC technique in fibre optic-based optical communication system has been carried out through an intensive numerical simulation. From the analysis, it is found that MSAC system performance is certainly better than DCDM-ADC, DCDM with DD-MZM and NMP-DCDM in terms of spectral width, spectral efficiency, receiver sensitivity, OSNR and CD tolerance. Besides that MSAC is also capable of achieving higher bandwidth or spectral efficiency compared to ETDM and AP-DCDM. Note that MSAC system is implemented using simple intensity modulation and direct detection scheme. As a result, the complexity of transmitter and receiver for this system is less than that of coherent scheme or optical time division multiplexing (OTDM) system. This issue is crucial for future high speed optical communication system in metropolitan region. This work has successfully demonstrated the possibility of implementing MSAC as advance multiplexing in the cost efficient high speed optical communication system for metropolitan application.

Acknowledgements

This work has been funded by the Ministry of Higher Education, Government of Malaysia under project FRGS/1/2015/TK04/UTHM/03/5.

Author details

Rahmat Talib and Mohammad Faiz Liew Abdullah*

*Address all correspondence to: faiz@uthm.edu.my

University Tun Hussein Onn Malaysia, Malaysia

References

[1] Lach E, Schuh K. Recent advances in ultrahigh bit rate ETDM transmission systems. Journal of Lightwave Technology. 2006;**23**(12):4455–4467. DOI: 10.1109/JLT.2006.886404

[2] Winzer PJ, Raybon G, Doerr CR, Duelk M, Dorrer C. 107-Gb/s optical signal generation using electronic time-division multiplexing. Journal of Lightwave Technology. 2006;**24**(8):3107–3113. DOI: 10.1109/JLT.2006.878016

[3] Mahdiraji GA, Abdullah MK, Mohammadi AM, Abas AF, Mokhtar M, Zahedi E. Duty-cycle division multiplexing (DCDM). Optics & Laser Technology. 2010;**42**(2):289–295. DOI: 10.1016/j.optlastec.2009.07.007

[4] Talib R, Abdullah MFL, Abdullah MK. Multi slot amplitude coding performance improvement with level spacing optimization. In: Conference Proceeding of the 4th International Conference on Photonics (ICP 2013); 28–30 October 2013; Melaka, Malaysia. New York: IEEE; 2013. pp. 206–208. DOI: 10.1109/ICP.2013.6687115

[5] Agrawal GP. Fiber-Optic Communication Systems. 4th ed. New York: John Wiley & Sons; 2010. DOI: 10.1002/9780470918524

[6] Winzer PJ, Gnauck AH, Doerr CR, Magarini M, Buhl LL. Spectrally efficient Long-Haul optical networking using 112-Gb/s polarization-multiplexed 16-QAM. Journal of Lightwave Technology. 2010;**28**(4): 547–556. DOI: 10.1109/JLT.2009.2031922

[7] Mahdiraji GA. Performance analysis of Duty-Cycle division multiplexing for optical fiber communication systems [thesis]. Serdang: Universiti Putra Malaysia; 2009

[8] Amouzad Mahdiraji G, Abas AF. Improving the performance of electrical duty-cycle division multiplexing with optimum signal level spacing. Optics Communications. 2012;**285**(7):1819–1824. DOI: 10.1016/j.optcom.2011.12.022

[9] Malekmohammadi A, Abdullah MK, Mahdiraji GA, Abas AF, Mokhtar M, Rasid MFA, Basir SM. Analysis of return-to-zero-on-off-keying over absolute polar duty cycle division multiplexing in dispersive transmission medium. Optoelectronics. IET. 2009;**3**(4):197–206. DOI: 10.1049/iet-opt.2008.0050

[10] Dahawü TH, Aljunid SA, Rashidi CBM, Rahman AK. Performance analysis of new multiplexed pattern for duty-cycle division multiplexing over optical communication system. In: 3rd International Conference on Electronic Design (ICED); 11–12 August 2015; Phuket. New York: IEEE; 2016. pp. 170–173. DOI: 10.1109/ICED.2016.7804630

[11] Malekmohammadi A, Mahdiraji GA, Abas AF, Abdullah MK, Mokhtar M, Rasid MFA. Realization of high capacity transmission in fiber optic communication systems using Absolute Polar Duty Cycle Division Multiplexing (AP-DCDM) technique. Optical Fiber Technology. 2009;**15**(4):337–343. DOI: 10.1016/j.yofte.2009.03.001

[12] Malekmohammadi A, Al-Mansoori MH, Mahdiraji GA, Abas AF, Abdullah MK. Performance enhancement of absolute polar duty cycle division multiplexing with dual-drive mach-zehnder-modulator in 40 Gbit/s optical fiber communication systems. Optics Communications. 2010;**283**(16):3145–3148. DOI: 10.1016/j.optcom.2010.04.026

[13] Miyamoto Y, Kataoka T, Tomizawa M, Yamane Y. 40-Gbit/s channel Technologies for optical Transport Network. Optical Fiber Technology. 2001;**7**(4):289–311. DOI: 10.1006/ofte.2001.0359

Chapter 5

OFDM Systems for Optical Communication with Intensity Modulation and Direct Detection

Jian Dang, Liang Wu and Zaichen Zhang

Additional information is available at the end of the chapter

http://dx.doi.org/10.5772/intechopen.68199

Abstract

Intensity modulation and direct detection (IM/DD) is a cost-effective optical communication strategy which finds wide applications in fiber communication, free-space optical communication, and indoor visible light communication. In IM/DD, orthogonal frequency division multiplexing (OFDM), originally employed in radio frequency communication, is considered as a strong candidate solution to combat with channel distortions. In this research, we investigate various potential OFDM forms that are suitable for IM/DD channel. We will elaborate the design principles of different OFDM transmitters and investigate different types of receivers including the proposed iterative receiver. In addition, we will analyze the spectral efficiency and decoding complexities of different OFDM systems to give a whole picture of their performance. Finally, simulation results are given to assess the detection performance of different receivers.

Keywords: OFDM, optical communication, IM/DD, modulation, detection

1. Introduction

Optical communication is an important part of modern communication techniques due to the excessive bandwidth of the light spectrum. Theoretically, optical communication has much higher system throughput than its radio frequency (RF) communication counterpart. Therefore, it finds many applications and facilitates our lives. Some typical optical communication scenarios include optical fiber communication, free-space optical communication, and visible light communication. In those communication scenarios, intensity modulation and direct detection (IM/DD) is a cost-effective communication scheme compared to coherent ones. In IM/DD, the intensity, or power, of the light beam from a laser or a light-emitting diode (LED) is modulated by the information bits and no phase information is needed. Due to this nature, no local oscillator is required for IM/DD communication, which greatly eases the cost of the hardware.

In IM/DD channel, there are still some non-ideal factors that may deteriorate the quality of communication. One key factor is the multipath effect. This effect is caused by several mechanisms. First, in wireless communications, the light could be reflected at multiple locations and by many times by the surroundings before arriving at the receiver side. Second, the modulation bandwidth of LED is limited, typically below 100 MHz. When the bandwidth of signal exceeds the modulation bandwidth of LED, multipath effect occurs. Third, in fiber communication, light components of different wavelength propagate through different paths, which also cause multipath effect. Therefore, effective means of mitigating the multipath effect are necessary in IM/DD optical communications.

In RF communication, orthogonal frequency division multiplexing (OFDM) is a powerful multi-carrier modulation scheme to combat the multipath effect. Compared to the single-carrier modulation schemes, OFDM avoids the usage of a complicated high-order time-domain equalizer. Instead, it employs frequency domain equalizer that only has a single tap. This greatly simplifies the equalization task and can perfectly resolves the multipath effects without any residual errors at high signal-to-noise ratio (SNR) region. Thus, introducing OFDM to IM/DD optical communication is a natural choice. However, different from RF communication, IM/DD requires that the transmitted signal must be real and positive, which imposes strict constraint on the modulation scheme and the original OFDM transceiver must be modified carefully to satisfy the new scenario. In addition, different applications may have diverse emphasis such as spectral efficiency, power efficiency, detection capability, as well as computational complexity.

Within such perspectives, the purpose of this chapter was to analyze the potential forms of OFDM that are suitable for IM/DD transmission as well as various receiver designs in optical communication. We first study the concepts and basic modulation schemes of OFDM systems in IM/DD optical communication. They can be generally classified into three categories: direct-current-biased optical OFDM (DCO-OFDM), non-DC-biased optical OFDM, and hybrid optical OFDM. We will elaborate the system models and explain the validity of some fancy designs in those systems through analysis. Second, we investigate the preliminary receivers of those OFDM systems. Besides, we will propose a new receiver that is capable of improving the detection performance based on the inherent signal structures of the specific transmitted signal. Third, the spectral efficiencies and computational complexities of different systems and receivers are analyzed and compared. Finally, the bit error rate (BER) performance of different systems is compared through computer simulations to give the reader a whole picture of different candidate OFDM systems in IM/DD optical communication.

2. OFDM principles

This section gives a brief introduction on the principles of OFDM in radio frequency communication which serves as a basis for further reading. The baseband diagram of OFDM is shown in **Figure 1**. At the transmitter side, coded information bits are first mapped to symbols through digital modulation such as pulse amplitude modulation (PAM), quadrature amplitude modulation (QAM), and phase shift keying (PSK). Typically, complex-valued QAM modulation is used

http://dx.doi.org/10.5772/intechopen.68199

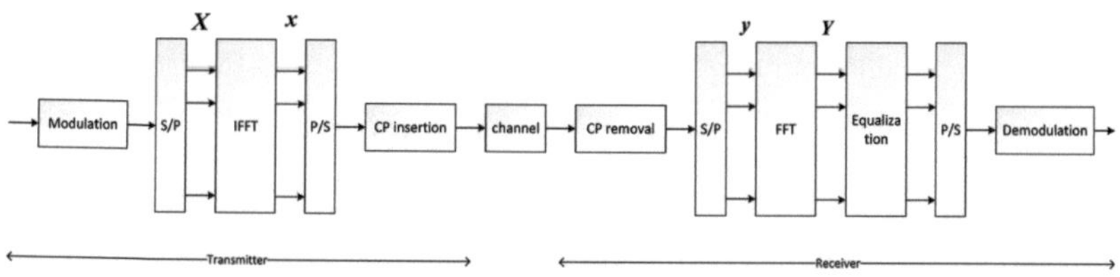

Figure 1. The "IFFT" module at the receiver side should be "FFT" module.

in OFDM. Then, the modulated symbols are divided into multiple groups and each group consists of N-modulated symbols, defined as $\boldsymbol{X} = [X(0), X(1), \ldots, X(N-1)]^T$, where the group index is omitted here for simplicity. In OFDM, each modulated symbol $X(k)$ is loaded on a *subcarrier* with center frequency $\frac{2\pi}{N}k$ and there are N subcarriers in total. All the symbols are transmitted on their subcarriers simultaneously. Mathematically, this is equivalent to transform the vector $\boldsymbol{X}$ by an N-point inverse fast Fourier transform (IFFT) module, resulting in a new vector $\boldsymbol{x} = [x(0), x(1), \ldots, x(N-1)]^T$, that is,

$$x(n) = \frac{1}{\sqrt{N}} \sum_{k=0}^{N-1} X(k) e^{j\frac{2\pi}{N}kn}. \tag{1}$$

In OFDM, $\boldsymbol{X}$ is generally considered as *frequency* domain signal and $\boldsymbol{x}$ is viewed as *time*-domain signal. In addition, $\boldsymbol{x}$ is typically called an *OFDM symbol*, which is different with the *modulated symbol* aforementioned. Finally, $\boldsymbol{x}$ is appended at its head with a cyclic prefix (CP), which is just the copy of the last few samples of $\boldsymbol{x}$. In general, the length of CP is no smaller than the length of transmission channel to avoid inter-symbol interference (ISI) between adjacent OFDM symbols.

Assuming the impulse response of the multipath channel is denoted by $\boldsymbol{h} = [h(0), h(1), \ldots, h(L-1)]^T$; then, the received signal at the receiver after channel transmission is given by

$$r(n) = h(n) * x_c(n) + z(n), \tag{2}$$

where $x_c(n)$ is the CP-appended version of the time-domain transmitted signal, $z(n)$ is the additive white Gaussian noise (AWGN) with zero mean, and * denotes linear convolution. The receiver first removes the CP parts of received signal, which results in a new vector $\boldsymbol{y}$ of length N, which can be rewritten as

$$y(n) = h(n) \otimes x(n) + z(n), \tag{3}$$

where $\otimes$ denotes cyclic convolution. We can see that due to the insertion and removal of CP, the linear convolution is now transformed to a cyclic one, which would be beneficial for equalization, as will be shown in the following text.

From signal processing theory, cyclic convolution in time domain is equivalent to product in frequency domain. Based on this fact, by defining $Y(k)$, $H(k)$, and $Z(k)$ as the N-point fast Fourier transform of $y(n)$, $h(n)$, and $z(n)$, respectively, one has

$$Y(k) = H(k)X(k) + Z(k), k = 0, 1, \ldots, N-1. \tag{4}$$

As we can see, each frequency-domain symbol $X(k)$ is transmitted as if in a *flat* channel of response $H(k)$ and different symbols transmit in different subchannels (subcarriers) without interfering with each other. This greatly simplifies the equalization task. For example, both zero forcing (ZF) and minimum mean square error (MMSE) equalization can be performed to recover $X(k)$ which only involves single-tap equalizer per subcarrier:

$$\widehat{X}(k) = \begin{cases} \dfrac{Y(k)}{H(k)}, & ZF \\ \dfrac{H^*(k)Y(k)}{|H(k)|^2 + {\sigma_n}^2}, & MMSE \end{cases} \tag{5}$$

where ${\sigma_n}^2$ is the variance of noise. The processing in Eqs. (4) and (5) can be realized by performing FFT and per-subcarrier equalization, as shown in **Figure 1**.

3. Optical OFDM systems for IM/DD channel

3.1. Preliminaries

In IM/DD channel, there is a key difference with RF channel: the transmitted signal must be real and positive. This results from the fact that the intensity of light must be a real and positive quantity. Therefore, the structure for OFDM shown in **Figure 1** cannot be directly used in IM/DD optical channel. Necessary changes must be made instead. A common approach is to generate a real time-domain signal first. This can be realized by imposing *Hermitian symmetry* on the frequency domain signal $\mathbf{X}$, which is defined as follows:

$$X(N-k) = X^*(k), k = 1, \ldots, N-1, X(0) = X(N/2) = 0. \tag{6}$$

It can be easily shown that the IFFT of $\mathbf{X}$ having property Eq. (6) is a pure *real*-valued signal $\boldsymbol{x}$. Based on this real signal, one can further generate a positive signal to drive the optical source by various means. Those resultant new OFDM systems are typically referred to as *optical OFDM* systems.

3.2. DC-biased optical OFDM

The most straightforward approach to generate a positive signal from a real signal is to impose a proper DC bias. In optical communication, the DC bias is typically chosen such that the mean value of the positive signal just lies on the center point of the linear range of the optical source. This system is called direct current-biased optical OFDM, or DCO-OFDM, whose transmitter is shown in **Figure 2**.

The clipping module shown in **Figure 2** is necessary. Since $x(n)$ is Gaussian distributed, it is possible that $x(n)$ plus a DC bias is still out of the linear range of the optical source. For example, if an LED accepts driving current within the range of $[a, b]$, where $0 \le a < b$, then the

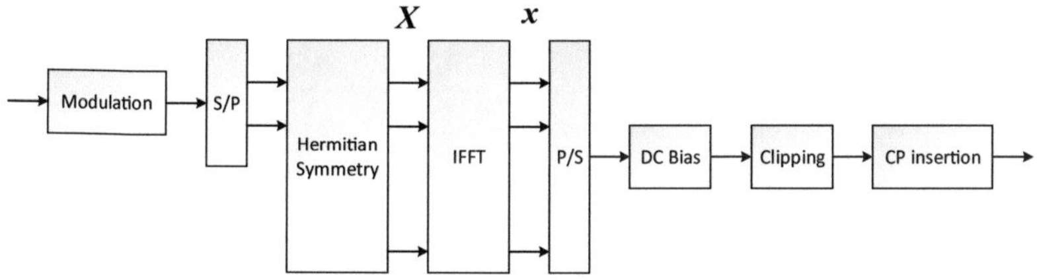

Figure 2. Diagram of DCO-OFDM transmitter.

clipping is needed to confine $x(n)$ plus a DC bias into this range. Otherwise, the LED could not be illumined due to under-driving or even be damaged due to over-driving.

3.3. Non-DC-biased optical OFDM

Besides DCO-OFDM, there are many forms of optical OFDM systems that are not relying on DC bias. The most famous ones are introduced in this subsection.

3.3.1. ACO-OFDM

Asymmetrically clipped optical OFDM (ACO-OFDM) is the most famous non-DC-biased optical OFDM system and has been extensively studied in literature [1]. The basic idea of ACO-OFDM is to generate an asymmetrically structured time-domain signal such that direct clipping at zero (without adding DC bias) is allowed without any information loss. To do so, the frequency-domain input symbol $\boldsymbol{X}$ has a special structure besides satisfying Hermitian symmetry. Specifically, the *odd* components of $\boldsymbol{X}$ contain useful information $\boldsymbol{U}$ but the *even* components of $\boldsymbol{X}$ are set to zeros. This is shown in **Figure 3**.

After IFFT, the time-domain signal $\boldsymbol{x}$ has an asymmetrical structure:

$$x(n) = -x\left(n + \frac{N}{2}\right), n = 0, 1, \ldots, N/2. \tag{7}$$

As shown in **Figure 4**, signal $\boldsymbol{x}$ can be directly clipped at zero without adding any DC bias, yet the information is kept after clipping thanks to the asymmetrical structure.

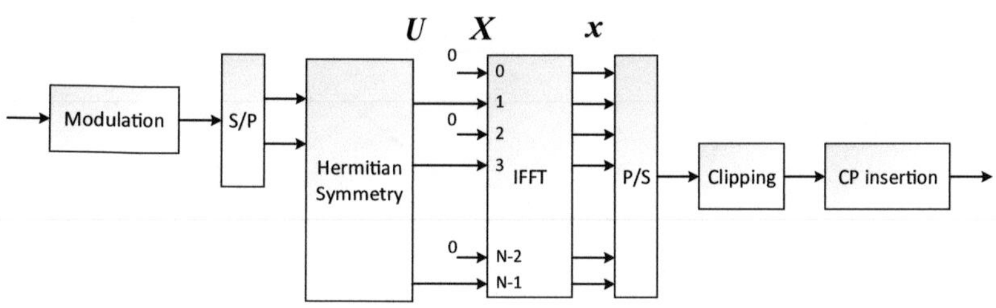

Figure 3. Diagram of ACO-OFDM transmitter.

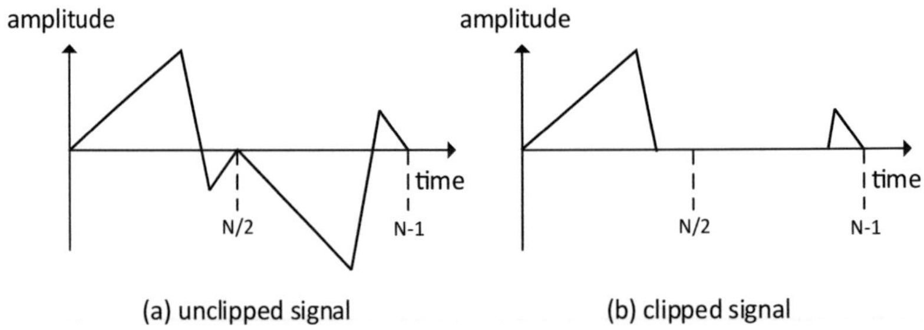

Figure 4. Asymmetrical structure before and after clipping at zero in ACO-OFDM.

3.3.2. PAM-DMT

Pulse amplitude modulation-discrete multi-tone (PAM-DMT) is another non-DC-biased optical OFDM system [2]. It is similar to ACO-OFDM, in that direct clipping is used. However, the difference is that in PAM-DMT, only the imaginary part of subcarrier input $\boldsymbol{X}$ carries useful information $\boldsymbol{U}$ while the real part is set to zero, as shown in **Figure 5**. Note that PAM modulation should be used in PAM-DMT rather than QAM in DCO-OFDM and ACO-OFDM.

It can be easily shown that the resultant time-domain signal $\boldsymbol{x}$ also has an asymmetric structure but is slightly different with that of ACO-OFDM, which is shown in Eq. (8) and **Figure 6**:

$$x(n) = -x(N-n), n = 1, 2, \ldots, \frac{N}{2}, x(0) = x\left(\frac{N}{2}\right) = 0. \tag{8}$$

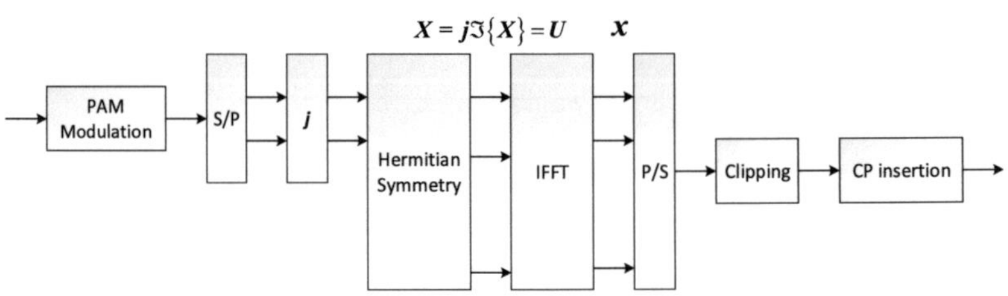

Figure 5. Diagram of PAM-DMT transmitter.

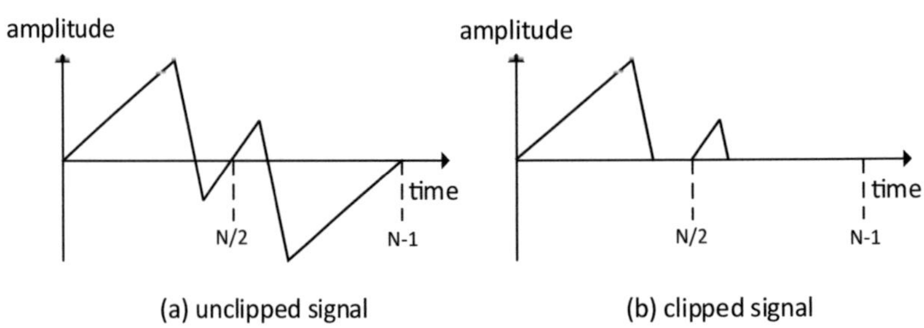

Figure 6. Asymmetrical structure before and after clipping at zero in PAM-DMT.

3.3.3. Flip-OFDM

Both ACO-OFDM and PAM-DMT rely on specially designed signal structures on the frequency- and time-domain signals. In contrast, flip-OFDM employs a simpler way such that a general frequency-domain signal $\boldsymbol{X}$ without any fancy structure is accepted [3]. Instead, the real time-domain signal $\boldsymbol{x}$, without any symmetry, is split into two parts: the first part only contains the samples of positive ones in $\boldsymbol{x}$, the negative ones are set to zeros; the second part only contains the samples of negative ones, but with flipped signs, and leaves the positive ones as zeros. This is shown in **Figure 7**.

Mathematically, the first and second parts of the flipped signal are given by

$$x_1(n) = \frac{x(n) + |x(n)|}{2}, x_2(n) = \frac{-x(n) + |x(n)|}{2}, n = 0, 1, \ldots, N-1. \tag{9}$$

After flip processing, the two signal parts are appended with CP, respectively, and are transmitted on channel consecutively. In some literature, flip-OFDM is also referred to as unipolar OFDM (U-OFDM).

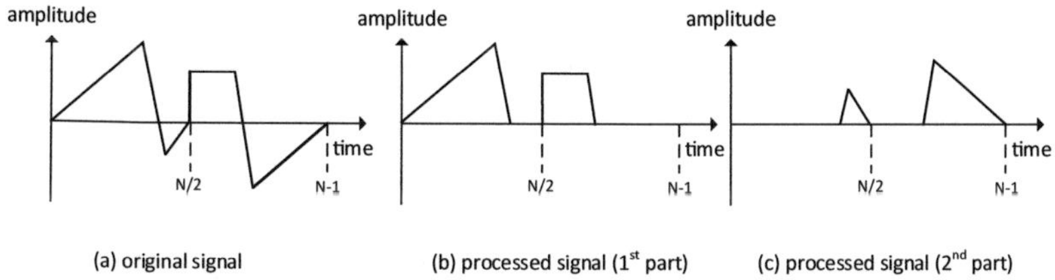

Figure 7. Signal structure before and after flipping processing in flip-OFDM.

3.4. Duality of non-DC-biased optical OFDM systems

This section gives a brief introduction on the duality of the non-DC-biased optical OFDM systems. Based on this duality, many receiver design methods could be easily extended from one system to other systems.

For ACO-OFDM, the transmitted signal is given by

$$x_c = (x + |x|)/2, \tag{10}$$

and $\boldsymbol{x}$ can be written as $\boldsymbol{x} = [\boldsymbol{x}_1; -\boldsymbol{x}_1]^T$, where $\boldsymbol{x}_1$ is the first half of $\boldsymbol{x}$. Therefore, the first and second halves of $\boldsymbol{x}_c$ can be written as

$$x_{c,1}(n) = \frac{x_1(n) + |x_1(n)|}{2}, x_{c,2}(n) = \frac{-x_1(n) + |x_1(n)|}{2}, \tag{11}$$

which is exactly the same as the model in Eq. (9) except that the size is changed from N to $N/2$.

For PAM-DMT, the transmitted signal is also given by Eq. (10). However, the unclipped signal $\boldsymbol{x}$ is slightly different, that is, $\boldsymbol{x} = [\boldsymbol{x}_1; \boldsymbol{x}_2]^T$, where $\boldsymbol{x}_2 = -\boldsymbol{J}\boldsymbol{x}_1$ with $\boldsymbol{J}$ being a matrix whose anti-diagonal elements are 1s and other elements are all 0s. Now, the first and second halves of the transmitted signal are given by

$$x_{c,1} = \frac{x_1 + |x_1|}{2}, x_{c,2} = \frac{-Jx_1 + J|x_1|}{2}. \tag{12}$$

However, $\boldsymbol{JJ} = \boldsymbol{I}$; therefore, by defining $\boldsymbol{x}_{c,3} = \boldsymbol{J}\boldsymbol{x}_{c,2}$, we have

$$x_{c,1} = \frac{x_1 + |x_1|}{2}, x_{c,3} = \frac{-x_1 + |x_1|}{2}. \tag{13}$$

Now, Eq. (13) is exactly the same as Eqs. (9) and (11).

Therefore, we can see that ACO-OFDM, PAM-DMT, and flip-OFDM essentially share the same signal structure and there is a duality between them. Based on this fact, the receivers designed for one system can be readily extended to other systems with simple substitution of variables.

3.5. Hybrid systems

Beside the basic forms of DC and non-DC-biased optical OFDM systems, there also exist some hybrid ones where multiple basic systems are superimposed in a specially designed fashion. In general, hybrid systems can be further classified into three categories.

3.5.1. Hybrid optical OFDM based on DCO-OFDM and a non-DC-biased one

A representative for this kind of system is ADO-OFDM, which combines DCO-OFDM and ACO-OFDM in a special way [4]. Specifically, in ACO-OFDM, the useful data are only loaded on the odd subcarriers, as illustrated in **Figure 3**, the even subcarriers are forced to be zero. After clipping in time domain, the clipping noise only falls onto even subcarriers and the odd subcarriers are not affected by the clipping noise. At the receiver side, the data could be recovered by using only the odd subcarriers. With the recovered data, one can further perfectly reconstruct the clipping noise on even subcarriers. Therefore, in ACO-OFDM, the even subcarriers can be exploited to load more data, which is the basic idea of ADO-OFDM.

In ADO-OFDM, the odd subcarriers are performed exactly the same as the ACO-OFDM. For the even subcarriers, a modified DCO-OFDM signal is generated, in which only the even subcarriers are used. Then, the signals generated from ACO-OFDM and DCO-OFDM are added together to obtain the ADO-OFDM signal. At the receiver side, ACO-OFDM signal, which is on odd subcarriers, are first detected. Then, the clipping noise on even subcarriers is estimated and subtracted. After that, the even subcarriers contain only DCO-OFDM signal, which is finally decoded.

3.5.2. Hybrid optical OFDM based on two different non-DC-biased ones

HACO-OFDM, or hybrid ACO-OFDM, combines ACO-OFDM and PAM-DMT in one system [5]. The basic idea is similar to that of ADO-OFDM, that is, the odd subcarriers are used for ACO-OFDM transmission while the even subcarriers are used for PAM-DMT transmission.

At the receiver side, interference cancellation is used for even subcarriers before decoding PAM-DMT signal. An alternative form for HACO-OFDM is also proposed [6].

3.5.3. *Hybrid optical OFDM based on a same non-DC-biased one*

Another form of hybrid optical OFDM is to superimpose multiple blocks of signals from a same non-DC-biased OFDM system. For example, enhanced unipolar OFDM (eU-OFDM) involves multiple blocks of signals from flip-OFDM [7]. **Figure 8** shows its time-domain structure with three layers [8], where the symbols $x_{i,j}^{+}$ and $x_{i,j}^{-}$ denote, respectively, the positive and flipped negative parts of the original j-th bipolar signal $x_{i,j}$ from layer-i. Each layer is just a repetition of flip-OFDM time-domain signal. We can see that for the first layer, four normal flip-OFDM symbols are used. For the second layer, two normal flip-OFDM symbols are repeated two times. For the third layer, one normal flip-OFDM symbol is repeated four times. Then, all the time symbols from three layers are added for transmission.

The receiver detection is very simple. The first layer is firstly decoded using normal subtraction. The second and third layers do not interfere in this procedure due to perfect self-cancellation. After the first layer is decoded, its impact is subtracted from the received signal. Then, the second layer is decoded subsequently. Then, the second layer signal is subtracted from the received signal and the third layer is decoded finally.

Besides eU-OFDM, the overlapping of ACO-OFDM or PAM-DMT is also proposed in literatures [9–11]. They all share similar idea with eU-OFDM and the receiver is based on layer-by-layer decoding.

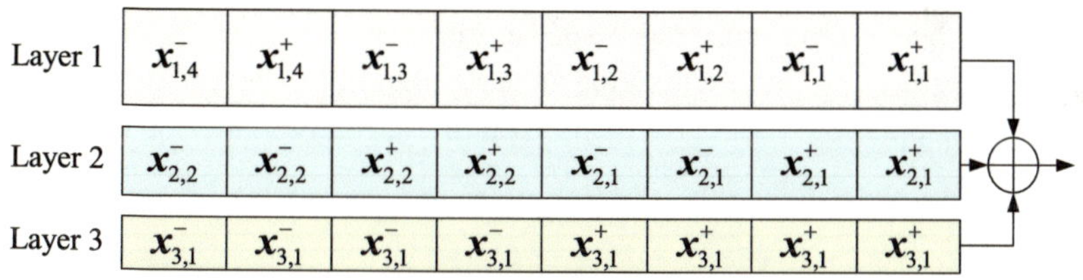

Figure 8. Illustration of a three-layer eU-OFDM time-domain signal components.

4. Receiver design for optical OFDM systems

In this section, we investigate the receiver design for optical OFDM systems. For DCO-OFDM, the receiver is straightforward. For non-DC-biased optical OFDM systems, there exist multiple candidate receivers which will be detailed later. For hybrid systems, since they are constructed mainly based on non-DC-biased ones, the receivers designed for non-DC-biased systems are also applicable for hybrid systems. Moreover, as the duality between non-DC-biased systems, in this chapter we focus on ACO-OFDM. We review the basic receiver design, diversity combining receiver design, and propose an iterative receiver design. All the formulations are based on an AWGN channel model but the results can be readily extended to multipath channels.

4.1. Basic receiver

The basic receiver for ACO-OFDM is very simple. The received signal after CP removal is given by

$$y(n) = x_c(n) + z(n). \tag{14}$$

In frequency domain, one has

$$Y(k) = X_c(k) + Z(k). \tag{15}$$

As proved by Ref. [1], the odd subcarriers of $X_c(k)$ is related to $X(k)$ by

$$X_c(k) = \frac{1}{2}X(k), \text{ for odd } k. \tag{16}$$

Therefore, the data $\boldsymbol{U}$, which is only on odd subcarriers of $\mathbf{X}$, can be recovered from $Y(k)$ by

$$\widehat{U}(k) = \begin{cases} 2Y(2k+1), & ZF \\ \dfrac{2Y(2k+1)}{1+{\sigma_n}^2}, & MMSE \end{cases} \quad k = 0, 1, \ldots, \frac{N}{2} - 1 \tag{17}$$

The receiver diagram is shown in **Figure 9**.

4.2. Diversity combining receiver

The basic receiver only utilizes the odd subcarriers for signal detection. The even subcarriers, bearing pure clipping noise, are simply discarded. Therefore, half of the received power is wasted. However, the clipping noise has a special inherent signal structure that is dependent on the unclipped signal. This inherent signal structure could be exploited for better detection performance. This is the basic idea of diversity combining receiver and the iterative receiver.

To unveil the relationship between the clipping noise and the unclipped signal, we rewrite the clipped signal as

$$x_c(n) = \frac{1}{2}[x(n) + |x(n)|], \forall n. \tag{18}$$

Based on Eqs. (16) and (18), one can see that the clipping noise falls only onto the even subcarriers and has a special form as

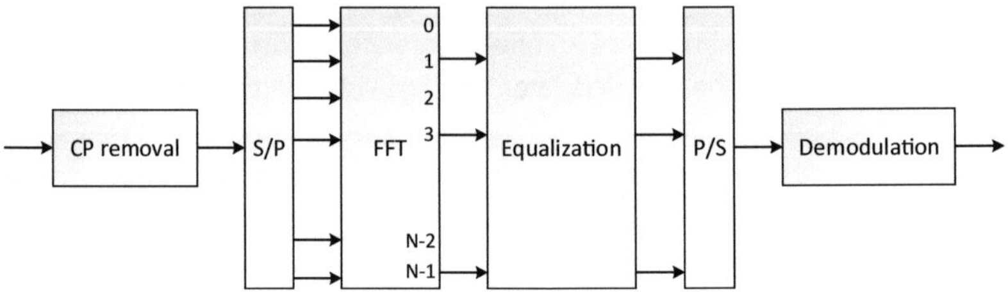

Figure 9. Diagram of the basic receiver for ACO-OFDM.

$$X_c(k) = \frac{1}{2} FFT\{|x(n)|\}, \text{ for even } k, \tag{19}$$

which says that the clipping noise on the even subcarriers is just the FFT of $|x(n)|$. Thus, we can generate two signals based on $Y(k)$: the first one is $\boldsymbol{Y}_0 = [0 \quad Y(1) \quad 0 \quad Y(3)\ldots 0 \quad Y(N-1)\ ]^T$, and the second one is $\boldsymbol{Y}_e = [Y(0) \quad 0 \quad Y(2) \quad 0\ldots Y(N-2) \quad 0]^T$. Denoting their time-domain signal by $\boldsymbol{y}_0$ and $\boldsymbol{y}_e$, respectively, we have

$$y_0 = \frac{1}{2} x + z_0, \tag{20}$$

$$y_e = \frac{1}{2} |x| + z_e. \tag{21}$$

Eq. (21) is obtained from Eqs. (16) and (19). A new signal $y_c(n)$ is generated based on Eqs. (20) and (21):

$$y_c(n) = \begin{cases} y_e(n), if & y_0(n) \geq 0, \\ -y_e(n), if & y_0(n) < 0. \end{cases} \tag{22}$$

Now, we get two branches of signals that are related to $x(n)$. Thus, diversity combining technique could be used to enhance the detection performance. The diversity combining is performed by

$$r(n) = a y_0(n) + (1-a) y_c(n), \forall n. \tag{23}$$

The combining coefficient a is usually a bit larger than 0.5 since $y_c(n)$ is not as accurate as $y_0(n)$ [12]. Based on $r(n)$, the data could be estimated just as in the basic receiver. The whole procedure of diversity combining receiver is shown in **Figure 10**.

A pairwise-ML receiver based on noise cancellation has been proposed in [13]. It has been proved that this receiver is in fact a special case of diversity combining receiver with $a = 0.5$ [14].

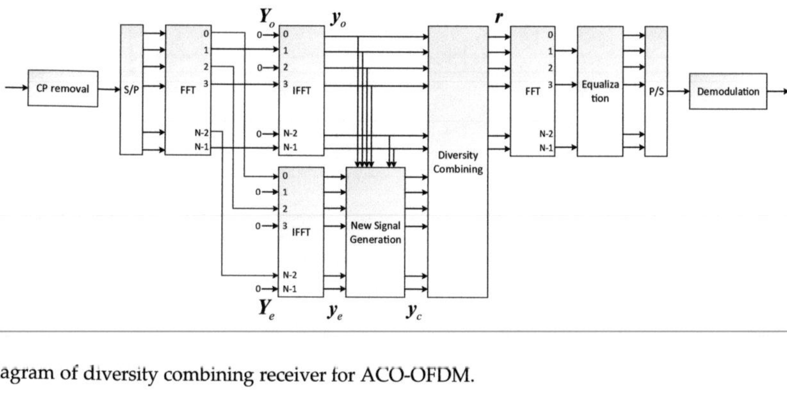

Figure 10. Diagram of diversity combining receiver for ACO-OFDM.

4.3. Proposed iterative receiver

Although the diversity combining receiver exploits the signal on even subcarriers, it is not performed in an optimal way, resulting in possible performance loss compared to optimal joint detection. Here, we propose an iterative receiver that has a better way to exploit the signal on even subcarriers [14]. The basic idea is to re-estimate the modulated data in a complete mathematical model at each iteration. At the very first iteration, the basic receiver is used for initialization. The details are given as follows.

Define an N by $N/2$ matrix $\boldsymbol{P}_0$ whose odd rows form an identity matrix and even rows are all zeros. Similarly, define another N by $N/2$ matrix $\boldsymbol{P}_e$ whose even rows form an identity matrix and odd rows are all zeros. Then, we have

$$\boldsymbol{X} = \boldsymbol{P}_0\boldsymbol{U}, \boldsymbol{U} = \boldsymbol{P}_0{}^T\boldsymbol{X} = \boldsymbol{P}_0{}^T\boldsymbol{P}_0\boldsymbol{U}. \tag{24}$$

In addition, based on Eqs. (16) and (19), we have

$$\boldsymbol{P}_0{}^T\boldsymbol{X}_c = \frac{1}{2}\boldsymbol{U}, \tag{25}$$

$$\boldsymbol{P}_e{}^T\boldsymbol{X}_c = \frac{1}{2}\boldsymbol{W}|\boldsymbol{x}| = \frac{1}{2}\boldsymbol{WSx} = \frac{1}{2}\boldsymbol{WSW}^H\boldsymbol{X} = \frac{1}{2}\boldsymbol{WSW}^H\boldsymbol{P}_0\boldsymbol{U}, \tag{26}$$

where $\boldsymbol{W}$ is the FFT matrix and $\boldsymbol{S}$ is a diagonal matrix whose entries on the main diagonal are the signs of $\boldsymbol{x}$. Then, Eq. (15) could be decomposed to

$$\boldsymbol{Y}_{odd} \triangleq \boldsymbol{P}_0{}^T\boldsymbol{Y} = \boldsymbol{P}_0{}^T\boldsymbol{X}_c + \boldsymbol{P}_0{}^T\boldsymbol{Z} = \frac{1}{2}\boldsymbol{U} + \boldsymbol{Z}_{odd}, \tag{27}$$

$$\boldsymbol{Y}_{even} \triangleq \boldsymbol{P}_e{}^T\boldsymbol{Y} = \boldsymbol{P}_e{}^T\boldsymbol{X}_c + \boldsymbol{P}_e{}^T\boldsymbol{Z} = \frac{1}{2}\boldsymbol{P}_e{}^T\boldsymbol{WSW}^H\boldsymbol{P}_0\boldsymbol{U} + \boldsymbol{Z}_{even}. \tag{28}$$

Collecting Eqs. (27) and (28) together, we have

$$\boldsymbol{R} = \boldsymbol{QU} + \boldsymbol{V}, \tag{29}$$

where

$$\boldsymbol{R} = \begin{bmatrix} \boldsymbol{Y}_{odd} \\ \boldsymbol{Y}_{even} \end{bmatrix}, \boldsymbol{Q} = \frac{1}{2}\begin{bmatrix} \boldsymbol{I} \\ \boldsymbol{P}_e{}^T\boldsymbol{WSW}^H\boldsymbol{P}_0\boldsymbol{U} \end{bmatrix}, \boldsymbol{V} = \begin{bmatrix} \boldsymbol{Z}_{odd} \\ \boldsymbol{Z}_{even} \end{bmatrix}, \tag{30}$$

where $\boldsymbol{I}$ denotes the identity matrix of proper size. Eq. (29) is a complete signal model of the received signal with respect to the information symbol. Therefore, based on Eq. (29), we can readily get the estimation of $\mathbf{U}$ by

$$\widehat{\boldsymbol{U}} = \begin{cases} (\boldsymbol{Q}^H\boldsymbol{Q})^{-1}\boldsymbol{Q}^H\boldsymbol{R}, & ZF \\ (\boldsymbol{Q}^H\boldsymbol{Q}+\sigma_n{}^2\boldsymbol{I})^{-1}\boldsymbol{Q}^H\boldsymbol{R}, & MMSE \end{cases} \tag{31}$$

Note that Q is in fact a function of $\mathbf{U}$ due to the component S. However, at each iteration, we assume Q is known by substituting $\widehat{U}$ from previous iteration to get Q. At the first iteration, the basic receiver is used to get the initial estimate of $\widehat{U}$.

5. Performance comparison

5.1. Spectral efficiency

In this section, we give a comparison on the spectral efficiencies of different optical OFDM systems.

For DCO-OFDM, each OFDM symbol only contains $N/2$ information-bearing complex-modulated symbols. Assuming the modulation order is M, then the spectral efficiency of DCO-OFDM is given by

$$\eta_{DCO-OFDM} = \frac{1}{2}\log_2 M \text{ bits/s/Hz}. \tag{32}$$

For all the non-DC-biased optical OFDM systems, as redundancy (zeros) is used in either frequency domain (ACO-OFDM and PAM-DMT) or time expansion is used (flip-OFDM), the spectral efficiencies are only half of DCO-OFDM:

$$\eta_{ACO-OFDM} = \eta_{PAM-DMT} = \eta_{Flip-OFDM} = \frac{1}{4}\log_2 M \text{ bits/s/Hz}. \tag{33}$$

For hybrid systems, things are a little bit complex. There is no general expression but one has to analyze the specific system.

For ADO-OFDM, in addition to a conventional ACO-OFDM transmission on odd subcarriers, a half-rate DCO-OFDM is used on even subcarriers. Therefore, its spectral efficiency is

$$\eta_{ADO-OFDM} = \eta_{ACO-OFDM} + \frac{1}{2}\eta_{DCO-OFDM} = \frac{1}{2}\log_2 M \text{ bits/s/Hz}. \tag{34}$$

For HACO-OFDM, a similar expression could be obtained:

$$\eta_{HACO-OFDM} = \eta_{ACO-OFDM} + \frac{1}{2}\eta_{PAM-DMT} = \frac{3}{8}\log_2 M \text{ bits/s/Hz}. \tag{35}$$

For eU-OFDM, the spectral efficiency depends on the number of layers. For an L-layer system, the spectral efficiency is given by

$$\eta_{eU-OFDM}(L) = \sum_{l=1}^{L}\left(\frac{1}{2}\right)^{l-1}\eta_{Flip-OFDM} = 2\left(1-\frac{1}{2^{L-1}}\right)\eta_{Flip-OFDM}. \tag{36}$$

When L approaches infinity, we have the upper bound of spectral efficiency:

$$\eta_{eU-OFDM} = 2\eta_{Flip-OFDM} = \frac{1}{2}\log_2 M \text{ bits/s/Hz}. \quad (37)$$

There is a tradeoff between the spectral efficiency and decoding complexity with respect to L: a larger L means the spectral efficiency is closer to its upper bound but the decoding complexity, mainly from signal cancellation for decoded layers, will increase linearly with L. In practice, a fairly small L is desired to achieve a balance between the spectral efficiency and decoding complexity, say, for example, $L = 5$ is a good choice. In fact, when $L = 5$, we have

$$\eta_{eU-OFDM}(5) = 2\left(1 - \frac{1}{2^{5-1}}\right)\eta_{Flip-OFDM} = 0.94\eta_{eU-OFDM'} \quad (38)$$

which shows that the spectral efficiency is very close to the upper bound.

On summarizing, we can see that DCO-OFDM has the highest spectral efficiency. However, its power efficiency is not very good due to the non-information-bearing DC. On the contrary, the non-DC-biased optical OFDM systems have better power efficiency due to the elimination of DC offset but their spectral efficiency is only half of that of DCO-OFDM. The hybrid systems, especially ADO-OFDM and eU-OFDM, have better balance between the spectral efficiency and power efficiency. With those facts, one can choose a proper implementation form in practice under specific communication requirement and constraint.

5.2. Receiver complexity

The computational complexity of different receivers is analyzed here using the order notation. For the basic receiver, the main computation burden is the FFT and equalization, which have complexities of $O(N\log_2 N)$ and $O(N)$, respectively. In total, it is just $O(N\log_2 N)$. For the diversity combining receiver, it involves finite number of FFT/IFFT and a final equalization. Therefore, although it is more complex than the basic receiver, there is no difference when considering the order notation, that is, it is still $O(N\log_2 N)$. For the proposed iterative receiver, its main computation burden is the matrix inversion of Eq. (31) and this operation should be repeated at each iteration. Thus, the total complexity is in the order of $O(TN^3)$, where T is the total number of iterations. As we can see, this receiver is the most complicated one among the three receivers. However, as will be shown later, its performance is the best and can be far better than the other two. Thus, it is acceptable considering the performance gains. In addition, with the rapid development of modern signal-processing hardware, the computation burden will not be a limiting factor for these small-scale computations.

6. Simulations

In this section, we compare the average uncoded BER performance of different receivers in VLC channels through simulations. The channels are generated using the method in [15] with the following configurations: an empty room of size 8 × 6 × 4 m with reflection coefficients 0.8, 0.8, and 0.3 for the ceiling, the walls, and the floor, respectively; LEDs are used as the optical source and they are attached 0.1-m below the ceiling. The photodetectors (PDs) are 1 m above the floor with an 80° of field of view (FOV). Both line-of-sight (LOS) and nonline-of-sight (NLOS) channels are tested (the LEDs point straight downward and upward, respectively). Multiple LEDs and PDs are used to enhance the performance and robustness of the communication link, resulting in a multiple-input multiple-output (MIMO) channel model. The receiver design methods described in Section 4 can be easily extended to this channel model by using vector notation. For each MIMO channel realization, the positions of the LEDs and the photodetectors are randomly drawn from their corresponding plains and the channels are normalized to have power $N_R N_T$, where N_R and N_T denote the number of PDs and LEDs, respectively. The ill-conditioned channels are rejected for fair comparison. The number of subcarriers is N = 64. In the legend of figures, "conventional" denotes the basic receiver, "pairwise ML" denotes the receiver in [13], which is a special case of the diversity combining receiver. "Lower bound" denotes the ideal curve of the proposed receiver with perfect estimation of matrix $\boldsymbol{Q}$. In all receivers, MMSE equalization is used.

Figure 11 shows a sample view of the impulse response of the LOS and NLOS channels with a sampling rate of 300 MHz. It can be seen that the LOS channel is more like a delta function but NLOS channel has relatively longer delay spread, which means it can be viewed as a multipath channel.

First, we compare the BER performance in single-input single-output (SISO) channel. **Figure 12** shows the performance with modulation order M = 64. It can be seen that in LOS channel, the

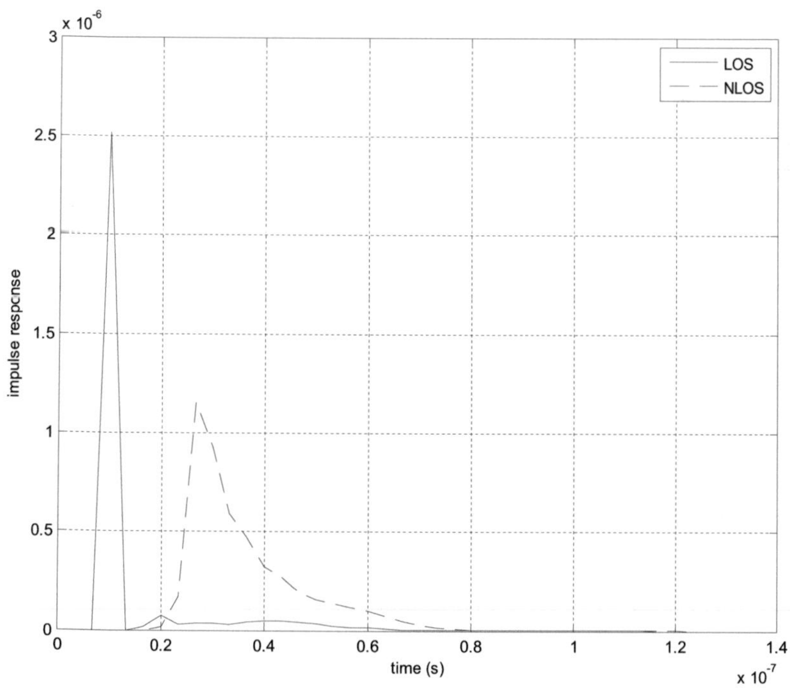

Figure 11. A sample view of LOS and NLOS channels with 300-MHz sampling rate.

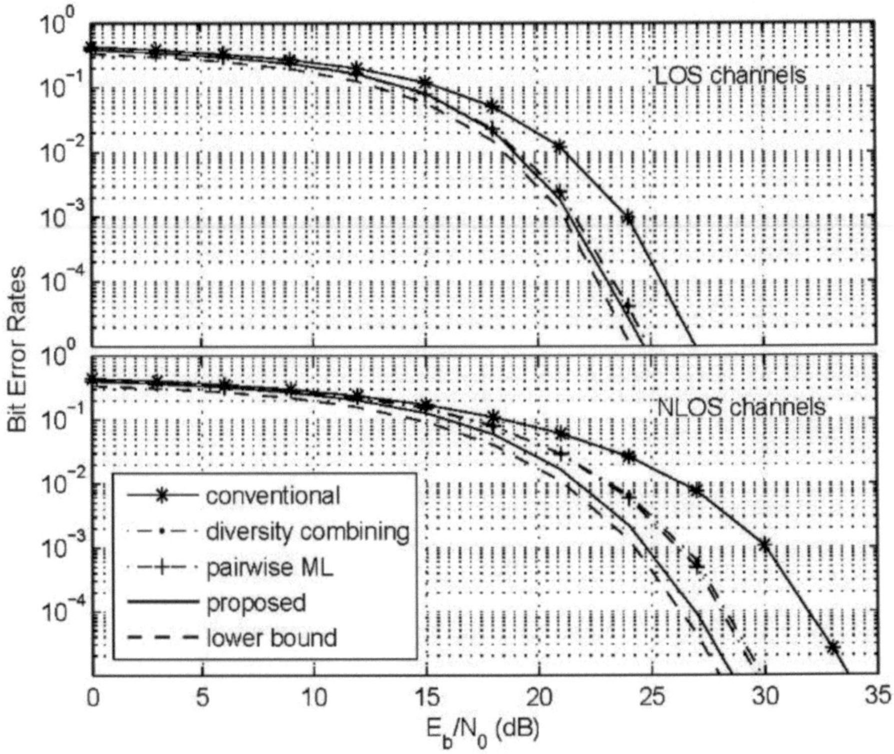

Figure 12. BER comparison of SISO ACO-OFDM using modulation sizes of $M = 64$.

diversity combining receiver and the proposed iterative receiver have similar performance and are much better than the basic receiver. In addition, their performance gap to the lower bound is limited. However, in NLOS channel, things are different: the proposed iterative receiver has the best performance. Compared to the basic receiver, its performance gain is more than 10 dB at high SNR range, which is significant. Even compared to the diversity combining receiver, 1-dB gain could be observed.

Now, we turn to MIMO channels. **Figure 13** shows the BER performance of a 4 × 4 MIMO ACO-OFDM using modulation sizes of $M = 16$. It can be seen that the performance of different receivers has similar behavior as the SISO case. However, compared to the SISO case, the performance gain of the proposed receiver is even larger: compared to the diversity combining receiver, it is more than 6 dB at high SNR regime; compared to the basic receiver, it is much more than 10 dB. Nonetheless, there is still a fairly large performance gap between the proposed receiver and its lower bound, which indicates that more advanced signal processing at the receiver side is desired in the future.

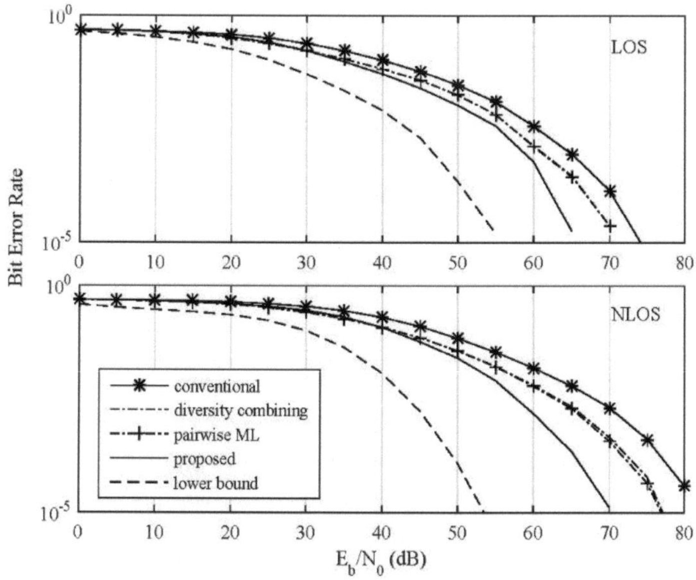

Figure 13. BER comparison of 4 × 4 MIMO ACO-OFDM using modulation sizes of $M = 16$.

7. Conclusions

In this research, we have investigated various forms of optical OFDM systems that are suitable for IM/DD optical channel. Different receivers are described with a proposed iterative one. Spectral efficiencies, computational complexities, as well as BER performance in LOS and NLOS channels of different systems and receivers are given. It is found that DCO-OFDM is more spectrally efficient than the non-DC-biased systems. The hybrid systems achieve a better tradeoff between the spectral efficiency and power efficiency. The proposed iterative receiver has the highest complexity but is far superior than other receivers, especially the basic receiver. Those results reveal the potential of OFDM systems in IM/DD channels for optical communication.

Acknowledgements

This work is supported by Southeast University 3-Category Academic Programs Project ("Optical Wireless Communication" and "Software-Defined Radio") and Top-notch Academic Programs Project of Jiangsu Higher Education Institutions (no. PPZY2015A035).

Author details

Jian Dang*, Liang Wu and Zaichen Zhang

*Address all correspondence to: newwanda@seu.edu.cn

School of Information Science and Engineering, Southeast University, Nanjing, P.R. China

References

[1] J. Armstrong, A. J. Lowery. Mint: Power efficient optical OFDM. Electronics Letters. 2006; **42**(6): 370–372. DOI: 10.1049/el:20063636.

[2] S. C. J. Lee, S. Randel, F. Breyer, A. M. J. Koonen. Mint: PAM-DMT for intensity-modulated and direct-detection optical communication systems. IEEE Photonics Technology Letters. 2009;**21**(23):1749–1751. DOI: 10.1109/LPT.2009.2032663.

[3] N. Fernando, Y. Hong, E. Viterbo. Mint: Flip-OFDM for unipolar communication systems. IEEE Transactions on Communications. 2012;**60**(12):3726–3733. DOI: 10.1109/TCOMM.2012.082712.110812.

[4] S. D. Dissanayake, K. Panta, J. Armstrong. A novel technique to simultaneously transmit ACO-OFDM and DCO-OFDM in IM/DD systems. In: IEEE GLOBECOM Workshops (GC Wkshps); 5–9 December 2011; Houston, Texas. IEEE; 2012. p. 782–786. DOI: 10.1109/GLOCOMW.2011.6162561.

[5] B. Ranjha, M. Kavehrad. Mint: Hybrid asymmetrically clipped OFDM-based IM/DD optical wireless system. Journal of Optical Communications and Networking. 2014;**6**(4):387–396. DOI: 10.1364/JOCN.6.000387.

[6] Q. Wang, Z. Wang, L. Dai. Mint: Asymmetrical hybrid optical OFDM for visible light communications with dimming control. IEEE Photonics Technology Letters. 2015;**27**(9):974–977. DOI: 10.1109/LPT.2015.2404972.

[7] D. Tsonev, S. Videv, H. Haas. Mint: Unlocking spectral efficiency in intensity modulation and direct detection systems. IEEE Journal on Selected Areas in Communications. 2015;**33**(9):1758–1770. DOI: 10.1109/JSAC.2015.2432530.

[8] J. Dang, Z. Zhang, L. Wu. Mint: Improving the power efficiency of enhanced unipolar OFDM for optical wireless communication. Electronics Letters. 2015;**51**(21):1681–1683. DOI: 10.1049/el.2015.2024.

[9] A. J. Lowery. Mint: Enhanced asymmetrically-clipped optical OFDM. Mathematics. 2015;**40**(1):36–40. DOI: 10.1364/OE.24.003950.

[10] M. S. Islim, D. Tsonev, H. Haas. On the superposition modulation for OFDM-based optical wireless communication. In: IEEE Global Conference on Signal and Information

Processing (Global SIP); 14–16 December 2015; Orlando, Florida, USA. IEEE; 2016. p. 1022–1026. DOI: 10.1109/GlobalSIP.2015.7418352.

[11] M. S. Islim, H. Haas. Mint: Augmenting the spectral efficiency of enhanced PAM-DMT-based optical wireless communications. Optics Express. 2016;**24**(11):11932–11949. DOI: 10.1364/OE.24.011932.

[12] L. Chen, B. Krongold, J. Evans. Diversity combining for asymmetrically clipped optical OFDM in IM/DD channels. In: IEEE Global Telecommunications Conference; 30 November–4 December 2009; Honolulu, Hawaii, USA. IEEE; 2010. p. 1–6. DOI: 10.1109/GLOCOM.2009.5425293.

[13] K. Asadzadeh, A. Dabbo, S. Hranilovic. Receiver design for asymmetrically clipped optical OFDM. In: IEEE GLOBECOM Workshops (GC Wkshps); 5–9 December 2011; Houston, Texas, USA. IEEE; 2012. p. 777–781. DOI: 10.1109/GLOCOMW.2011.6162559.

[14] J. Dang, Z. Zhang, L. Wu. Mint: A novel receiver for ACO-OFDM in visible light communication. IEEE Communications Letters. 2013;**17**(12):2320–2323. DOI: 10.1109/LCOMM.2013.111113.132223.

[15] J. B. Carruthers, P. Kannan. Mint: Iterative site-based modeling for wireless infrared channels. IEEE Transactions on Antennas Propagation. 2002;**50**(5):759–765. DOI: 10.1109/TAP.2002.1011244.

Chapter 6

Selective Mode Excitation: A Technique for Advanced Fiber Systems

Dmitry V. Svistunov

Additional information is available at the end of the chapter

http://dx.doi.org/10.5772/intechopen.68409

Abstract

Actual problems arising in development of fiber optical systems are increasing the information capacity and enhancing data security. Different encoding methods and data compression techniques have been developed to meet these requirements. The presented materials emphasize advantages of application of selective mode excitation in fiber systems that lead to both increasing the system information capacity and enhancing data security. Three-stage hierarchical scheme of data compression [where time division multiplexing (TDM) method is the content of the first stage, the second stage utilizes wavelength division multiplexing, and mode division multiplexing (MDM) is applied at the last stage] is discussed. Furthermore, it is highlighted that selective mode excitation is able to embarrass eavesdropping. It is shown that just application of the mentioned technique allows enhancing data security, while designing of special system architectures provides additional increase of data protection level. The examples of such system schemes are presented. Thus, application of selective mode excitation could improve the performances of fiber systems significantly, at least the ones such as short- and middle-haul communication lines and local area networks (LANs).

Keywords: selective mode excitation, information capacity, data compression, mode division multiplexing, data security

1. Introduction

The demands on increasing the information capacity of fiber systems and also on higher secrecy are declared the whole time since optical communication came to the practical stage of commercial systems, and intensive investigations have been performed in order to meet the mentioned requirements.

Regarding improvement of data security, conducted research resulted in a number of developed encoding algorithms and the specific methods as quantum cryptography that can embarrass decoding of eavesdropped information significantly.

As for information capacity, substantial progress has been achieved by employment of data compression methods. Various developed techniques can be combined in several groups by operating functions. The first group—time division multiplexing (TDM) method—had been developed for wire and radio communications and then was implemented also in optical fiber communication on appearance of fiber systems. That method bases on managing the launch of different input signal streams in turn to the same trunk line. In TDM method, time domains (frames) are defined, and each one is divided into several time slots which are filled with the data blocks of transmitted signal streams. While former applications of the method use managing the electric signals (ETDM), the latter application employs also extra techniques performing switching of optical signal streams (OTDM). Whereas a number of variants of TDM method have been developed, further works on TDM techniques and algorithms are still performed actively.

The next group—wavelength division multiplexing (WDM) method, was implemented later upon achieving appropriate performances and lower costs of required fiber system units. The method uses single optical fiber for parallel independent transmission of various light signals of different wavelengths. Performed standardization resulted in defining several wavelength windows and also wavelength channel spacing into the window. Maximal system information capacity can be reached under employment of dense WDM (DWDM) technique capable of providing minimal channel spacing defined by the standardized spectral grid for 12.5 and 25 GHz separations. Presently, WDM techniques have wide application in current fiber systems.

Evident way to get further increase of system information capacity is cooperative employment of different data compression techniques in advanced systems. Conducted research resulted in application of WDM method in combinations with ETDM or OTDM techniques mainly in single-mode fiber (SMF) systems and also in short-haul multimode systems. Developments of the latter systems have led to elaborating a promising approach to significant increasing the bandwidth-distant products of those systems: selective excitation of sole mode within multimode fiber that allows obtaining a regime of quasi-single-mode data transmission. The way to get further substantial rise of information capacity of those systems is employment of the specific data compression method named mode division multiplexing (MDM). Selective mode excitation is a crucial technique of that method, and the interest to research on that theme rises year-by-year.

Here, our subject is to consider possibility of complex employment of noted data compression methods. Furthermore, we shall show that application of selective mode excitation in the form of MDM technique allows enhancing data security due to just application of the method, while designing of special system architectures provides additional increase of data protection level.

2. Hierarchic scheme of data compression

The essence of MDM technique is the use of different fiber modes or mode groups of the same multimode fiber as independent information channels. High level of intermodal coupling in

http://dx.doi.org/10.5772/intechopen.68409

former commercial fibers delayed implementation of this method for a long time. However, achieved progress in fabrication technology reduced drastically the intermodal coupling (as for example, the average level of intermodal coupling coefficient as 0.007 km^{-1} for the case of neighbor-order modes had been reached already in the past decade [1]). Due to that progress, propagation of sole modes has been obtained in multimode commercial fibers, and further investigations concentrated just on realization of MDM technique. First of all, different selective mode couplers had been developed, and designing of experimental systems started. One group of developments was directed to application of MDM to coherent systems in order to reach long-haul trunk lines. However, these systems are complicated and high cost. On the other hand, a number of short- and middle-haul current fiber systems are based on multimode fibers. Bearing in mind, the possibility to reconstruct those systems by application of MDM technique; further, we consider low-cost direct detection systems that are quite appropriate for mentioned trunk lengths. By present, experimental MDM-based systems are developed demonstrating the available distances being sufficient for fiber systems such as metropolitan or toll lines and local area networks (LANs) (for example, [2, 3]). As feasibility of MDM systems is proven practically, the goal of the next stage is complex combination of this technique with other data compression methods.

Cascade scheme of data compression is shown in **Figure 1**. The scheme is capable of providing hierarchical three-level compression of data streams. Primary stage of data processing in the scheme bases on TDM techniques issuing the set of compressed data streams each intended for separate spectral channel. The second stage uses WDM method that joins a number of spectral channels in the same fiber. Application of MDM technique is the content of the third stage. At this stage, each compressed optical signal resulting from previous stages launches certain separate mode of multimode trunk fiber. The scheme has a symmetric structure, and reverse sequence of those stages at the receiver part provides recovering the initial data streams.

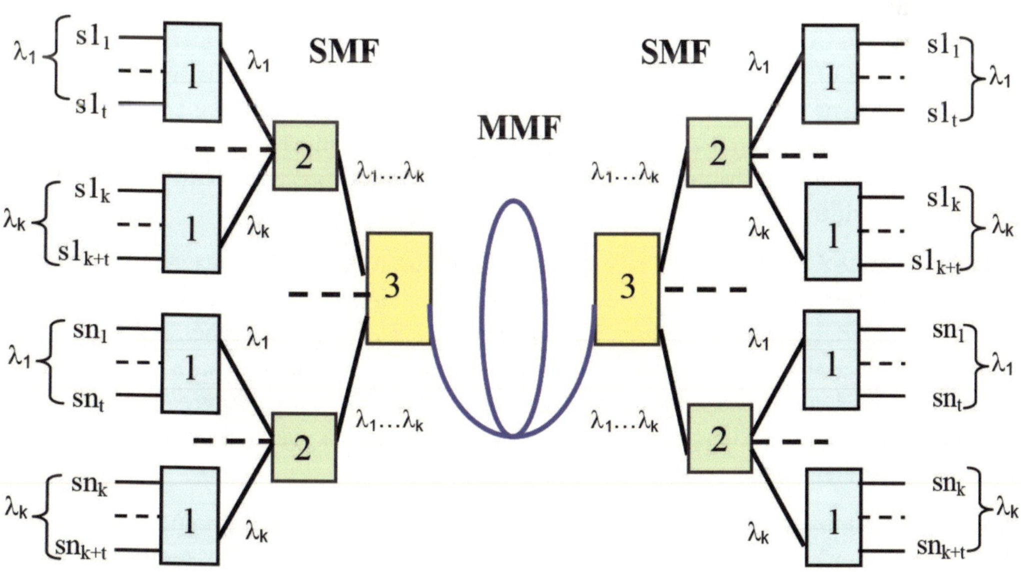

Figure 1. Hierarchical cascade scheme of data compression in multimode fiber systems. The blocks denote the devices of following types: 1 is TDM; 2 is WDM; 3 is MDM.

Multimode fiber (MMF) is utilized as a trunk fiber, while single-mode fibers (SMFs) required for optimal operating of mode multiplexers are employed in other optical interconnections into the scheme.

Every mode channel in multimode fiber can be considered as an analogue to separate single-mode fiber for which cooperative implementation of TDM and WDM techniques is already proven. So, the question concerning the proposed scheme is whether mutual compatibility of developed WDM and MDM operating units is limited restricting common application of the techniques.

Feasibility of the considered scheme depends on ability of mode multiplexers/demultiplexers to operate effectively in some spectral range. Those units are based on selective mode couplers which excite/detect independently fiber modes of different orders. A set of such elements basing on different operating principles have been developed (for example, [4–16]). In most cases, selective excitation of fiber modes is performed with external units that are built as waveguide (fiber or integrated optical) elements or bulk devices based on traditional optical elements. Optical matching of those units with the fibers is performed by traditional or GRIN lenses, and also butt joining is used for waveguide elements. Waveguide selective mode couplers require special consideration. Particularly, the attention should be paid to selective units whose operating principles lead to dependence of directions of intrinsic light propagation on the light wavelength. Waveguide mode multiplexer/demultiplexer described in Refs. [17, 18] is just the unit of this kind, and experimental samples of the unit were examined to determine the noted dependence.

Scheme of the considered unit is shown in **Figure 2**. Planar selective element operates as known input/output prism coupler and matches optically each mode of the multimode channel waveguide with the corresponding waveguide beam in single-mode planar region joined with the set of single-mode channel guides by horn transition structures. The angle α between the axis of certain planar beam and multimode channel guide depends on the ratio of mode indices of that beam and corresponding channel mode. As planar selective coupler is the key element of the unit, experimental samples of this element have been studied by measurement of directional diagrams of planar beams [19] in order to estimate whether those beams become superimposed if the channel waveguide is excited with light of certain spectral bandwidth.

Reliable accurate measurement of beam directivity diagram into planar waveguide could be difficult; therefore, extra prism input/output coupler having cylindrical base was applied, and directional diagrams of output beams were measured. **Figure 3** presents the results of examinations of the trial sample. Numerical angular values at the diagram axis correspond to goniometric readouts under arbitrary benchmark position. Obtained patterns are typical for all couplers of this kind. Every peak corresponds to the separate planar beam associated with the certain channel mode.

Study of the sample intended for operating at light wavelength as 1.3 μm and simulation of coupler excitation with light of whole standard O bandwidth showed that angular widths of three neighbor planar beams become as 22, 20, and 16 arcmin, while the angles between the

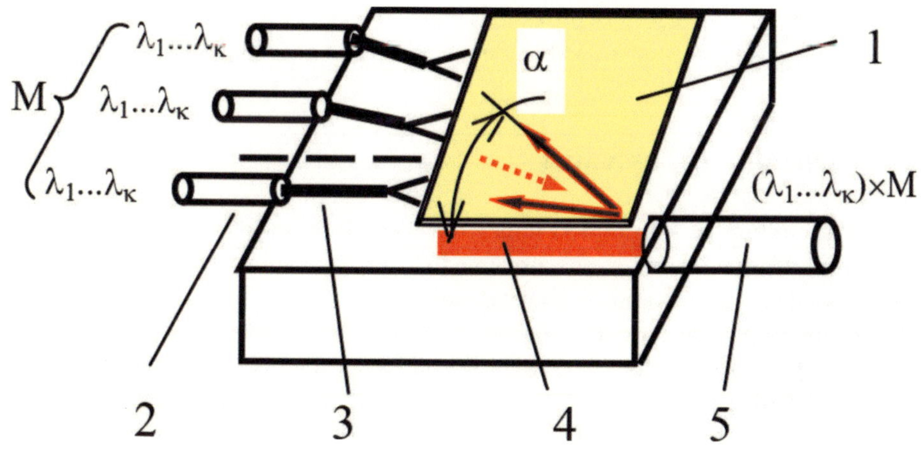

Figure 2. Scheme of waveguide unit for selective mode excitation/detection. 1 is single-mode planar section; 2 is SMF; 3 is single-mode channel waveguides with horn transitions; 4 is multimode channel waveguide; 5 is MMF.

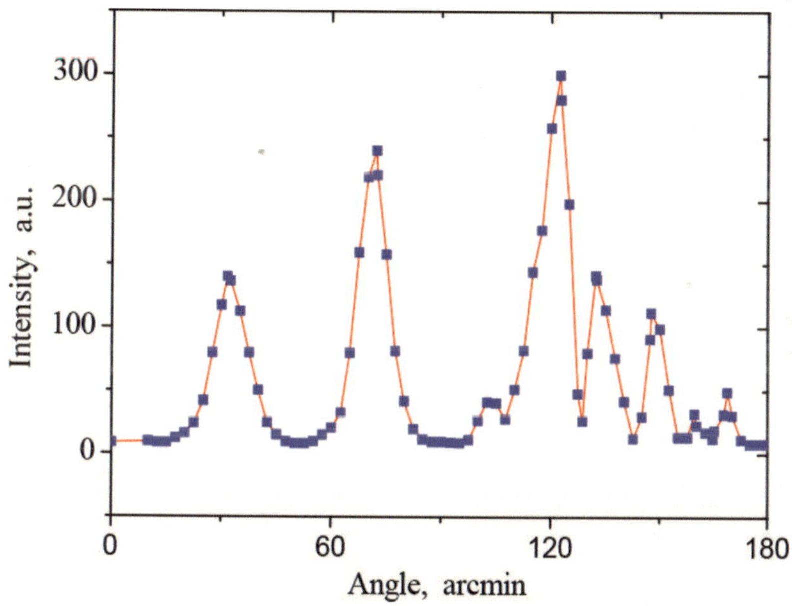

Figure 3. Angular distribution of light beams associated with the set of excited channel modes.

axes of adjacent beams are 64 and 33 arcmin [19]. These results indicate that the beams are spatially separated and can be directed to corresponding horn structures without appearance of significant crosstalk. So, the experimental results confirm that this mode multiplexer is capable of operating with optical signals compressed by WDM.

Considering that the mentioned mode multiplexer is assumed to be among the ones having perceptible spectral sensitivity, we can note compatibility of MDM and WDM units and conclude that the obtained results prove feasibility of cooperative application of MDM and WDM techniques. Thus, the mentioned hierarchic scheme of data compression can be realized.

3. Enhancement of data security

Nowadays, protection of transmitted data from eavesdropping is considered as a crucial problem, and intensive investigations are aimed to development of new approaches leading to effective solutions. One branch of developments is counteracting to eavesdropping under intrusion to the trunk fiber line. In order to understand the challenge, let us estimate what techniques could be applied for illegal data reading from the fiber bearing in mind that the essence of eavesdropping means that only minor part or optical power is spit off, otherwise alarm signals are issued by the control system blocks. Therefore, partial mode extraction to lateral direction is to be performed in some intrusion procedure, while the main part of optical power propagates further into the fiber. Some possible variants of the mentioned impact on the fiber are shown in **Figure 4**.

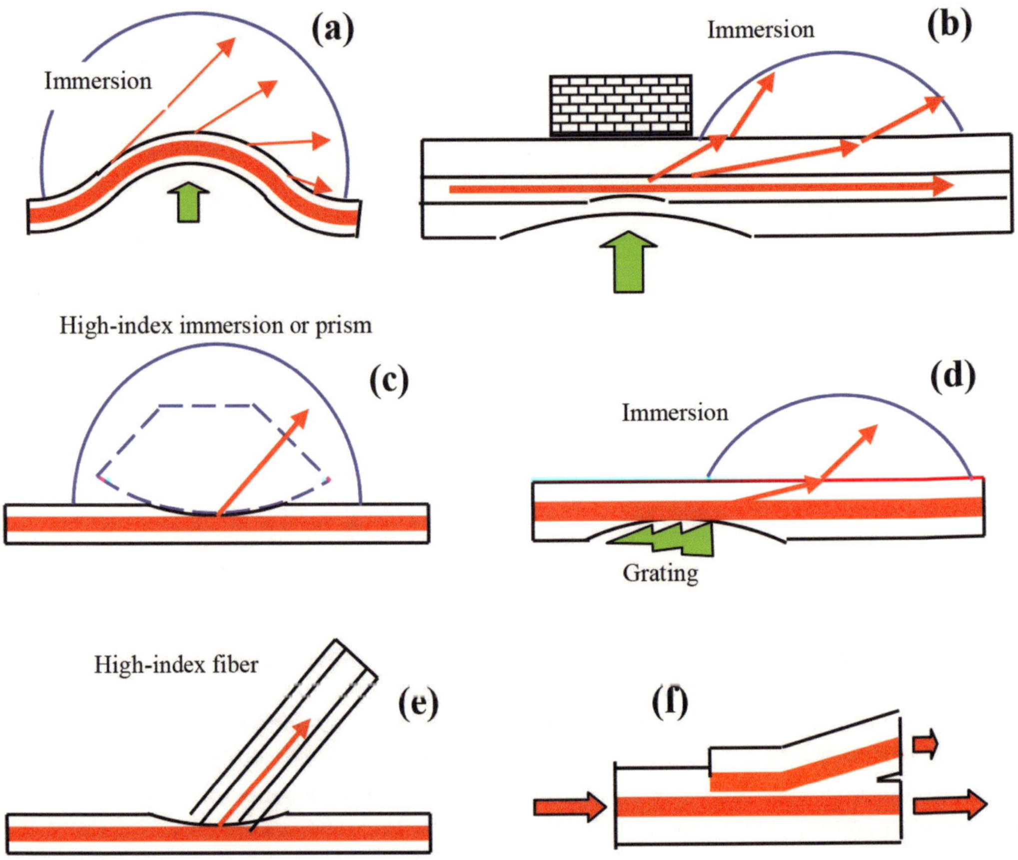

Figure 4. Variants of impact on the fiber resulting in partial extraction of the optical power from the fiber modes: fiber bending (a), local pressing (b), prism-like (c), and diffractive (d) couplers, higher-index fiber (e) and as all-fiber coupler (f).

Rather simple way to perform mentioned eavesdropping is to bend the fiber. Then, bent losses occur because the part of the tail of mode field cross distribution is cut off due to fiber bending and launches separate light beam (namely cladding mode) propagating into the fiber cladding. Those cladding modes form light beams in the adjacent medium due to scattering on cladding imperfections or refraction if the medium is higher index. Mode leakage increases as the bend radius decreases. As the leakage reasons remain the same along the bend, mode transformation occurs at every point of bent fiber, and the whole radiated beam is high-divergent as it is shown in **Figure 4a**. Similar technique uses microbendings (obtained with pressed special tools as it is performed in some pressure sensors) and/or deviations of the core radius caused by local impact to the fiber (see **Figure 4b**). In those cases, a part of fiber mode power is also split off being proportional to the pressure strength and forms cladding beams that can be outcoupled with immersion drop and registered.

The noted splitting techniques do not require preliminary treatment of fiber inner cladding, only outer protective jackets must be removed. The next variants exploit optical tunneling effect so they need affecting the cladding adjacent to fiber core. Appropriate gap between the applied tool and the fiber core can be obtained by partial remove of cladding layer, for example with etching by dropped chemical reagent or pressing under heating. Upon preparing the gap, higher index immersion is placed to the formed contact region or a bulk prism is pressed, and one obtain splitting tool operating as known longitudinal input/output prism couplers where the level of extracted optical power depends on combination of refractive indices and also on the gap length and thickness. Spatial extracted beams are formed directly with such tool drawn in **Figure 4c**. That variant of impact on the trunk fiber under illegal data reading seems more difficult in use but more dangerous with relation to the mentioned manner of data protection because longitudinal couplers can spatially separate output light beams associated with the modes of different orders. Similar tool can use external diffraction grating placed to the contact area with grating strokes normal to the fiber axis as shown in **Figure 4d**. The effective regime of that diffractive tool is realized when a long-period grating transfers the mode power to cladding modes. A higher index extra fiber can also replace the bulk prism as shown in **Figure 4e**. End face of the extra fiber is placed to the contact region, while its axis forms a certain angle with the trunk fiber. If inclination of the extra fiber provides meeting the condition of phase matching of the modes in both fibers, mode launching occurs in the extra fiber due to optical tunneling through the cladding gap. And of course, optical tunneling can be performed by exciting the extra fiber located along the trunk fiber as in all-fiber directional couplers (see **Figure 4f**). Treatment of fiber claddings is also necessary in this case in order to provide appropriate coupler structure.

It is shown in Ref. [20] that selective mode excitation has a valuable advantageous feature: being simply employed in the form of MDM technique, selective mode launching can resist seriously to eavesdropping performed with noted intrusion techniques. Indeed, fiber bending results in light irradiation in all points along the formed fiber curve. So, every fiber mode is associated with the certain highly divergent output beam having angular width equal to the central angle of the fiber bend arc. Evidently, these output beams superimpose when different fiber modes propagate simultaneously. Then, the information in the registered eavesdropped signal becomes mixed.

Similar situation occurs in cases of microbendings and core radius variations where every core mode excites a group of cladding modes of adjacent orders, and information carried by neighbor fiber modes becomes mixed just at this stage because the same cladding mode groups appear from different fiber modes. When these cladding mode groups are split off with placed immersion or fiber macrobending, or if scattering at cladding imperfections is registered, a degree of information mixing rises.

As for the scheme employing optical tunneling to high-index immersion drop or pressed bulk prism, consideration performed in Ref. [20] for commercial MMF showed that one can also expect a superposition of output space beams accompanied with information mixing if neighbor fiber mode groups are used for data transmission according to the MDM scheme. Similar assumption concerns the external grating tool because of likeness between performances of space beams formed by prism and grating directional couplers operating as longitudinal directional input/output units. Mixed information can also be expected in case of application of extra higher index fiber. Difference between mode spectra of trunk and mentioned fibers does not allow choosing the fiber inclination angle that could provide meeting the matching conditions for the set of modes simultaneously, and parasitic mode launching occurs in the extra fiber leading to cross-talk and mixing the data read from different mode channels.

The last noted splitting scheme is tunneling to the extra fiber along with the fiber length. In this case, particularly if both fibers are of the same type, simple application of MDM technique can fail in data protecting, and special system schemes should be built to counteract eavesdropping.

First of all, the fiber system must be capable of detecting the intrusion attempt and issue the alarm signal. This important feature is already realized in some fiber systems, for example, in the experimental system described in Ref. [21]. That system exploits selectively excited fundamental mode of graded-index MMF for high-bit-rate data transmission, and that mode is launched with external strip waveguide by on-axis butt joint. Another strip waveguide of that external chip excite several higher order fiber modes by off-axis butt joint, and these modes form together the monitor channel. Due to the character of mode field cross distributions, higher order modes are more sensitive to fiber bending and other variants of impact on the fiber than the fundamental mode. Therefore, monitor signal decreases immediately at the beginning of intrusion, and the control block issues the alarm signal timely. Because of the used mode coupler, only selectively excited lowest order fiber mode can be exploited for high-bit-rate data transmission in the built system. However, application of the mode coupler of another type could enable employing the MDM technique and providing parallel transmission of additional data streams and/or random signals, while the monitor channel still controls intrusion attempts. Besides the increase of the system information capacity that could allow achieving additional enhancement of data security in the system due to the reasons noted above. Data transmission part of the system of that kind is shown schematically in **Figure 5**.

Of course, the presented scheme can be reorganized to bidirectional transmission by adding the appropriate system blocks. The features to be remained in the scheme revisions are application of MDM technique for low-order modes and building the monitor channel for higher order modes.

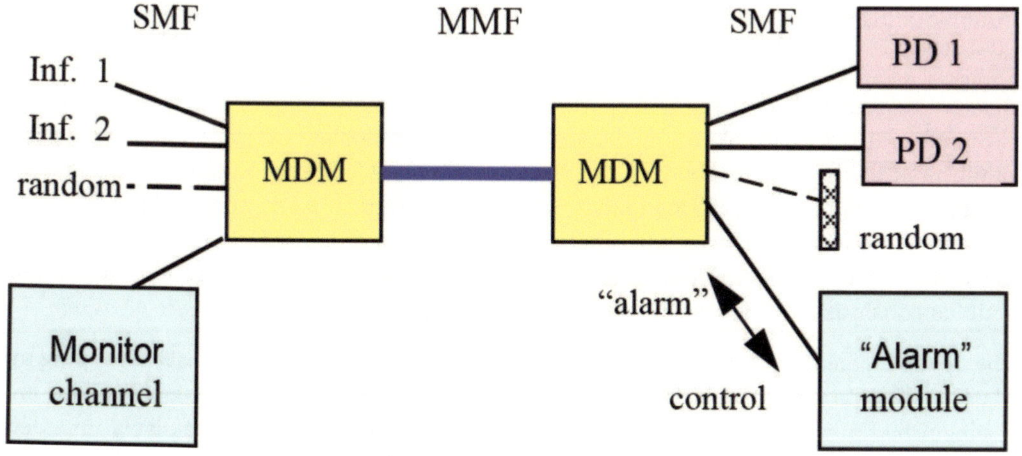

Figure 5. Scheme of transmission part of MDM system with monitor/alarm line. PD is photodetector.

Another possible system scheme could use a specific technique of data transmission. As in TDM technique, the data stream can be divided in time periods each containing data blocks. Each block of the same data period is directed individually to the certain optical delay line. The kit of delay lines provides time synchronization of blocks at the line outputs. Then, the optical signal from each line is transmitted over the certain mode of the trunk fiber. So, we obtain simultaneous transmission of originally sequential data blocks. At the end of trunk fiber, outcoupled mode signals pass via the output kit of delay lines performing reversal time shifts of data blocks and thus providing restoration of their original sequence. **Figure 6** shows the

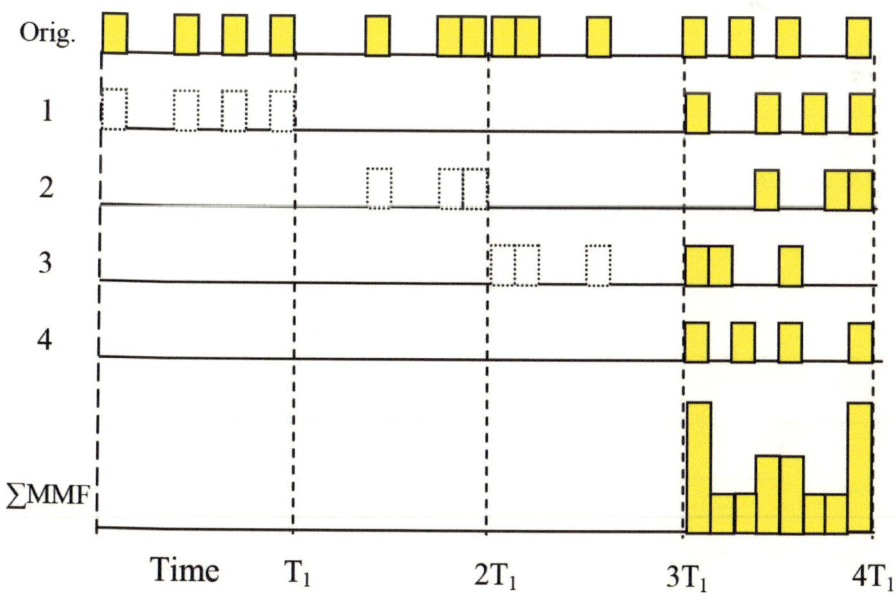

Figure 6. Time distribution of data bit blocks by the modes of the trunk fiber.

example of distribution of data stream by four trunk fiber modes. Here, one data time period is demonstrated as a bit stream in NRZ format.

In order to simplify the diagram, the period length is extremely shortened, and only 8 bits are included into each data block. T_1 denotes the time duration of one data block, while the numbers at the lines correspond to different operating trunk fiber modes. Four fiber modes having different principal mode numbers can be really excited independently (it follows, for example, from **Figure 3** where experimental characteristics of the selective mode coupler are presented). One can see that individually delayed data blocks come to the trunk fiber input simultaneously as they are shown within the time range $3T_1 \ldots 4T_1$.

The signal denoted as ∑MMF represents the superposed data resulted from eavesdropping by fiber bending or similar nonselective technique. However, even if different fiber modes are distinguished in some intrusion procedure, decoding of the read data is massively impeded because the eavesdropper must determine right sequence of registered data blocks.

Figure 7 presents the scheme capable of performing the proposed technique for the noted case of four mode channels into the trunk fiber. The circles at SMF lines denote optical delay lines which can be represented by fiber loops of certain lengths. The extra line is reserved for random signal that could be transmitted over fifth selectively excited fiber mode filling the "empty" time windows $0 \ldots 3T_1$ appeared in the trunk fiber.

The data distributor can be built as the integrated optical chip combining known waveguide switches as it is shown in **Figure 8** together with the time diagram of tuning electric signals. For simplicity of the pattern, only one of every pair of switch electrodes is plotted.

Distributor provides division of data period to data blocks and individual directing each block to corresponding output channel. The input signal is the original data bit sequence. Every switch is activated with tuning electric impulse whose duration equals to the chosen duration of the data block. Consecutive turning on the switches performs directing the consecutive data blocks to different outputs of the distributor, and time distribution of those data blocks by device outputs is depicted in **Figure 6** as dotted bit images. According to the scheme shown in **Figure 7**, each output of the distributor is joined with the corresponding SMF delay line. Performing individual delaying, those lines synchronize data bit blocks at the input of the MDM unit as it is seen in **Figure 6** in the time range $3T_1 \ldots 4T_1$. The MDM unit matches each

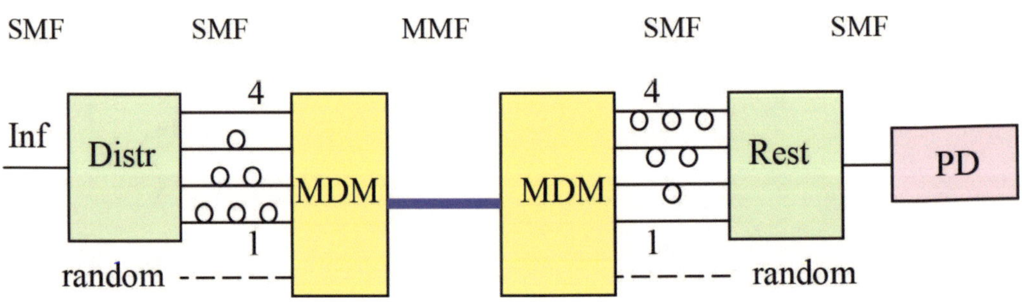

Figure 7. Scheme of transmission part of MDM system performing distribution of information data sequence by trunk modes.

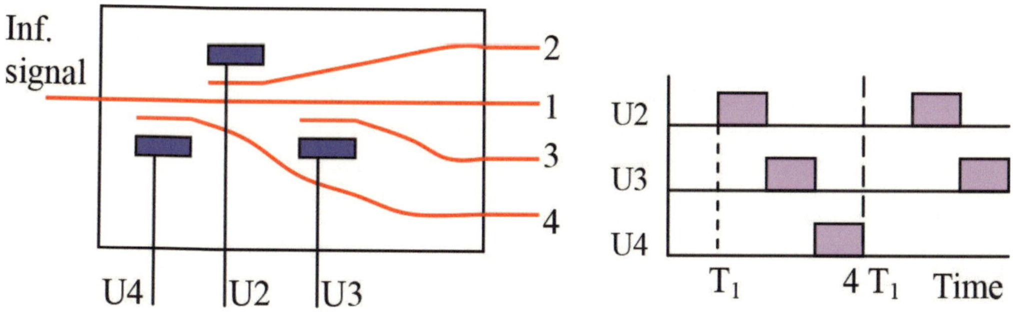

Figure 8. Scheme of data distributor/restorer together with the diagram of tuning voltage.

delay line with the certain trunk mode, and shown distribution of delayed data bit blocks represents time distribution of light pulses by different trunk fiber mode channels at the entire end of trunk line.

Let us evaluate the required lengths of SMF delay loops at the transmitter for the scheme with four mode information channels into the trunk fiber. The length of the certain SMF at the input of the MDM unit can be written as

$$L_i^{inp} = L_4^{inp} + (4 - i) \cdot \delta L_i \tag{1}$$

where i is the number of SMF (let us define that the signal from the first SMF is to be coupled to the trunk fiber of lower order); $\delta L_i = cT_1/N_{ms} = cK_b/fN_{ms}$ is the difference between the lengths of two neighbor SMFs; c is the light speed in vacuum; K_b is the amount of bits in the data block; N_{ms} is the mode index of SMF; f is the data stream frequency. L_4^{inp} corresponds to the SMF line whose signals are not delayed specially, and this length is minimal among four SMFs being defined by constructive reasons only. For $K_b = 8$, $f = 800$ MHz, $N_{ms} \approx 1.47$ (for $\lambda = 1.3$ μm), we obtain from Eq. (1) $\delta L_i \approx 2$ m. Although the used data block length is too small for practice, this rough evaluation resulted in the reasonable meaning of δL_i which could be still appropriate for block lengths of more than ten times longer. The frequency of tuning electrical signal applied to data distributor is $f_{el} = f/K_b K_{ic}$, where K_{ic} is the amount of mode informational channels. In the considered example $K_{ic} = 4$, and then $f_{el} = 25$ MHz that is also the reasonable value which will decrease inversely to the rise of data block length.

Regarding SMF loops in the receiver system part, corresponding SMF lengths should be in reversal relation in order to compensate data block delays performed in the transmitter and to recover original data sequence. Furthermore, here the differences in propagation times of different trunk fiber modes must be considered additionally. So, the length of the certain SMF in the receiver becomes as

$$L_i^{out} = L_1^{out} + (i - 1) \cdot \delta L_i + \delta L_i^{dmd} \tag{2}$$

where L_1^{out} is the length of shortest output SMF whose optical signal is not delayed specially in processing at the receiver. δL_i^{dmd} is the increment of SMF length defined by intermodal dispersion in the trunk MMF. That increment can be determined as

$$\delta L_i^{dmd} = c\delta t_i / N_{ms} \tag{3}$$

where δt_i is the differential delay time of the corresponding trunk mode. Depending on the types of employed trunk fibers, this delay time can vary in a wide range having significantly larger values in step-index fibers than the ones in graded-index fibers.

Maximal delay time caused by intermodal dispersion can be determined from optical pulse broadening when the whole mode spectrum is launched. Then, the specific delay related to the trunk fiber length is

$$\delta t_{sf} / L_{tr} = n_1 \Delta / c \tag{4}$$

where $\Delta = (n_1 - n_2)/n_1$, L_{tr} is the fiber length; n_1 is the maximal index in the core cross-section (uniform value in a step-index fiber); n_2 is the cladding index [22]. Substituting to Eq. (4), the values $n_1 = 1.48$ and $\Delta = 0.01$ for the case of step-index fiber, we obtain $\delta t_{sf}/L_{tr} \approx 50$ ns/km that seems too big despite delay times for several low-order neighbor modes are evidently less. The first scheme with control/alarm line and independent data channels seems more appropriate for step-index fiber systems.

Unlike that fiber type, graded-index fibers demonstrate much less intermodal delay due to the nature of mode propagation via the fiber with radial refractive index cross distribution. The intermodal dispersion reaches minimal values when the index cross profile is close to a parabolic form. Optimization of profile decrement parameters for the chosen wavelength allows reaching the values of maximal delay less than 100 ps/km for $\Delta = 0.01$, but then, the real index profile must correspond precisely to the theoretical one. In practice, the value as about $\delta t_{gf}/L_{tr} \approx 1$ ns/km can be chosen in approximate estimations as a maximal delay time in commercial graded-index fibers. Now, let us evaluate a differential delay time in such fibers. Considering the performances of MMFs having parabolic-like index profile in circular cross-section, one finds equidistant dependence of mode propagation constant β in terms of β^2 on the principal number M (the value characterizing separate mode groups each containing the set of degenerate fiber modes; just those groups can be exploited as information channels), and group mode indices can also be assumed approximately equidistant. Total amount of mode groups is evaluated as

$$M_0 = kn_1 a(\Delta/2)^{1/2} \tag{5}$$

where a is the core radius; $k = 2\pi/\lambda$; λ is the light wavelength in vacuum [23]. Equidistant mode indices mean equidistant distribution of mode group velocities that in turn lead to equal increments of delay time counted between adjacent mode groups. Therefore, for our case of four mode channels, we obtain following term that determines the specific delay time with respect to propagation of the fastest mode:

$$\delta t_i / L_{tr} \approx \delta t_{gf} (4 - i) / M_0 L_{tr} \tag{6}$$

Then, the desired SMF length increment for the case of graded-index trunk fiber is

$$\delta L_i^{dmd} \approx c(i-1) \cdot \delta t_{gf} / M_0 N_{ms} \tag{7}$$

(the higher-order mode propagates faster and must be delayed in more degree). Substituting to Eqs. (5) and (7), the values $a = 25$ μm, $\Delta = 0.01$, $n_1 = 1.486$, $N_{ms} = 1.47$, $\lambda = 1.33$ μm, we obtain (for trunk line as $L_{tr} = 1$ km) the following set of SMF elongations δL_i^{dmd}: 0, 16.5, 33, and 49.5 mm placed here in priority according to the rise of the channel number. For longer trunk lines, one must multiply these values by the factor of the distance in km. Regarding the reasonable trunk distances (no more than 100 km), the SMF lengths required for recovering the original data sequence evidently dominate over these additional elongations in case of graded-index trunk fiber. Additional precise tuning of data block time shifting can be achieved if active fibers with longitudinal electrodes exploiting an electrooptical effect are employed as SMFs in the receiver. Appropriate levels of uniform voltage biases are to be determined in set-up procedures when installing the system.

The functions of restorer and distributor are reversal, and their constructions are symmetric. The restorer has four inputs (in our four-channel example) and one output. Input signals are the data blocks issued from delay lines at the receiver part where they have got individual time shifts that provide right time sequence of these blocks, whereas they propagate in parallel lines yet. These four fiber lines are joined with restorer inputs, and time distribution of the input blocks corresponds to location of dotted data bit images in **Figure 6**. Sequential turning on the restorer electrooptical switches by the tuning voltages (whose diagram is shown in **Figure 8**) enables transferring light pulses of all data blocks to one waveguide (one can imagine reversal light tracing in the scheme plotted in **Figure 8**). Then, the output restorer signal becomes as the original data bit sequence shown in **Figure 6**.

Simple filling the "empty" time windows by random signals in the fifth trunk mode can be replaced with different random bit streams (whose frequencies equal to the data stream frequency) each filling the certain local "empty" window in different trunk fiber mode. Then, the scheme transforms into the variant shown in **Figure 9**.

Every random sequence is divided to random bit blocks by the same manner as described above for the data sequence, and SMF loops are also employed to synchronize random blocks at the input of the MDM unit (those loops are plotted as corresponding circles at the scheme lines). That random blocks are directed from distributors to multiplexing connectors where the channel signals are formed containing each a continuous sequence of one data block and three different random blocks as it is depicted in **Figure 10**.

These signals are multiplexed by the MDM unit providing here filling each trunk mode channel with continuous bit sequences. That circumstance impedes significantly separation of random and data blocks in eavesdropped signal even if trunk modes are distinguished in intrusion procedure.

Upon passage the distance and executing a mode demultiplexing, bit sequences come to the extractor where random signals are omitted, while extracted data blocks are directed further to the restorer and processed there as described above for recovering the original data sequence. **Figure 11** demonstrates the schemes of the multiplexing connector and the extractor together with the diagram of tuning voltage. The devices are built on widespread elements—known 3dB-splitters and electrooptical switches.

Extractor includes four (for our example) switches each having one output joined with further system scheme line, while another output is empty. The signals that come to every extractor input are the sequences of data and random blocks shown in **Figure 10**. Corresponding blocks come to parallel extractor inputs simultaneously. Therefore, extractor switches are tuned with the same voltage signal (for simplicity, only one electrode of every pair is plotted at all switches). Activating the switches during the time range $T_1 \ldots T_4$ every data period, we direct

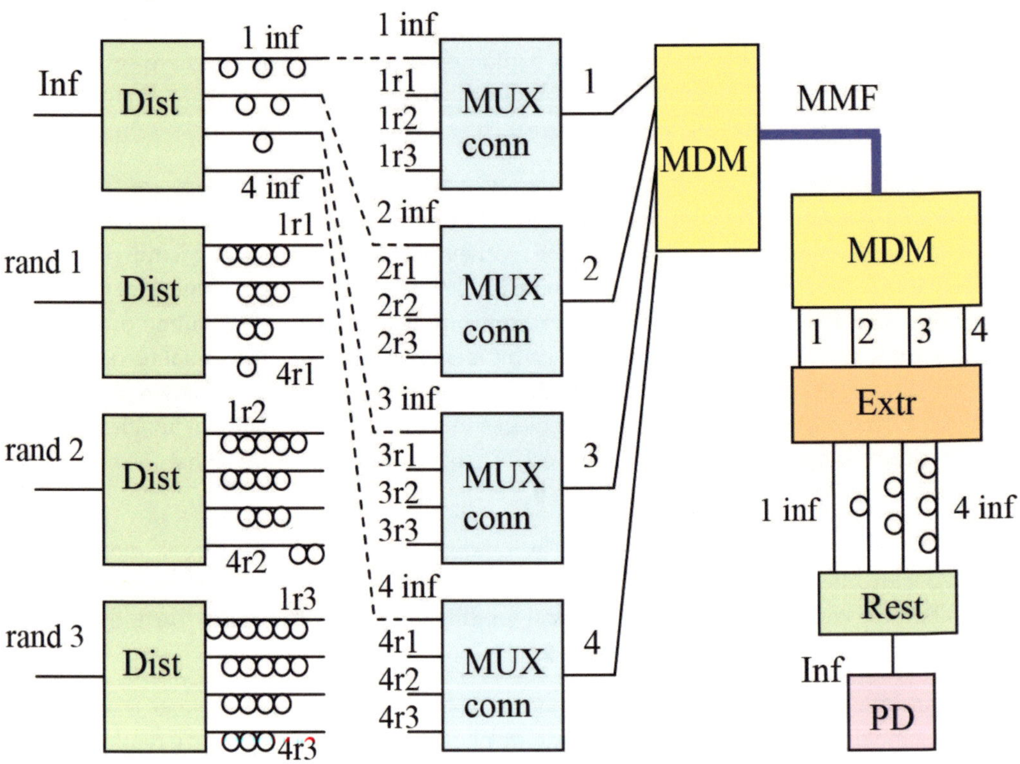

Figure 9. Scheme of transmission part of MDM system performing distribution of information data sequence by trunk modes and filling the empty time windows with random signals.

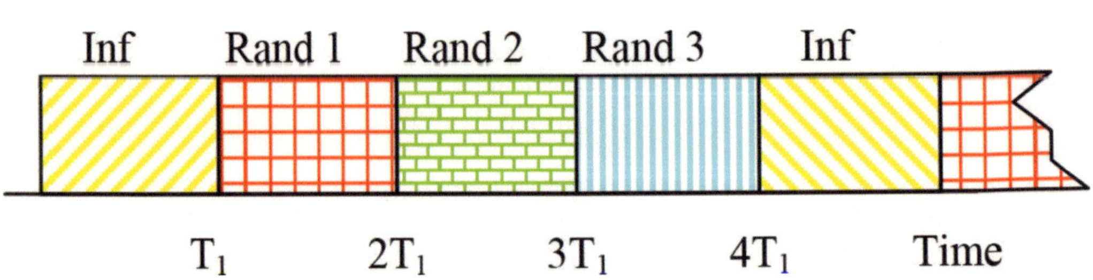

Figure 10. Sequence of data and random bit blocks in each mode of the trunk fiber.

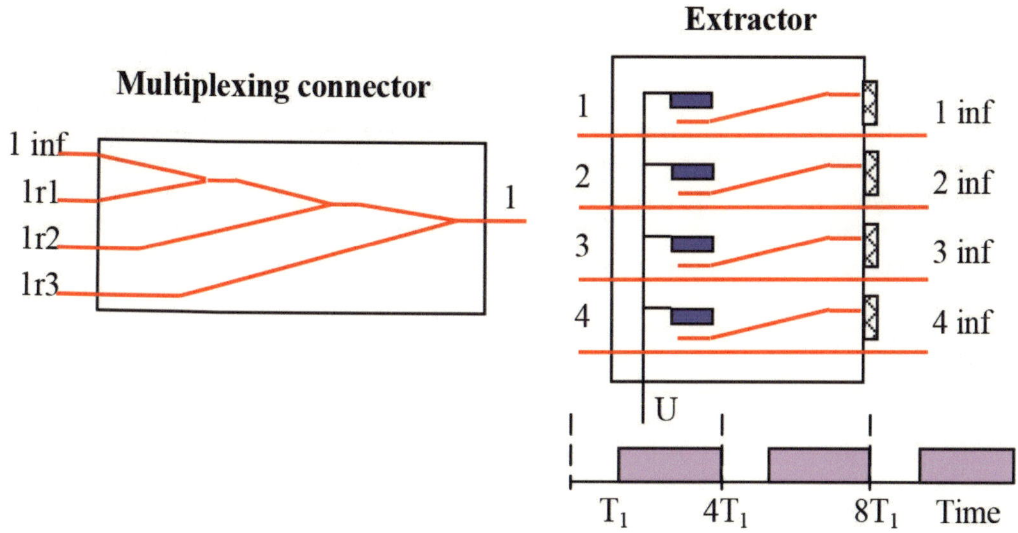

Figure 11. Schemes of the connector and the extractor depicted in **Figure 9** together with the diagram of tuning voltage.

the random bit blocks to empty outputs, while the data bit blocks pass through the extractor during times 0 … T_1. Then, the signal at the extractor output is represented by data bit blocks issued simultaneously from four parallel channels, and their distribution by those channels is the same as the one plotted in **Figure 6** within time limits $3T_1 \ldots 4T_1$. Synchronization of tuning voltages with data blocks can be executed with repeated timely service signals transmitted instead of data bit blocks.

To provide simultaneous coming of the data blocks to extractor inputs, differential mode delay in the trunk fiber must be considered preliminary. Therefore, in this scheme, the lengths of delaying SMFs at the input of MDM unit in the transmitter are determined as

$$l_i^{inp} = l_4^{inp} + (4 - i) \cdot \delta L_i + \delta L_i^{dmd} \tag{8}$$

where l_4^{inp} is the length of shortest SMF line, and other values are determined as above using Eqs. (1) and (7). The lengths of SMF delay lines that follow the extractor become as

$$l_i^{out} = l_1^{out} + (i - 1) \cdot \delta L_i \tag{9}$$

where l_1^{out} is the length of shortest SMF corresponding to the lowest order operating trunk mode. Elongation of delaying SMFs for random bit blocks is conducted by the same manner except that the required amount of SMF sections (each of δL_i length) for certain line can be determined from the scheme in **Figure 9** as a number of circles plotted at the corresponding line. The delayed signals are directed to the restorer where they are processed as described above. As the result, original data bit sequence becomes recovered.

Considering all above, one can conclude that application of selective mode excitation could really lead to substantial improvement of system data security.

4. Conclusion

The presented materials emphasize advantages of application of selective mode excitation that lead to both increasing the system information capacity and enhancing data security. Realizability of three-stage hierarchical scheme of data compression is shown, and also ability of the mentioned technique to embarrass eavesdropping is highlighted. Application of selective mode excitation in the form of MDM technique could improve significantly performances of fiber systems, at least the ones like short- and middle-haul communication lines and LANs.

Author details

Dmitry V. Svistunov

Address all correspondence to: svistunov@mail.ru

Peter-the-Great St.-Petersburg Polytechnic University, St.-Petersburg, Russia

References

[1] Tur M, Menashe D, Japha Y, Danziger Y. High-order mode based dispersion compensating modules using spatial mode conversion. Journal of Optical Fiber Communications Reports. 2007;**4**(2):110-172. DOI: 10.1007/s10297-007-0081-0

[2] Salsi M, Koebele C, Sperti D, Tran P, Mardoyan H, Brindel P, Bigo S, Boutin A, Verluise F, Sillard P, Bigot-Astruc M, Provost L, Charlet G. Mode-division multiplexing of 2 × 100 Gb/s channels using an LCOS based spatial modulator. Journal of Ligthwave Technology. 2012;**30**(4):618-623. DOI: 10.1109/JLT.2011.2178394

[3] Hu T, Li J, Ren F, Tang R, Yu J, Mo Q, Ke Y, Du C, Liu Z, He Y, Li Z, Chen Z. Demonstration of bidirectional PON based on mode division multiplexing. IEEE Photonics Technology Letters. 2016;**28**(11):1201-1204. DOI: 10.1109/LPT.2016.2530663

[4] Kawaguchi Y, Tsutsumi K. Mode multiplexing and demultiplexing devices using multimode interference couplers. Electronic Letters. 2002;**38**(25):1701-1702. DOI: 10.1049/el:20021154

[5] Lee BT, Shin SY. Mode-order converter in a multimode waveguide. Optical Letters. 2003;**28**(18):1660-1662. DOI: 10.1364/OL.28.001660

[6] Karpeev SV, Pavelyev VS, Khonina SN, Kazanskiy NL, Gavrilov AV, Eropolov VA. Fibre sensors based on transverse mode selection. Journal of Modern Optics. 2007;**54**(6):833-844. DOI: 10.1080/09500340601066125

[7] Carpenter J, Thomsen BC, Wilkinson TD. Degenerate mode-group division multiplexing. Journal of Lightwave Technology. 2012;**30**(24):3946-3952. DOI: 10.1109/JLT.2012.2206562

[8] Koonen AMJ, Chen H, Boom HPA, Raz O. Silicon photonic integrated mode multiplexer and demultiplexer. IEEE Photonics Technology Letters. 2012;**24**(21):1961-1964. DOI: 10.1109/LPT.2012.2219304

[9] Love JD, Riesen N. Mode-selective couplers for few-mode optical fiber networks. Optical Letters. 2012;**37**(19):3990-3992. DOI: 10.1364/OL.37.003990

[10] Li A, Chen X, Al Almin A, Shieh W. Fused fiber mode couplers for few-mode transmission. Photonics Technology Letters. 2012;**24**(21):1953-1956. DOI: 10.1109/LPT.2012.2218803

[11] Vaissie L, Johnson EG. Selective mode excitation by non-axial evanescent coupling for bandwidth enhancement of multimode fiber links. Optical Engineering. 2002;**41**(8):1821-1828. DOI: 10.1117/1.1488607

[12] Hoyningen-Huene J, Ryf R, Winzer P. LCoS-based mode shaper for few-mode fiber. Optics Express. 2013;**21**(15):18097-18110. DOI: 10.1364/OE.21.018097

[13] Chen W, Wang W, Yang J. Mode multi/demultiplexer based on cascaded asymmetric Y junctions. Optics Express. 2013;**21**(21):25113-25119. DOI: 10.1364/OE.21.025113

[14] Tsekrekos CP, Syvridis D. Symmetric few-mode fiber couplers as the key component for broadband mode multiplexing. Journal of Lightwave Technology. 2014;**32**(14):2461-2467. DOI: 10.1109/JLT.2014.2327479

[15] Igarashi K, Souma D, Tsuritani T, Morita I. Selective mode multiplexer based on phase plates and Mach–Zehnder interferometer with image inversion function. Optics Express. 2015;**23**(1):183-194. DOI: 10.1364/OE.23.000183

[16] Dong J, Chiang KS, Jin W. Compact three-dimensional polymer waveguide mode multiplexer. Journal of Lightwave Technology. 2015;**33**(22):4580-4588. DOI: 10.1109/JLT.2015.2478961

[17] Svistunov DV. A planar prism for detection and selective excitation of modes in a multimode channel waveguide. Technical Physics Letters. 2004;**30**(4):332-334. DOI: 10.1134/1.1748616

[18] Svistunov DV. Selective mode launching in a multimode channel waveguide by planar coupler. Journal of Optics A Pure Applied Optics. 2004;**6**(9):859-861. DOI: 10.1088/1464-4258/6/9/007

[19] Svistunov DV. Evaluating the possibility of using mode compression in a cascade system for data compression in multimode fiber-optic communication lines. Journal of Optics Technology. 2012;**79**(1):29-35. DOI: 10.1364/JOT.79.000029

[20] Svistunov DV. Cascade scheme of data compression in multimode fiber systems. In: Simmons M, editor. Network Coding and Data Compression: Theory, Applications and Challenges. New York: Nova Science; 2015. pp. 151-161. ISBN: 978-1-63483-185-7

[21] Asawa CK. Intrusion-alarmed fiber optic communication link using a planar waveguide bimodal launcher. Journal of Lightwave Technology. 2002;**20**(1):10-18. DOI: 10.1109/50.974813

[22] Lecoy P. Fiber-Optic Communications. London-Hoboken: ISTE-Wiley; 2008. p. 333. ISBN: 978-1-84821-049-3

[23] Olshansky R, Keck DB. Pulse broadening in graded-index optical fibers. Applied Optics. 1976;**15**(2):483-491. DOI: 10.1364/AO.15.000483

Section 2

Optical Wireless Systems

Chapter 7

Holograms in Optical Wireless Communications

Mohammed T. Alresheedi,
Ahmed Taha Hussein and Jaafar M.H. Elmirghani

Additional information is available at the end of the chapter

http://dx.doi.org/10.5772/intechopen.68408

Abstract

Adaptive beam steering in optical wireless communication (OWC) system has been shown to offer performance enhancements over traditional OWC systems. However, an increase in the computational cost is incurred. In this chapter, we introduce a fast hologram selection technique to speed up the adaptation process. We propose a fast delay, angle and power adaptive holograms (FDAPA-Holograms) approach based on a divide and conquer methodology and evaluate it with angle diversity receivers in a mobile optical wireless (OW) system. The fast and efficient fully adaptive FDAPA-Holograms system can improve the receiver signal to noise ratio (SNR) and reduce the required time to estimate the position of the receiver. The adaptation techniques (angle, power and delay) offer a degree of freedom in the system design. The proposed system FDAPA-Holograms is able to achieve high data rate of 5 Gb/s with full mobility. Simulation results show that the proposed 5 Gb/s FDAPA-Holograms achieves around 13 dB SNR under mobility and under eye safety regulations. Furthermore, a fast divide and conquer search algorithm is introduced to find the optimum hologram as well as to reduce the computation time. The proposed system (FDAPA-Holograms) reduces the computation time required to find the best hologram location from 64 ms using conventional adaptive system to around 14 ms.

Keywords: adaptive hologram, delay spread, SNR

1. Introduction

In general, holography is the storage of the phase and amplitude information of a wavefront. It is usually used as an approach to create three-dimensional (3D) images of objects through interference between a wavefront diffracted and a coherent reference beam. In optical wireless communications (OWC), the hologram is a transparent or reflective device that is used to

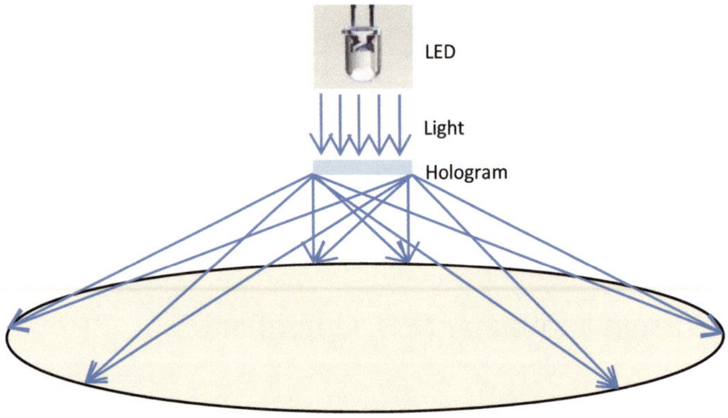

Figure 1. Holographic diffuser with uniform intensities that cover desired area.

spatially modulate the phase or amplitude of the energy passing through it. **Figure 1** illustrates the effect of a diffusing hologram on a set of rays from a light source. Here, the light source (light emitting diode (LED)) is split into a number of beams that cover the desired area.

Holograms can be produced from mathematical description or physical object. In mathematical approach, any wavefront can be generated. If mathematical method is implemented with a computer, the hologram is called a computer generated hologram (CGH). A ground glass diffuser can be given as a simple example of physical object diffuser, where ground glass can be placed at the output of a laser to change it from a point source to a large area source. However, this type of hologram cannot be controlled.

Beam steering has been widely studied in wireless communication systems to maximise the signal to noise (SNR) at the receiver [1, 2]. It is also considered as an attractive option in optical wireless communication (OWC) systems to enhance the system performance [3, 4]. New adaptive technique using beam steering is introduced in visible light communication (VLC) links in Ref. [5]. The goal is to maximise the SNR at the receiver in all possible locations within an indoor environment. Simulations results have shown that high data rate up to 20 Gb/s can be achieved by partially steering some of the beams towards the receiver location. Multi-input-multi-output (MIMO) infrared (IR) links employing beam steering method has been introduced in Ref. [6]. Furthermore, demonstration of IR-linked energy transmission using beam forming along with a spatial light modulator (SLM) is shown in Ref. [7]. An efficient power and angle adaptation technique is proposed in Refs. [1–4] in order to help the IR optical wireless (IROW) transmitter to optimise the diffusing spots distribution. These methods (power and angle adaptations) are able to enhance the received signal strength level, regardless of the receiver's location, the receiver's field of view (FOV) and transmitter's position. A significant performance improvement using beam angle and beam power adaptation in a line strip multi-beam system (APA-LSMS) is shown in Refs. [1, 2]. However, a cost has to be paid due to the complex adaptation requirements. The adaptive APA-LSMS transmitter needs to generate a single spot and scanning with all the possible locations (around 8000 locations) in the room in order to find the receiver and then generate the hologram with optimum powers and angles. This makes APA-LSMS system design very challenging.

In this chapter, we aim to point out the impairments of IROW links and propose new efficient solutions beyond those reported in Refs. [1–4]. We report an adaptive hologram selection method employing simulated annealing (SA) to generate diffusing spots (multi-beam). The proposed system is pre-calculated and stored all the holograms in memory. Each stored hologram is suited for a given transmitter and receiver location. Hence, it eliminates the need to calculate holograms real time at each transmitter-receiver location. We model fast angle and power adaptive holograms (FAPA-Holograms) and fast delay, angle and power adaptive holograms (FDAPA-Holograms) mobile OW systems, in conjugation with angle diversity receivers [8]. The conventional diffuse system (CDS) and line strip multi-beam systems (LSMS) are studied for comparison purposes. The ultimate goal of the proposed systems: FAPA-Holograms and FDAPA-Holograms is to reduce the time required to generate hologram at optimum transmitter and receiver location as well as to enhance the overall system performance such as SNR and channel bandwidth in a typical indoor environment. A significant improvement can be obtained by increasing the scanned stored holograms in our systems to approach the original power and angle adaptive methods proposed previously in Refs. [1, 2]. However, increasing the number of scanned stored holograms leads to an increase in the computation time needed to find the best hologram. To overcome this issue, we introduce a divide and conquer (D&C) algorithm to select the best hologram among a finite vocabulary of holograms, hence speed up the adaptation process associated with these adaptive systems [8]. High data rates of 2.5 and 5 Gb/s are considered for the FAPA-Holograms and FDAPA-Holograms systems.

The remainder of this chapter is organised into the following sections: Section 2 presents The IROW room setup and channel characteristics. Section 3 presents the proposed systems' configurations. Section 4 introduces the simulation results and discussion of the IROW systems. Finally, conclusions are drawn in Section 5.

2. The IROW room setup and channel characteristics

In order to study the impact of directive ambient light noise sources and multipath dispersion, as well as their effect on the received data flow, consideration was given to an unoccupied rectangular room that had no furnishings, with dimensions of 8 m × 4 m × 3 m (length × width × height). Researchers in Ref. [9] have studied and investigated the power reflected in indoor IROW system. The study found that the light reflected on either ceiling or wall is Lambertian in nature (mode $n = 1$). They also found that the wall reflected power by 80% whereas ceiling by only 30%. In this chapter, we consider that the rays reflected from door and windows are similar to those coming from walls. The reflecting elements from walls and ceiling can be modelled by dividing the surfaces into a small square shape, which can operate as secondary Lambertian transmitter with $n = 1$.

The accuracy of the impulse response profile is controlled by the size of the reflecting elements. Therefore, element sizes of 5 cm × 5 cm for the first order reflections and 20 cm × 20 cm for the second order reflections are employed for all arrangements. Previous work studied the received optical power within an indoor environment. They found that most of the received optical power is located within the two first order reflections (1st and 2nd). Third order and higher reflections are highly attenuated [10, 11]. Hence, two bounces are considered in our calculations. All the proposed systems use an upright transmitter with 1 W optical power. Furthermore, the significant signal to noise ratio (SNR) improvement of the hologram-proposed systems is used to reduce the transmit power to 80 mW reducing the power density on the adaptive hologram and helping eye safety.

In OW communication links, intensity modulation with direct detection (IM/DD) is considered the most viable approach. The indoor OW IM/DD channel can be fully specified by its impulse response $h(t)$, and it can be modelled as a baseband linear system given by

$$I(t, Az,\ El) = \sum_{m=1}^{M} Rx(t) \otimes h_m(t,\ Az,\ El) + \sum_{m=1}^{M} Rn(t,\ Az,\ El). \tag{1}$$

where $I(t,\ Az,\ El)$ is the current instantaneous due to m reflecting elements, El and Az are the directions of arrival in the elevation and azimuth angles, t is the absolute time, $x(t)$ is the optical power transmitted, ⊗ denotes convolution, M is the total number of receiving elements, R is the photodetector responsivity and $n(t,\ Az,\ El)$ is the background noise. The delay spread is a good tool to measure signal spread due to multipath propagation. The delay spread can be written as [12, 13]:

$$DS = \sqrt{\left(\sum_{\forall i} (t_i - \mu)^2 P_{r_i}^2\right) \Big/ \sum_{\forall i} P_{r_i}^2} \tag{2}$$

where the time delay t_i is associated with the received power P_{r_i} (P_{r_i} reflects the impulse response $h(t)$ behaviour) and μ is the mean delay given by

$$\mu = \sum_{\forall i} t_i P_{r_i}^2 \Big/ \sum_{\forall i} P_{r_i}^2. \tag{3}$$

The delay spared is deterministic for a given stationary transmitter-receiver and reflecting elements' positions. The delay spread can change for a given transmitter-receiver location when the reflecting elements moves or an object is entering and leaving the environment. However, the impact of such a change is not considered in this work and has not been investigated by other researchers.

The SNR of the received signal can be calculated by taking into account the powers associated with logic 0 and logic 1 (P_{S0} and P_{S1}), respectively. The SNR is given by [14]:

$$SNR = \left(\frac{R(P_{s1} - P_{s0})}{\sigma_0 + \sigma_1}\right)^2 \tag{4}$$

$$\sigma_0 = \sqrt{\sigma_{pr}^2 + \sigma_{bn}^2 + \sigma_{s0}^2} \text{ and } \sigma_1 = \sqrt{\sigma_{pr}^2 + \sigma_{bn}^2 + \sigma_{s1}^2} \tag{5}$$

where σ_{pr}^2 represents the receiver noise, which is a function of the design used for the preamplifier; σ_{bn}^2 represents the background shot noise component and σ_{s0}^2 and σ_{s1}^2 represent the shot noise associated with the received signal (P_{S0} and P_{S1}), respectively. The signal-dependent noise (σ_{si}^2) is very small due to the weak received optical signal, see the experimental results reported in Ref. [15]. In this study, we used the PIN-FET transimpedance preamplifier proposed in Ref. [16]. The background shot noise calculations can be found in Ref. [3]. Nine branches angle diversity receiver is used to reduce the impact of multipath dispersion. In this work, we employed the non-imaging angle diversity receiver design proposed in Ref. [8]. We considered maximum ratio combining (MRC) scheme. Calculations of MRC method can be found in our previous work in Refs. [1, 19].

In order to consider the impact of background noise, we use eight light bulbs in the room. These lights are used for illuminations. However, in OW receiver, the signals arrived from each light are considered as undesired signals which can be modelled as background shot noise. In this study, we assumed that Philips PAR 38 Economic' (PAR38) was used as spotlight in which each unit of light radiates 65 W within a narrow beam width. The light from these units can be modelled as a Lambertian radiant intensity with order nl = 33.1 [11]. Additional simulation parameters are given in **Table 1**.

3. Proposed systems' configurations

A CDS and non-adaptive LSMS are two widely studied configurations in the literature; therefore, they are modelled and used for comparison purposes in order to evaluate the improvements offered through the proposed configurations. More information about CDS and LSMS system can be found in Refs. [9–11].

3.1. FAPA-hologram

Power adaptation and beam angle can be considered as an effective approach, which helps to have an optimum power allocation and distribution of the diffusing spots. A single spot is produced by the adaptive transmitter to scan the ceiling and walls at approximately 8000 possible locations (2.86° beam angle increment [3]) in order to identify the best location. A liquid crystal device (IR hologram) is used to change the spot location at each step. Power adaptation technique can be used with angle adaptation to further enhance signal to noise ratio. The transmitter switches spots one by one and the receiver calculates the SNR (weight) associated with each spot. Then the feedback signal is sent by receiver at a low data rate to inform the transmitter about the SNR associated with each spot. The transmitter re-distributes the power among the spots based on their SNR weights [1] .The transmitter generates the hologram after finding optimum angles and power levels of spots. Intensive calculations and time are required from a digital signal processor (DSP). An adaptation approach is proposed where a finite vocabulary of stored holograms is used in order to get rid of computing the holograms at each step to identify the best location. The ceiling is divided into 80 regions (0.4 m × 1m per region), see **Figure 2**. This large number of regions has been selected based on our recent optimisation in Ref. [17].

Parameter	Configuration								
Width	4 m								
Height	3 m								
ρ x z wall	0.8								
y-z wall	0.8								
x-z op wall	0.8								
y-z op wall	0.8								
Floor	0.3								
	Transmitter								
Quantity				1					
Location (x,y,z)	(1 m, 1 m, 1 m)						(2 m, 7 m,1 m)		
Elevation				90°					
Azimuth				0°					
	Receiver								
Quantity	9								
Photodetector's area	10 mm^z								
Acceptance semi-angle	12°								
Location (x, y, z)	(1,1,1),(1,2,1),(1,3,1,),(1,4,1),(1,5,1),(1,6,1),(1,7,1)								
Elevation	90°	65°	65°	65°	65°	65°	65°	65°	65°
Azimuth	0°	0°	45°	90°	135°	180°	225°	270°	315°
	Resolution								
Time bin duration			0.5 ns				0.01 ns		
Bounces			1				2		
Surface elements			32,000				2000		
Number-of-spot lamps					8				
Locations				(1,1,1), (1,3,1), (1,5,1), (1,7,1) (3,1,1), (3,3,1), (3,5,1), (3,7,1)					
Wavelength					850 nm				
Preamplifier design	PIN-BJT						PIN-FET		
Bandwidth (BW)	50 MHz				2.5 GHz				5 GHz
Bit rate	50 Mbit/s				2.5 Gbit/s				5 Gbit/s

Table 1. Parameters used in simulation.

Holograms generated by means of a computer can produce spots with any prescribed amplitude and phase distribution. For FAPA-Holograms, all the spots have different weights (powers) and different phases. CGHs have many useful properties. Spot distributions can be computed on the

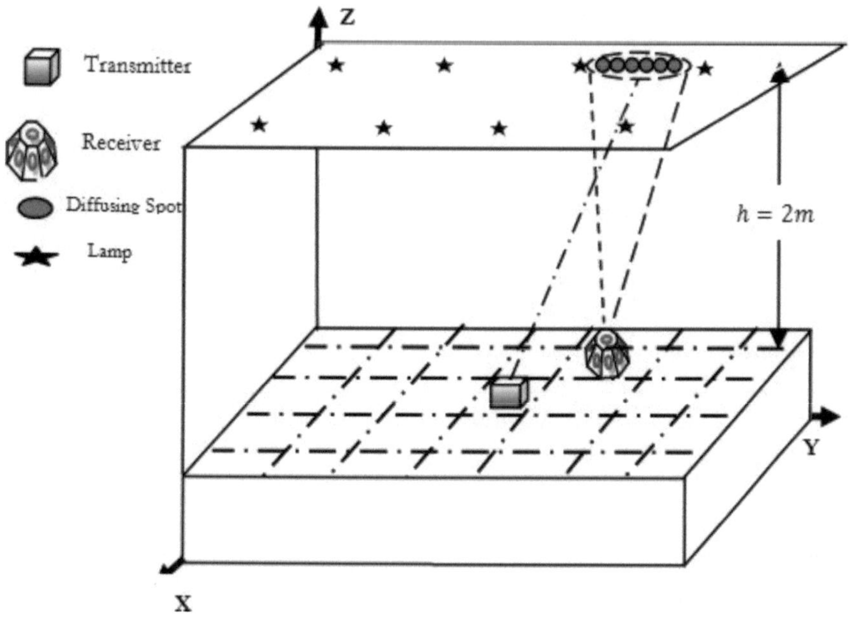

Figure 2. OW communication architecture of FAPA-Hologram system.

basis of diffraction theory and encoded into a hologram. Calculating a CGH means the calculation of its complex transmittance. The transmittance is expressed as follows:

$$H(u,v) = A(v,u).exp[j\phi(u,v)] \tag{6}$$

where $H(u, v)$ is complex transmittance function, $A(u, v)$ and $\phi(u, v)$ are amplitude and phase distribution, respectively. The parameters (u, v) are coordinates in the frequency space. The phase of incoming wavefront is modulated by hologram, whereas the transmittance amplitude is equal to unity. The analysis used in Refs. [18–20] has been employed for the design of the CGHs. The hologram $H(u, v)$ is considered to be in the frequency domain and the observed diffraction pattern $h(x, y)$ in the spatial domain. They are related by the continuous Fourier transform:

$$h(x,y) = \iint H(u,v)exp[-i2\pi(ux+vy)]dudv \tag{7}$$

The diffraction pattern of the hologram when it is placed in the frequency plane is given by

$$h(x,y) = RSsinc(Rx,Sy)\sum_{k=-\frac{M}{2}}^{\frac{M}{2}-1}\sum_{l=-\frac{N}{2}}^{\frac{N}{2}-1} H_{kl}exp[i2\pi(Rkx+Syl)] \tag{8}$$

where $sinc(a,b) = sin(\pi a)\, sin(\pi b)/\pi^2 ab$. The complex amplitude of the spots is proportional to some value of interest. But, the reconstruction will be in error because of the finite resolution of the output device and the complex transmittance of the resulting hologram. This error can be considered to be a cost function. Simulated annealing (SA) is used to minimise the cost function [21]. The phases and amplitudes of every spot are determined by the hologram pixel pattern and are given by its Fourier transform. The constraints considered in the hologram plane are to discretise the phase from 0 *to* 2π and a constant unit amplitude for the phase only CGH.

Let the desired spots in the far field be $f(x\ ,y) = |f(x\ ,y)|\exp(i\varphi(x\ ,y))$. The main target is to find the CGH distribution $g(v,\ u)$ that produces optimum reconstruction $g(x,\ y)$ that is very close to the desired distribution $f(x,\ y)$. The cost function (CF) is defined as a mean squared error which can be interpreted as the difference between the normalised desired object energy $f''(x,\ y)$ and the scaled reconstruction energy $g''(x,y)$:

$$CF_k = \sqrt{\sum_{i=1}^{M} \sum_{j=1}^{N} \left(|f''(i,j)|^2 - |g''_k(i,j)|^2\right)^2}, \tag{9}$$

where $f''(x,\ y)$ represents the normalised desired object energy and $g''_k(i,j)$ represents the scaled reconstruction energy of the kth iteration. Simulated annealing was used to optimise the phase of the holograms offline in order to minimise the cost function. The simulating annealing algorithm can help jump from local optima to close to a global optimum (minimising the cost function close to zero). The transition out of a local minima to global one is accomplished by accepting hologram phases that increase the mean squared error of the reconstruction with a given probability. The probability of accepting these phases is $exp(-\Delta CF/T)$, where $\Delta CF$is the change in error and T is a control parameter (the temperature of the annealing process). First, we start with a high value of T so that all the change in the hologram phases are accepted and then slowly lower T at each iteration until the number of accepted changes is small. This method is similar to melting a metal at a high value of T and then reducing the T slowly until the metal crystals freezes at a minimum energy. The changes of hologram phases relate to a small perturbation of the physical system, and the resulting change in the mean squared error of the reconstruction corresponds to the resulting change in the energy of the system. Therefore, this technique finds a hologram configuration, which has a minimum mean squared error (CF). For phase only CGHs, the constraints are constant amplitude and a random phase distribution φ_0 ($M \times N$). In the first iteration, a random phase is applied to help in the convergence of the algorithm.

For a large room of 8 m × 4 m, the floor is divided into 80 regions. A library that contains 6400 holograms is optimised offline using SA. In order to accurately identify the receiver location, a large number of holograms are required [17]. The optimum diffusing spots were pre-calculated based on the power and angle techniques shown in Ref. [1]. A total of 80 holograms are stored in a library and allocated for each region, the transmitter should cover the 80 possible receiver positions in the room, which means 6400 holograms are required to cover the entire room. The total number of holograms required is N^2, where N represents the number of regions into which the floor/ceiling is divided. **Figure 2** illustrates one hologram when the receiver is present at (1 m, 6 m and 1 m) and the transmitter is placed in the middle of room at (2 m, 4 m and 1 m). SA is used to optimise the phase of the CGH. **Figure 3** shows

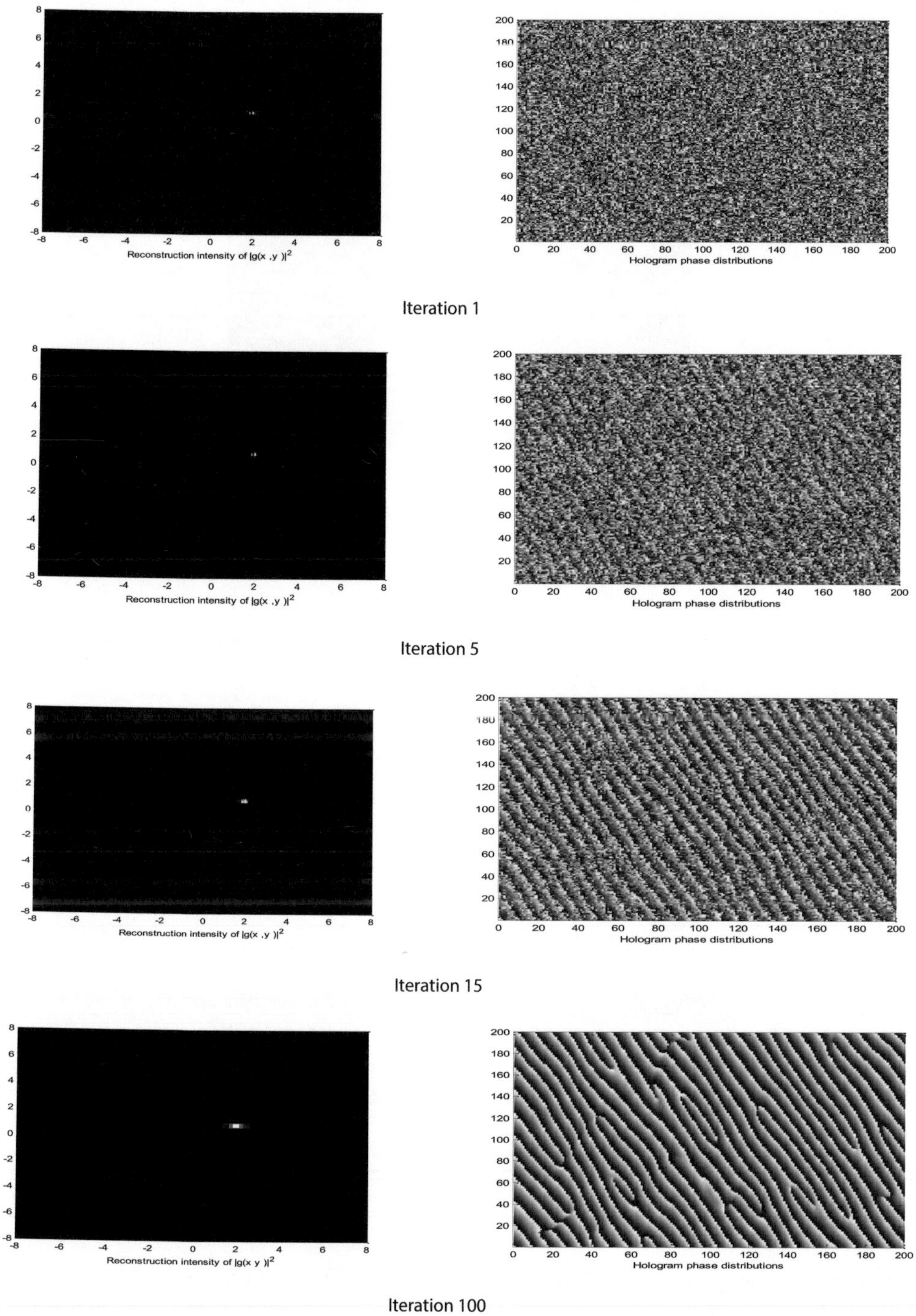

Figure 3. The reconstruction intensity at the far field and hologram phase pattern at iterations 1, 5, 15, and 100 using simulated annealing optimisation. Different grey levels represent different phase levels ranging from 0 (black) to 2π (white).

phase distributions and hologram reconstruction intensity at the far field in four snapshots. When the number of iterations increases, the reconstruction intensities are improved. The desired spot intensity in the far field is shown in **Figure 4**. **Figure 5** shows the number of iterations versus cost function.

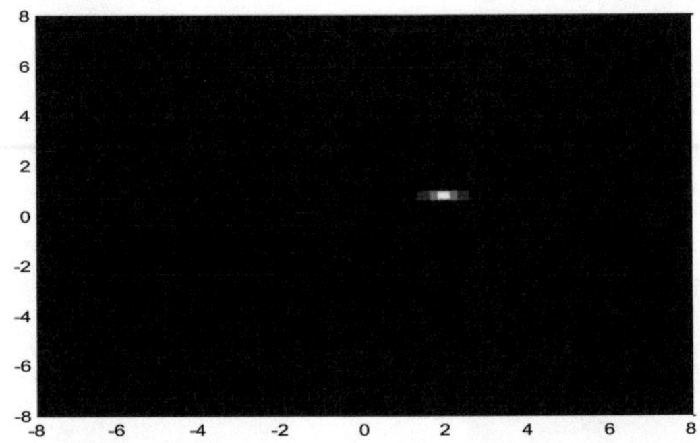

Figure 4. The desired spots intensity in the far field.

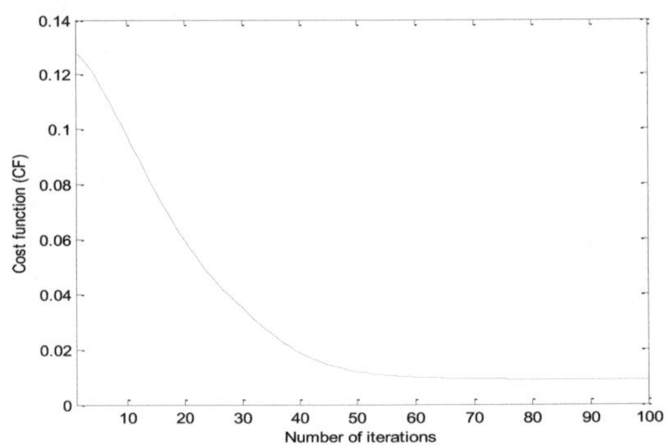

Figure 5. Cost function versus the number of iterations.

Scanning 6400 stored holograms in the system required a full search among all stored holograms, in order to select the best hologram. However, high complexity in term of the computation time is introduced. To solve this issue, a fast search technique is proposed to enhance the SNR via selecting the best hologram while reducing the computational time. The proposed search technique is based on a divide and conquer (D&C) method. Using the D&C algorithm, the transmitter is able to select the best hologram that can achieve the optimum receiver SNR. The fast search algorithm of our proposed system is applied for a single transmitter/receiver scenario as follows:

1. The stored holograms in the transmitter are first divided into four quadrants based on their transmission angles. The transmission angles associated with each quadrant are δ_{max-x} to δ_{min-x} and δ_{max-y} to δ_{min-y} in the x-axis and y-axis, respectively.

2. A single middle hologram is used in each group (quadrant) in order to find the first sub-optimum hologram.

3. The receiver calculates the SNR for each hologram and sends a pilot signal at low rate (feedback channel) to inform the transmitter about the SNR associated with each scanned hologram.

4. The transmitter selects the hologram that achieves the best SNR and identifies the new optimal quadrant (first sub-optimum quadrant) for next iteration.

5. The transmitter divides the sub-optimum quadrant into four sub-quadrants and applies steps 2–4 in order to find the second sub-optimal quadrant.

6. The D&C process is carried out and steps 2–5 are repeated to find the best location that has the highest SNR level at the receiver.

The proposed system reduces the computation time from 64 ms (each hologram required 1 ms to scan) taken by the classic beam steering system to 13 ms (13 possible locations should be scanned in all iterations × 1 ms).

3.2. FDAPA-hologram

To additionally enhance the performance of the IROW system, we introduce beam delay adaptation technique coupled with beam angle and power adaptation using hologram selection approach. The delay spread in a multi-beam spot IROW system is influenced by the spots' numbers and locations seen by each detector's FOV [3]. Therefore, switching all the beams simultaneously can introduce a differential delay between the neighbouring signals received at the receiver, hence spreading the pulse and limiting the 3-dB channel bandwidth. Instead of transmitting all the beams simultaneously, the delay adaptation algorithm sends the signal that has the longest journey first, and then sends the other signals with different differential delays (Δt) so that all the signals reach the receiver at the same time. A total of 10 μs time delay computation is carried out for each beam [8]. Therefore, a total of 1ms delay adaptation time is required for a beam. Our FDAPA-Hologram (delay, power and angle methods) requires only 14 ms adaptation time in order to select the best hologram. Assume the receiver initiates the adaptation every 1 s. This time is associated with a pedestrian speed of 1 m/s. Therefore, employing FDAPA-Hologram with a total 14 ms adaptation time introduces only 1.4%. Array element delayed switching is used to implement the delay adaptation method. The delay adaptation algorithm is explained as follows:

1. A transmitter first switches only the first beam (spot).

2. The receiver estimates the mean delay (μ) associated with first beam in the branch.

3. The receiver calculates the mean delay for all other branches.

4. The transmitter repeats steps 1–3 for other beams.

5. The receiver sends a feedback channel to update the transmitter about the delay associated with each beam.

6. The transmitter calculates the delay difference (Δt) between the beams as follows:

$$\Delta t_i = \max(t_{i\_max}) - t_i \; 1 \leq i \leq N_{spot} \quad (10)$$

7. The transmitter sends the beams with different times associated with their differential delay estimated in Eq. (10) to help the beam arrive at the receiver at the same time

The differential delay depends on the distances among the beams. If all the beams touch each other on the ceiling, then the differential delay will be few nano seconds, hence requiring timing control to switch such a beam within the required time.

4. Simulation results and discussion

In this section, we evaluate the performance of the proposed support systems in an empty room in the presence of multipath dispersion, receiver noise, back ground noise (light units) and mobility. The results are presented in terms of delay spread and SNR.

4.1. Delay spread

The delay spread of our adaptive proposed systems (FAPA-Holograms and FDAPA-Holograms) compared with CDS and LSMS system is shown in **Figure 6**. The results are presented when the transmitter is located near the room corner while the receiver moves along x = 2 m. The conventional pure diffuse system (CDS) has the largest delay compared with other systems. This is due to the diffuse transmission along with wide FOV at the receiver. Moreover, the delay spread of non-adaptive system LSMS is increased as the distance between the transmitter and receiver increases. The delay spread is almost independent of the distance between the transmitter-receiver in our hologram configurations, FAPA-Holograms and FDAPA-Hologram. This is due to beam angle adaptation technique where the proposed systems choose the hologram that has the best SNR. A significant reduction in the delay spread to 0.04 ns is achieved in our FAPA-Hologram system. Moreover, our delay adaptation method in FDAPA-Hologram reduces the delay spread of FDPA-Hologram by factor of 8. This improvement enhances the 3 dB channel bandwidth and increases the SNR at high transmission rates.

4.2. SNR

The SNR results of our proposed systems are shown in **Figure 7**. The proposed systems are tested under the influence of background noise and transmitter/receiver mobility. The pro posed adaptive holograms are compared with conventional CDS and multi-beam angle diversity LSMS system to facilitate the results with previous work published in Refs. [9–11]. The results are shown when transmitters operate at 50 Mb/s. High data rates of 2.5 and 5 Gb/s will be also considered in the next section. The transmitter is located near the room corner while the receiver moves 1 m step along x = 1 m and x = 2 m. The LSMS system provides better results than CDS with wide FOV receiver. This is due to providing direct link though spots and using non-imaging angle diversity receiver. Although the improvement has been achieved, there is

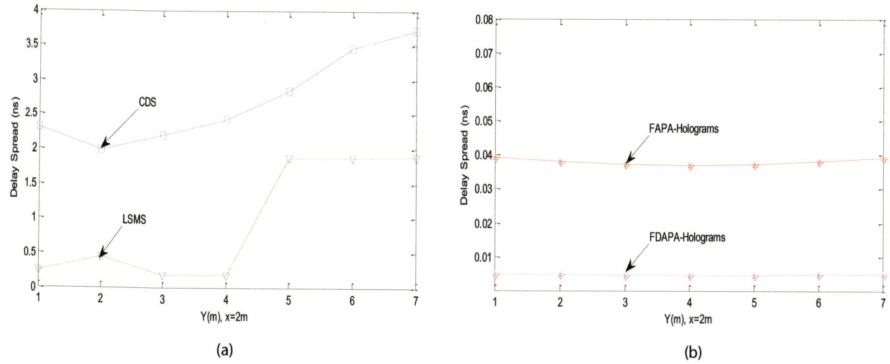

Figure 6. Delay spread of four configurations (a) CDS and LSMS, (b) FAPA-Holograms and APA-LSMS with angle diversity receiver, when the transmitter is placed at (1 m, 1 m, 1 m) and the receiver moves along $x = 2$ m line.

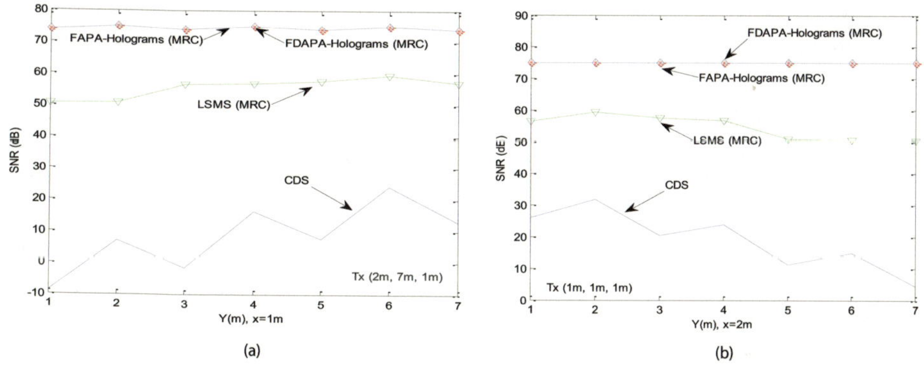

Figure 7. SNR of OW CDS, LSMS, FAPA-Holograms and FDAPA-Hologram at 50 Mbit/s, when the transmitter is located at (2 m, 7 m, 1 m) and (1 m, 1 m, 1 m) and the receiver mobiles along $x = 1$ m and $x = 2$ m lines.

degradation in the SNR results as LSMS transmitter move away from the receiver. For example, this is observed when the transmitter is moved towards the edge or the corner of the room at (2 m, 7 m and 1 m) and (1 m, 1 m and 1 m) while the receiver moves along $x = 1$ m and $x = 2$ m lines, respectively, as seen in **Figure 7(a)** and **(b)**. In order to overcome this significant reduction as well as improve the system performance, fast adaptive hologram (FAPA-Hologram and FDAPA-Hologram) systems are employed. Our proposed FAPA-Hologram achieves around 24 dB over the traditional LSMS system; see **Figure 7(b)**.

4.3. High data rate mobile IROW system

The results of the SNR achieved with our proposed FAPA-Hologram and FDAPA-Hologram allow the systems to reduce the total optical power transmit while operating at high data rates of 2.5 and 5 Gb/s. At high data rates, we considered a small photodetector area of 10 mm^2, in order to reduce the impact of high capacitance and improving receiver bandwidth. To the best of our knowledge, commercial photodetectors with a 10 mm^2 area and operating at high data rate are not common. However, researchers in Refs. [16, 22] have shown that the use of a small

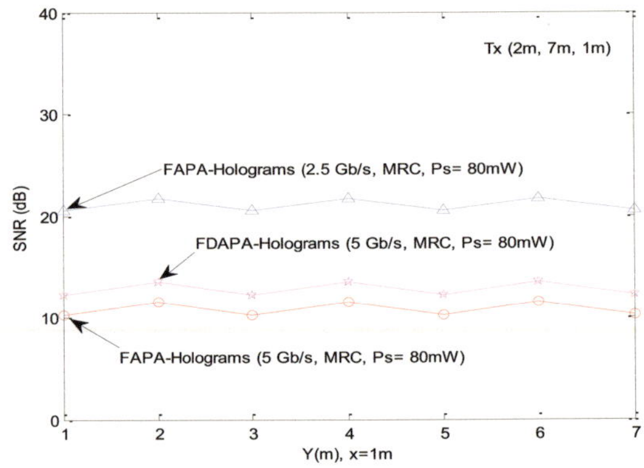

Figure 8. The SNR of proposed FDAPA-Holograms and FAPA-Hologram systems when operated at 2.5 and 5Gb/s, with a total transmit power of 80 mW.

detector area reduces the impact of high capacitance. In large commercial area, high speed detectors are starting to be used in free space optical systems. For example, Ref. [23] indicates that areas as large as 10 mm^2 and rise time as low as 10 ps are starting to emerge; however, the combination of large areas and fast response remains a challenge in photodetectors design. A 1 mW per beam is used to address the eye safety requirement in our proposed systems. Furthermore, we limit the power adaptation method where each beam cannot increase the power beyond 0.5 mW. The beams travel from the transmitter as group and spread until it reaches the object (reach to the ceiling in our case). At 10 cm distance each beam travel with different angle which can help in eye safety where the human eye cannot see more than one beam at time. Therefore, we propose that the transmitter is contained within a 10 cm deep enclosure to ensure that the human eye cannot be placed next to the transmitter. This can be achieved for example by placing the transmitter at the bottom of a laptop back cover (screen) and letting the beams emerge from the top of the back cover (screen). The proposed FAPA-Hologram achieves 20.5 dB at 2.5 Gb/s, see **Figure 8**. Moreover, at 5 Gb/s, the proposed FDAPA-Hologram offers around 2 dB SNR improvement over the FAPA-Hologram. This improvement is due to the use of beam delay adaptation method which helps to reduce the delay spread and improve 3dB channel bandwidth, hence increasing the SNR at the receiver. In terms of practical implementation, it should be noted that the diffraction limit has to be considered when considering commercially available spatial light modulators as the smallest pixel size that can be manufactured and operating wavelength to determine the maximum range of angles over which the beam can be steered [24]. This warrants further study.

5. Conclusions

The performance evaluation of the conventional CDS and non-adaptive LSMS can be significantly degraded by the transmitter/receiver mobility. In this chapter, the finite adaptive hologram using beam angle, power and delay adaptation techniques is introduced. All holograms are stored and pre-calculated in our adaptive system. A fast search algorithm based on the divide and conquer is reported. The fast algorithm reduced the time needed to generate hologram to select the best stored hologram in the system. The adaptive proposed system is combined with an angle diversity receiver. Nine beaches non-imaging angle diversity receiver was used to further improve the received optical signal in the presence of background noise and transmitter mobility. At 50 Mb/s, our simulation results show that the adaptive FAPA-Holograms system provides around 35 dB SNR gain over non-imaging diversity LSMS system. The proposed FAPA-Holograms system using beam angle and power adaptation methods is able to guide the spots nearer to the receiver location at each given transmitter-receiver location. The angles and powers associated with each hologram stored in the system are pre-calculated without adding any complexity at the transmitter to recomputed holograms. In order to further improve system performance and reduce the effect of multipath dispersion, beam delay adaptation method coupled was introduced when the system operated at high data rates. The proposed FDAPA-Holograms system was examined under eye safety regulations. A total transmit power of 80 mW was used. The SNR results of our 5 Gb/s FDAPA-Holograms system were around 13 dB, under the impact of mobility as well as background noise. A fast search algorithm based on the divide and conquer was proposed to reduce the time needed to generate a real hologram and select the best stored hologram in the system.

Acknowledgements

The authors extend their appreciation to the International Scientific Partnership Program ISPP at King Saud University for funding this research work through ISPP# 0093.

Author details

Mohammed T. Alresheedi[1]*, Ahmed Taha Hussein[2] and Jaafar M.H. Elmirghani[2]

*Address all correspondence to: malresheedi@ksu.edu.sa

1 Department of Electrical Engineering, King Saud University, Riyadh, Saudi Arabia

2 School of Electronic and Electrical Engineering, University of Leeds, Leeds, UK

References

[1] Alresheedi MT, Elmirghani JMH. Performance evaluation of 5 Gbit/s and 10 Gbit/s mobile optical wireless systems employing beam angle and power adaptation with diversity receivers. IEEE Journal on Selected Areas in Communications. 2011;**29**(6):1328-1340. DOI: 10.1109/JSAC.2011.110620

[2] Alsaadi FE, Elmirghani JMH. High-speed spot diffusing mobile optical wireless system employing beam angle and power adaptation and imaging receivers. Journal of Lightwave Technology. 2010. **28**(16):2191-2206. DOI: 10.1109/JLT.2010.2042140

[3] Alresheedi MT, Elmirghani JMH. 10 Gbit/s indoor optical wireless systems employing beam delay, angle and power adaptation methods with imaging detection. IEEE Journal of Lightwave Technology. 2012;**30**(12):1843-1856. DOI: 10.1109/JLT.2012.2190970

[4] Alsaadi FE, Alhartomi MA, Elmirghani JMH. Fast and efficient adaptation algorithms for multi-gigabit wireless infrared systems. Journal of Lightwave Technology. 2013;**31** (23):3735-3751. DOI: 10.1109/JLT.2013.2286743

[5] Hussein AT, Alresheedi MT, Elmirghani JMH. 20 Gbps mobile indoor visible light communication system employing beam steering and computer generated holograms. Journal of Lightwave Technology. 2015;**33**(24):5242-5260. DOI: 10.1109/JLT.2015.2495165

[6] Wu L, Zhang Z, Liu H. Transmit beamforming for MIMO optical wireless communication systems. Wireless Personal Communications. 2014;**78**(1):615-628. DOI: 10.1007/s11277-014-1774-3

[7] Kim S, Kim S. Wireless optical energy transmission using optical beamforming. Optical Engineering. 2013;2(4):205-210. DOI: 10.1117/1.OE.52.4.043205

[8] Alresheedi MT, Elmirghani JMH. Hologram selection in realistic indoor optical wireless systems with angle diversity receivers. IEEE Journal on Optical and Networking (JOCN). 2015;**7**(8):797-813. DOI: 10.1364/JOCN.7.000797

[9] Gfeller FR, Bapst UH. Wireless in-house data communication via diffuse infrared radiation. Proceedings of the IEEE. 1979;**67**(11):1474-1486.DOI: 10.1109/PROC.1979.11508

[10] Kahn JM, Barry JR. Wireless infrared communications. Proceedings of the IEEE. 1997;**85** (2):265-298

[11] AI-Ghamdi AG, Elmirghani JMH. Line strip spot-diffusing transmitter configuration for optical wireless systems influenced by background noise and multipath dispersion. IEEE Transactions on Communications. 2004;**52**(1):37-45. DOI: 10.1109/TCOMM.2003.822160

[12] Yun G, Kavehrad M. Spot diffusing and fly-eye receivers for indoor infrared wireless communications. In: Proceedings 1992 IEEE Conference. Selected Topics in Wireless Communications; Vancouver, BC, Canada; 1992. pp. 286-292 DOI: 10.1109/ICWC.1992.200761

[13] Fadlullah J, Kavehrad M. Indoor high-bandwidth optical wireless links for sensor networks. Journal of Lightwave Technology. 2010;**28**:3086-3094. DOI: 10.1109/JLT.2010.2076775

[14] Desurvire E. Erbium-Doped Fiber Amplifiers: Principles and Applications. New York; A Wiley-Interscience Publication; 1994, ISBN: 978-0-471-58977-8

[15] Moreira A, Valadas R, Duarte AO. Optical interference produced by artificial light. Wireless Networks. 1997;**3**(2):131-140. DOI: 10.1023/A:1019140814049

[16] Leskovar B. Optical receivers for wide band data transmission systems. IEEE Transactions on Nuclear Science. 1989;**36**(1):787-793. DOI: 10.1109/23.34550

[17] Alresheedi MT, Elmirghani JMH. High-speed indoor optical wireless links employing fast angle and power adaptive computer-generated holograms with imaging receivers. IEEE Transactions on Communications. 2016;**64**:1699-1710. DOI: 10.1109/TCOMM.2016.2519415

[18] Jivkova S, Kavehard M. Multispot diffusing configuration for wireless infrared access. IEEE Transactions on Communications. 2000;**48**(6):970-978. DOI: 10.1109/26.848558

[19] Seldowitz MA, Allebach JP, Sweeney DE. Synthesis of digital holograms by direct binary search. Applied Optics. 1987;**26**:2788-2798. DOI: org/10.1364/AO.26.002788

[20] Ramirez FA. Holography—Different Fields of Application. InTech. 2011. p. 158. ISBN 978-953-307-635-5. DOI: 10.5772/750

[21] Carnevali P, Coletti L, Patarnello S. Image processing by simulated annealing. IBM Journal of Research and Development. 1985;**29**(6):569-579

[22] Elmirghani JMH, Chan HH, Cryan RA. Sensitivity evaluation of optical wireless PPM systems utilizing PIN-BJT receivers. IEE Proceedings—Optoelectronics. 1996;**143**(6):355-359. 10.1049/ip-opt:19960880

[23] Available from: http://www.cablefree.co.uk/faqs/pin-photodiodes-faqs/. [Accessed 20 June 2015]

[24] McManamon PF et al. A review of phased array steering for narrow-band electrooptical systems. Proceedings of the IEEE. 2009;**97**(6):1078-1096. DOI: 10.1109/JPROC.2009.2017218

Chapter 8

Fundamental Analysis for Visible Light Communication with Input-Dependent Noise

Jin-Yuan Wang, Jun-Bo Wang and Yongjin Wang

Additional information is available at the end of the chapter

http://dx.doi.org/10.5772/68019

Abstract

Recently, visible light communication (VLC) has drawn much attention. In literature, the noise in VLC is often assumed to be independent of the input signal. This assumption neglects a fundamental issue of VLC: due to the random nature of photon emission in the lighting source, the strength of the noise depends on the signal itself. Therefore, the input-dependent noise in VLC should be considered. Given this, the fundamental analysis for the VLC with input-dependent noise is presented in this chapter. Based on the information theory, the theoretical expression of the mutual information is derived. However, the expression of the mutual information is not in a closed form. Furthermore, the lower bound of the mutual information is derived in a closed form. Moreover, the theoretical expression of the bit error rate is also derived. Numerical results verify the accuracy of the derived theoretical expressions in this chapter.

Keywords: visible light communication, input-dependent noise, mutual information, bit error rate

1. Introduction

As one of the emerging optical wireless communication techniques, the visible light communication (VLC) has drawn considerable attention recently from both the academy and industry [1–3]. Compared to the traditional radio frequency (RF) wireless communication, VLC has many advantages, such as freedom from hazardous electromagnetic radiation, no licensing requirements, low-cost frontends, large spectrum bandwidth (as shown in **Figure 1**), large channel capacity, and so on. In VLC, both illumination and communication are simultaneously implemented. Moreover, the transmitted optical signal is non-negative. Therefore, the developed theory and analysis results in traditional RF wireless communication are not directly applicable to VLC.

Up to now, the research on VLC can be divided into two categories: the demo system design and theoretical analysis. As research continues, a variety of demo platforms arise. **Table 1** shows the development of the VLC demo systems. As can be seen in **Table 1**, the transmit rate of the VLC system increases from several Mbps to several Gbps in the last decade, which indicates that the VLC has attractive prospects of development. Specifically, the transmit rates of the early demo systems are low, but the transmit distances are long and the data are processed in real time. With the development of communication techniques, more and more

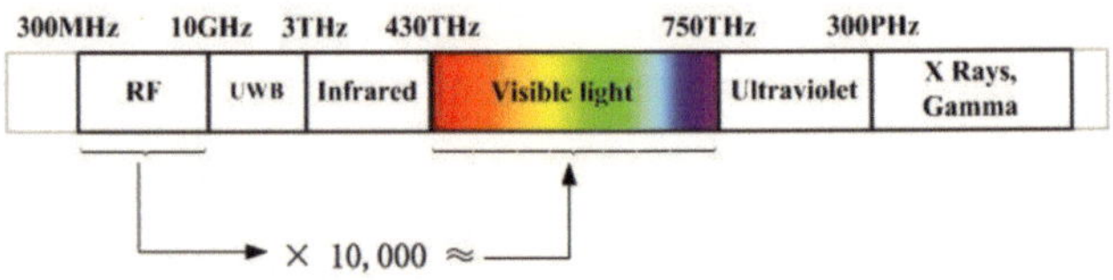

Figure 1. The electromagnetic spectrum.

Time (year)	Research & Development Group	Transmit rate (bit/s)	Transmit distance (m)	Data processing mode	
				Offline	Online
2000	Keio University, Japan	10 M	5	√	
2002	Keio University, Japan	87 M	1.65	√	
2008	Taiyo Yuden Co., Ltd, Japan	100 M	0.2	√	
	Jinan University, China	4 M	2.5	√	
2009	University of Oxford, UK, et al.	100 M	0.1	√	
	Heinrich Hertz Institute, Germany	125 M	5		√
2011	Heinrich Hertz Institute, Germany	803 M	0.12		√
2012	Kinki University, Japan	614 M			√
	National Chiao Tung University, Taiwan	1.1 G	0.23		√
	Santa Ana school for Advanced Studies, Italy	3.4 G	0.3		√
2013	University of Strathclyde, UK	1.5 G			√
	National Chiao Tung University, Taiwan	3.22 G	0.25		√
	The University of Edinburgh, UK	10 G			√
	Southeast University, China	480 M	3	√	
2014	Fudan University, China	3.25 G			√
2015	Pknuyong National University, Korea	3 G	2.15		√
	The PLA Information Engineering University, China	50 G			√

Table 1. The development of the VLC demo systems.

VLC testbeds with high transmit rates are developed successfully, but the real-time processing becomes very hard. Therefore, more advanced processing techniques are needed for VLC.

In the aspect of theoretical analysis, much work has been done on VLC. In Ref. [4], the channel capacity for VLC using inverse source coding is investigated. However, the theoretical expression of the capacity is not presented. Under the non-negative and average optical intensity constraints, the closed expression of capacity bounds is derived in Ref. [5]. Based on Ref. [5], a tight upper bound on the capacity is derived in Ref. [6]. By adding a peak optical intensity constraint, tight capacity bounds are further derived in Ref. [7]. In Ref. [8], the capacity bounds for multiple-input-multiple-output VLC are derived. In Ref. [9], the capacity and outage probability for the parallel optical wireless channels are analysed. Furthermore, low signal-to-noise ratio (SNR) capacity for the parallel optical wireless channels is obtained in Ref. [10]. It should be noted that the noises in Refs. [4–10] are all assumed to be independent with the input signal. This assumption is reasonable if the ambient light is strong or if the receiver suffers from intensive thermal noise. However, in practical VLC systems, typical illumination scenarios offer very high SNR [11, 12]. For high power, this assumption neglects a fundamental issue of VLC: due to the random nature of photon emission in the light emitting diode (LED), the strength of noise depends on the signal itself [13]. Up to now, the performance of the VLC with input-dependent noise has not been discussed completely.

In this chapter, we consider a VLC system with input-dependent Gaussian noise and investigate the fundamental performance of the VLC system. The main contributions of this chapter are given as follows:

1. A channel model with input-dependent Gaussian noise for the VLC is considered. In existing literature, the noise is generally assumed to be independent of the signal. However, this assumption is not applicable to the VLC system in some cases. In this chapter, a more general channel model is established which is corrupted by an additive Gaussian noise, however, with noise variance depending on the signal itself.

2. The mutual information of the VLC system is analysed. Based on the channel model, the exact expression of the mutual information is derived. However, the exact expression of the mutual information is not in a closed form. After that, a closed-form expression of the lower bound on the mutual information is derived.

3. The bit error rate (BER) of the VLC system is obtained. By employing the on-off keying (OOK), the theoretical expression of the BER for the VLC system is derived. Moreover, some asymptotic behaviour for the BER is also presented.

4. To show the accuracy of the derived theoretical expressions, the theoretical results are thoroughly confirmed by Monte-Carlo simulations.

The remainder of this chapter is organized as follows. The system model is described in Section 2. Section 3 presents the exact expression and the lower bound of the mutual information. In Section 4, the theoretical expression of the BER is derived. Numerical results are given in Section 5 before conclusions are drawn in Section 6.

2. System model

Consider a point-to-point VLC system, as shown in **Figure 2**. At the transmitter, an LED is employed as the lighting source, which performs the electrical-to-optical conversion. Then, the optical signal is propagated through the VLC channel. At the receiver, a PIN photodiode (PD) is used to perform the optical-to-electrical conversion. To amplify the derived electrical signal, a high impedance amplifier is employed. In this chapter, the main noise sources include thermal noise, shot noise and amplifier noise. The thermal noise and the amplifier noise are independent of the signal, and each of the two noise sources can be well modelled by Gaussian distribution [14]. Although its distribution can also be assumed to be Gaussian, the strength of the shot noise depends on the signal itself. Mathematically, the received electrical signal Y at the receiver can be written as [13]

$$Y = rGX + \sqrt{rGX}Z_1 + Z_0 \tag{1}$$

where r denotes the optoelectronic conversion factor of the PD. $Z_0 \sim N(0, \sigma^2)$ denotes the input-independent Gaussian noise. $Z_1 \sim N(0, \varsigma^2\sigma^2)$ denotes the input-dependent Gaussian noise, where $\varsigma^2 \geq 0$ denotes the ratio of the input-dependent noise variance to the input-independent noise variance. Z_0 and Z_1 are independent with each other.

In Eq. (1), G denotes the channel gain between the LED and the PD, which can be expressed as [15]

$$G = \frac{(m+1)A}{2\pi d^2} \cos^m(\varphi)T(\psi)g(\psi)\cos(\psi) \tag{2}$$

where m denotes the order of the Lambertian emission, A is the physical area of the PD and d, φ and ψ are the distance, the angle of irradiance and the angle of incidence from the LED to the PD, respectively. $T(\psi)$ is the gain of an optical filter and $g(\psi)$ is the gain of an optical concentrator.

Note that the channel gain in Eq. (2) is a constant, where the positions of the LED and the PD are given. Moreover, r in Eq. (1) is a constant for a fixed PD. Without loss of generality, the values of both G and r are set to be one. Therefore, Eq. (1) can be simplified as [16]

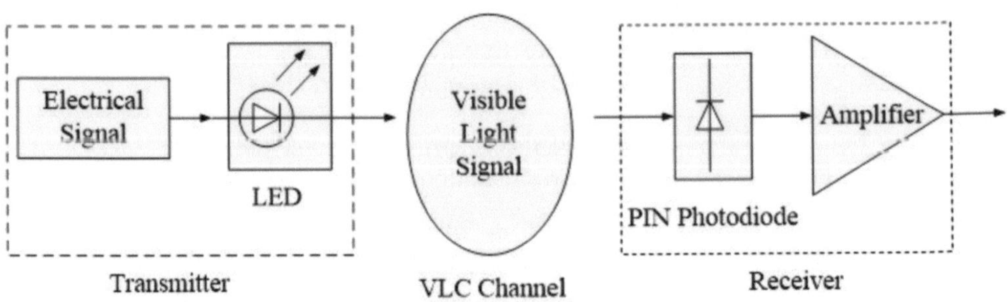

Figure 2. The point-to-point VLC system.

$$Y = X + \underbrace{\sqrt{X}Z_1 + Z_0}_{\triangleq Z}. \tag{3}$$

In VLC, information is transmitted by modulating the instantaneous optical intensity [17], and thus, X should be non-negative, that is,

$$X \geq 0. \tag{4}$$

Due to the eye and skin safety regulations, the peak optical intensity of the LED is limited [17], that is,

$$X \leq A \tag{5}$$

where A is the peak optical intensity of the LED.

Considering the illumination requirement in VLC, the average optical intensity cannot be changed but can be adjusted according to the users' requirement (dimming target) [18]. Therefore, the average optical intensity constraint is given by

$$E(X) = \xi P \tag{6}$$

where $E(\cdot)$ denotes the expectation operator and $\xi \in (0, 1]$ denotes the dimming target. $P \leq A$ is the normal optical intensity of the LED.

3. Mutual information analysis

Mutual information is an important performance indicator for wireless communication systems. In this section, the exact expression of the mutual information and the closed-form expression of the lower bound on the mutual information for the VLC will be derived, respectively.

3.1. Exact expression of mutual information

Assume that N-ary intensity modulation is employed. Let $X \in \{x_1, x_2, \cdots, x_N\}$ be the optical intensity symbol drawn from the equiprobable modulation constellation, that is,

$$\Pr(X = x_i) = \frac{1}{N}. \tag{7}$$

According to Eq. (3), the conditional probability density function (PDF) of Y when given $X = x_i$ can be written as [19]

$$f_{Y|X}(y|x_i) = \frac{1}{\sqrt{2\pi(1 + x_i\varsigma^2)}\sigma} \exp\left(-\frac{(y - x_i)^2}{2(1 + x_i\varsigma^2)\sigma^2}\right). \tag{8}$$

Furthermore, the PDF of Y can be expressed as

$$\begin{aligned} f_Y(y) &= \sum_{i=1}^{N} \Pr(X = x_i) f_{Y|X}(y|x_i) \\ &= \frac{1}{N}\sum_{i=1}^{N} \frac{1}{\sqrt{2\pi(1+x_i\varsigma^2)}\sigma} \exp\left(-\frac{(y-x_i)^2}{2(1+x_i\varsigma^2)\sigma^2}\right). \end{aligned} \tag{9}$$

The mutual information between X and Y is given by

$$\begin{aligned} I(X;Y) &= H(X) - H(X|Y) \\ &= \sum_{i=1}^{N} \frac{1}{N}\log_2 N - \sum_{i=1}^{N}\int_{-\infty}^{\infty} \frac{1}{N} f_{Y|X}(y|x_i) \log_2\left(\frac{f_Y(y)}{\Pr(X=x_i) f_{Y|X}(y|x_i)}\right) dy \\ &= \log_2 N - \frac{1}{N}\sum_{i=1}^{N} \underbrace{\int_{-\infty}^{\infty} \frac{\exp\left(-\frac{(y-x_i)^2}{2(1+x_i\varsigma^2)\sigma^2}\right)}{\sqrt{2\pi(1+x_i\varsigma^2)}\sigma} \log_2\left(\frac{\sum_{t=1}^{N}\frac{\exp\left(-\frac{(y-x_t)^2}{2(1+x_t\varsigma^2)\sigma^2}\right)}{\sqrt{2\pi(1+x_t\varsigma^2)}\sigma}}{\frac{\exp\left(-\frac{(y-x_i)^2}{2(1+x_i\varsigma^2)\sigma^2}\right)}{\sqrt{2\pi(1+x_i\varsigma^2)}\sigma}}\right) dy}_{\triangleq I_1} \end{aligned} \tag{10}$$

where $H(\cdot)$ denotes the entropy.

From Eq. (3), we have $Z = Y - X$. Therefore, let $z = y - x_i$, and thus, I_1 in Eq. (10) can be further written as

$$\begin{aligned} I_1 &= \int_{-\infty}^{\infty} \frac{\exp\left(-\frac{z^2}{2(1+x_i\varsigma^2)\sigma^2}\right)}{\sqrt{2\pi(1+x_i\varsigma^2)}\sigma} \log_2\left[\frac{\sum_{t=1}^{N}\frac{\exp\left(-\frac{(z+x_i-x_t)^2}{2(1+x_t\varsigma^2)\sigma^2}\right)}{\sqrt{2\pi(1+x_t\varsigma^2)}\sigma}}{\frac{\exp\left(-\frac{z^2}{2(1+x_i\varsigma^2)\sigma^2}\right)}{\sqrt{2\pi(1+x_i\varsigma^2)}\sigma}}\right] dy \\ &= E_Z\left\{\log_2\left[\frac{\sum_{t=1}^{N}\frac{\exp\left(-\frac{(z+x_i-x_t)^2}{2(1+x_t\varsigma^2)\sigma^2}\right)}{\sqrt{2\pi(1+x_t\varsigma^2)}\sigma}}{\frac{\exp\left(-\frac{z^2}{2(1+x_i\varsigma^2)\sigma^2}\right)}{\sqrt{2\pi(1+x_i\varsigma^2)}\sigma}}\right]\right\} \\ &= E_Z\left\{\log_2\left[\sum_{t=1}^{N}\frac{\sqrt{1+x_i\varsigma^2}}{\sqrt{1+x_t\varsigma^2}} \exp\left(\frac{z^2}{2(1+x_i\varsigma^2)\sigma^2} - \frac{(z+x_i-x_t)^2}{2(1+x_t\varsigma^2)\sigma^2}\right)\right]\right\} \end{aligned} \tag{11}$$

Therefore, Eq. (10) can be further written as

$$I(X;Y) = \log_2 N - \frac{1}{N}\sum_{i=1}^{N} E_Z\left\{\log_2\left[1+\sum_{\substack{t=1\\t\neq i}}^{N}\frac{\sqrt{1+x_i\varsigma^2}}{\sqrt{1+x_t\varsigma^2}}\exp\left(\frac{z^2}{2(1+x_i\varsigma^2)\sigma^2}-\frac{(z+x_i-x_t)^2}{2(1+x_t\varsigma^2)\sigma^2}\right)\right]\right\} \tag{12}$$

Remark 1: Let the average SNR be $\gamma = \xi P/[(1+\xi P\varsigma^2)\sigma^2]$. Because ξ, P and ς are non-negative and finite numbers, $\gamma \to \infty$ (or 0) is equivalent to $\sigma^2 \to 0$ (or ∞). Apparently, $I(X;Y)$ in Eq. (12) is a monotonic increasing function with respect to γ. Therefore, we have

$$\lim_{\gamma\to\infty} I(X;Y) = \log_2 N \tag{13}$$

which indicates that the maximum value of $I(X;Y)$ is $\log_2 N$.

Moreover, we have

$$\lim_{\gamma\to 0} I(X;Y) = \log_2 N - \frac{1}{N}\sum_{i=1}^{N}\log_2\left(1+\sum_{\substack{t=1\\t\neq i}}^{N}\frac{\sqrt{1+x_i\varsigma^2}}{\sqrt{1+x_t\varsigma^2}}\right) \tag{14}$$

Remark 2: When $\varsigma = 0$, Eq. (3) reduces to $Y = X + Z_0$. Therefore, the mutual information can be simplified as

$$I(X;Y)|_{\varsigma=0} = \log_2 N - \frac{1}{N}\sum_{i=1}^{N} E_Z\left\{\log_2\left[1+\sum_{\substack{t=1\\t\neq i}}^{N}\exp\left(\frac{z^2-(z+x_i-x_t)^2}{2\sigma^2}\right)\right]\right\} \tag{15}$$

3.2. Lower bound on mutual information

It should be noted that it is very hard to derive a closed-form expression of Eq. (12). In this subsection, a lower bound on the mutual information will be derived.

To facilitate the description, Eq. (12) can be further expressed as

$$\begin{aligned} I(X;Y) = \log_2 N &- \underbrace{\frac{1}{N}\sum_{i=1}^{N} E_Z\left\{\log_2\left[\exp\left(\frac{z^2}{2(1+x_i\varsigma^2)\sigma^2}\right)\right]\right\}}_{I_2} \\ &- \underbrace{\frac{1}{N}\sum_{i=1}^{N} E_Z\left\{\log_2\left[\sum_{t=1}^{N}\frac{\sqrt{1+x_i\varsigma^2}}{\sqrt{1+x_t\varsigma^2}}\exp\left(-\frac{(z+x_i-x_t)^2}{2(1+x_t\varsigma^2)\sigma^2}\right)\right]\right\}}_{I_3} \end{aligned} \tag{16}$$

For I_2 in Eq. (16), we have

$$\begin{aligned} I_2 &= \frac{\log_2(e)}{2(1+x_i\varsigma^2)\sigma^2}\int_{-\infty}^{+\infty} z^2 \frac{\exp\left(-\frac{z^2}{2(1+x_i\varsigma^2)\sigma^2}\right)}{\sqrt{2\pi(1+x_i\varsigma^2)\sigma^2}}\mathrm{d}z \\ &= \frac{\log_2(e)}{2(1+x_i\varsigma^2)\sigma^2}(1+x_i\varsigma^2)\sigma^2 \\ &= \frac{1}{2}\log_2(e). \end{aligned} \tag{17}$$

Using the Jensen's inequality for concave function, an upper bound of I_3 in Eq. (16) can be written as

$$\begin{aligned} I_3 &= E_Z\left\{\log_2\left[\sum_{t=1}^{N}\frac{\sqrt{1+x_i\varsigma^2}}{\sqrt{1+x_t\varsigma^2}}\exp\left(-\frac{(z+x_i-x_t)^2}{2(1+x_t\varsigma^2)\sigma^2}\right)\right]\right\} \\ &\leq \log_2\left\{\sum_{t=1}^{N}\frac{\sqrt{1+x_i\varsigma^2}}{\sqrt{1+x_t\varsigma^2}}E_Z\left[\exp\left(-\frac{(z+x_i-x_t)^2}{2(1+x_t\varsigma^2)\sigma^2}\right)\right]\right\} \\ &= \log_2\left\{\sum_{t=1}^{N}\frac{\sqrt{1+x_i\varsigma^2}}{\sqrt{1+x_t\varsigma^2}}\int_{-\infty}^{+\infty}\frac{\exp\left(-\frac{(z+x_i-x_t)^2+z^2}{2(1+x_t\varsigma^2)\sigma^2}\right)}{\sqrt{2\pi(1+x_t\varsigma^2)\sigma^2}}\mathrm{d}z\right\} \\ &= \log_2\left(\sum_{t=1}^{N}\frac{\sqrt{1+x_i\varsigma^2}}{\sqrt{2(1+x_t\varsigma^2)}}\exp\left(-\frac{(x_i-x_t)^2}{4(1+x_t\varsigma^2)\sigma^2}\right)\right). \end{aligned} \tag{18}$$

Substituting Eqs. (17) and (18) into Eq. (16), a lower bound of $I(X;Y)$ can be derived as

$$\begin{aligned} I_{\mathrm{Low}}(X;Y) &= \log_2 N - \frac{1}{2}\log_2(e) + \frac{1}{2} \\ &- \frac{1}{N}\sum_{i=1}^{N}\log_2\left(1+\sum_{\substack{t=1\\t\neq i}}^{N}\frac{\sqrt{1+x_i\varsigma^2}}{\sqrt{1+x_t\varsigma^2}}\exp\left(-\frac{(x_i-x_t)^2}{4(1+x_t\varsigma^2)\sigma^2}\right)\right). \end{aligned} \tag{19}$$

***Remark* 3**: Obviously, $I_{\mathrm{Low}}(X;Y)$ in Eq. (19) is a monotonic increasing function with respect to γ. Therefore, we have

$$\lim_{\gamma\to\infty} I_{\mathrm{Low}}(X;Y) = \log_2 N - \frac{1}{2}\log_2(e) + \frac{1}{2} \tag{20}$$

which indicates that the maximum value of $I_{\mathrm{Low}}(X;Y)$ is $\log_2 N - \log_2(e)/2 - 1/2$.

Moreover, we have

$$\lim_{\gamma\to 0} I_{\mathrm{Low}}(X;Y) = \log_2 N - \frac{1}{2}\log_2(e) + \frac{1}{2} - \frac{1}{N}\sum_{i=1}^{N}\log_2\left(1+\sum_{\substack{t=1\\t\neq i}}^{N}\frac{\sqrt{1+x_i\varsigma^2}}{\sqrt{1+x_t\varsigma^2}}\right) \tag{21}$$

Remark 4: According to Eqs. (13) and (20), we have

$$\lim_{\gamma\to\infty} I(X;Y) - \lim_{\gamma\to\infty} I_{\text{Low}}(X;Y) = \frac{1}{2}[\log_2(e) - 1]. \tag{22}$$

Similarly, from Eqs. (14) and (21), we have

$$\lim_{\gamma\to 0} I(X;Y) - \lim_{\gamma\to 0} I_{\text{Low}}(X;Y) = \frac{1}{2}[\log_2(e) - 1]. \tag{23}$$

From Eqs. (22) and (23), it can be concluded that a constant performance gap $[\log_2(e) - 1]/2$ exists between $I(X;Y)$ and $I_{\text{Low}}(X;Y)$ at low and high SNR regions.

Remark 5: When $\varsigma = 0$, $I_{\text{Low}}(X;Y)$ can be simplified as

$$\begin{aligned} I_{\text{Low}}(X;Y)|_{\varsigma=0} &= \log_2 N - \frac{1}{2}\log_2(e) + \frac{1}{2} \\ &- \frac{1}{N}\sum_{i=1}^{N}\log_2\left(1 + \sum_{\substack{t=1 \\ t\neq i}}^{N} \exp\left(-\frac{(x_i - x_t)^2}{4\sigma^2}\right)\right). \end{aligned} \tag{24}$$

4. BER analysis

In this section, the BER of the VLC with input-dependent noise is analysed. To facilitate the analysis, OOK is employed as the modulation scheme. Suppose that the transmitted optical signal is drawn equiprobably from the OOK constellation and $2\xi P \leq A$ always holds, we have

$$X \in \{0, 2\xi P\}. \tag{25}$$

Therefore, the BER for the VLC with OOK can be written as

$$BER = \Pr(\text{off})\Pr(\text{on}|\text{off}) + \Pr(\text{on})\Pr(\text{off}|\text{on}) \tag{26}$$

where $\Pr(\text{on})$ and $\Pr(\text{off})$ are the probabilities of sending "on" and "off" bits, respectively. Because the transmitted signal is taken as symbols drawn equiprobably, thus $\Pr(\text{on}) = \Pr(\text{off}) = 0.5$. $Pr(\text{on}|\text{off})$ and $Pr(\text{off}|\text{on})$ are the conditional bit error probabilities when the transmitted bit is "off" and "on," respectively.

According to Eq. (8), $\Pr(\text{off}|\text{on})$ can be written as

$$\begin{aligned} \Pr(\text{off}|\text{on}) &= \Pr(y < \xi P|\text{on}) \\ &= \int_{-\infty}^{\xi P} \frac{1}{\sqrt{2\pi(1+2\xi P\varsigma^2)}\sigma} e^{-\frac{(y-2\xi P)^2}{2(1+2\xi P\varsigma^2)\sigma^2}} dy \\ &= \mathcal{Q}\left(\frac{\xi P}{\sqrt{1+2\xi P\varsigma^2}\sigma}\right) \end{aligned} \tag{27}$$

where $\mathcal{Q}(x)$ is the Gaussian Q-function.

Moreover, Pr(off|on) can be similarly written as

$$\begin{aligned}\Pr(\text{off}|\text{on}) &= \Pr(y > \xi P|\text{off}) \\ &= \int_{\xi P}^{\infty} \frac{1}{\sqrt{2\pi}\sigma} e^{-\frac{y^2}{2\sigma^2}} dy \\ &= \mathcal{Q}\left(\frac{\xi P}{\sigma}\right).\end{aligned} \tag{28}$$

Therefore, the BER can be finally written as

$$BER = \frac{1}{2}\left[\mathcal{Q}\left(\frac{\xi P}{\sqrt{1+2\xi P \varsigma^2}\sigma}\right) + \mathcal{Q}\left(\frac{\xi P}{\sigma}\right)\right]. \tag{29}$$

Remark 6: Let the average SNR be $\gamma = \xi P/[(1+\xi P \varsigma^2)\sigma^2]$. Because ξ, P and ς are non-negative and finite numbers, $\gamma \to \infty$ (or 0) is equivalent to $\sigma^2 \to 0$ (or ∞). Apparently, BER in Eq. (29) is a monotonic decreasing function with respect to γ. Therefore, we have

$$\lim_{\gamma\to\infty} BER = 0 \tag{30}$$

$$\lim_{\gamma\to 0} BER = \frac{1}{2} \tag{31}$$

This indicates that the minimum BER and the maximum BER are 0 and 0.5, respectively.

Remark 7: When $\varsigma = 0$, BER can be simplified as

$$BER|_{\varsigma=0} = \mathcal{Q}\left(\frac{\xi P}{\sigma}\right). \tag{32}$$

5. Numerical results

In this section, some classical numerical results will be presented. The derived theoretical expressions of the mutual information, the lower bound of mutual information and the BER will be verified.

5.1. Results of mutual information

Figure 3 shows the mutual information (i.e., $I(X;Y)$ in Eq. (12)) and its lower bound (i.e., $I_{\text{Low}}(X;Y)$ in Eq. (19)) versus SNR with different modulation orders N. In the simulation, without loss of generality, ξ, P and ς are set to be one. In **Figure 3**, it can be seen that $I(X;Y)$ and $I_{\text{Low}}(X;Y)$ are monotonic increasing functions with respect to SNR. Moreover, with the increase of N, $I(X;Y)$ and $I_{\text{Low}}(X;Y)$ also increase. It can also be found that the maximum value of $I(X;Y)$ is $\log_2 N$, and the maximum value of $I_{\text{Low}}(X;Y)$ is $\log_2 N - \log_2(e)/2 - 1/2$, which coincides with *Remark 1*. Furthermore, the gap between $I(X;Y)$ and $I_{\text{Low}}(X;Y)$ is $(\log_2 e - 1)/2$ bits at low and high SNR regions, which coincides with *Remark 4*.

Figure 4 shows the mutual information (i.e., $I(X;Y)$ in Eq. (12)) and its lower bound (i.e., $I_{\text{Low}}(X;Y)$ in Eq. (19)) versus dimming targets ξ with different ς. In the simulation, P is set to be one, $\gamma = 20$dB and $N = 4$.As can be seen, when $\varsigma = 1$ and $\varsigma = 10$, $I(X;Y)$ and $I_{\text{Low}}(X;Y)$ increase with the increase of ξ, while $I(X;Y)$ and $I_{\text{Low}}(X;Y)$ do not change with the increase of ξ when $\varsigma = 0$. Moreover, it can be seen that $I(X;Y)$ and $I_{\text{Low}}(X;Y)$ are both the monotonic increasing functions with respect to ς.

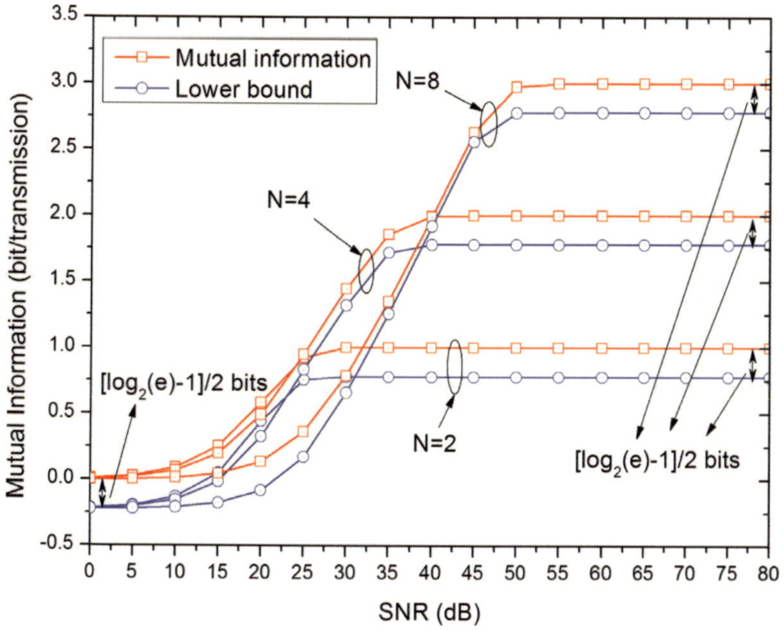

Figure 3. Mutual information and its lower bound versus SNR with different N.

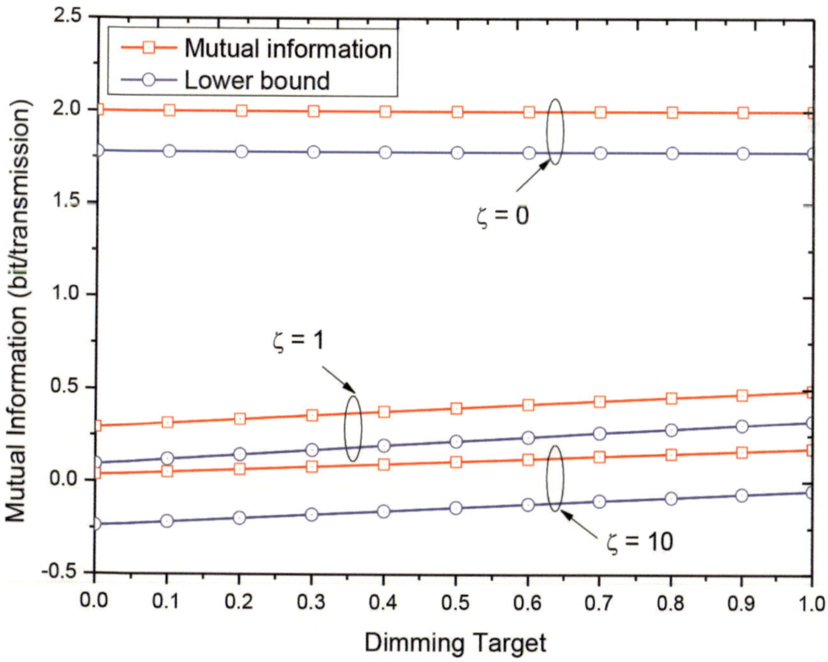

Figure 4. Mutual information and its lower bound versus dimming target ξ with different ς.

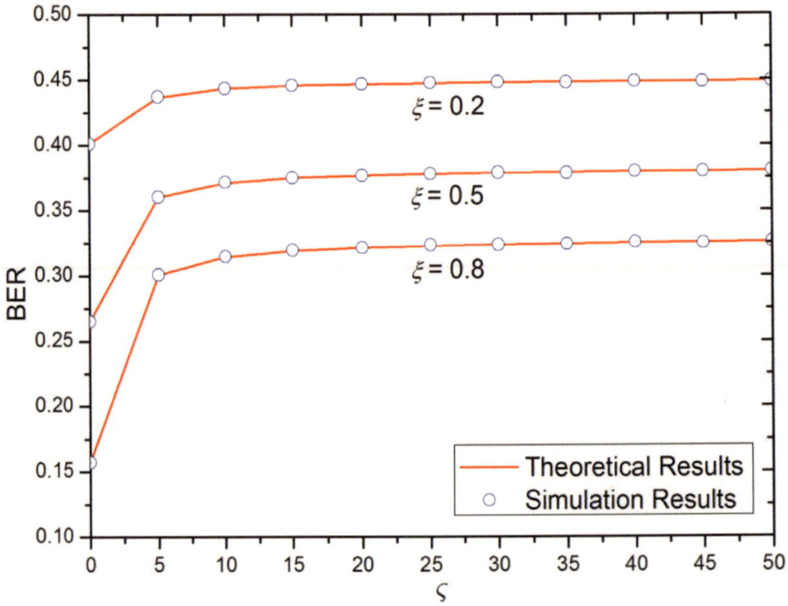

Figure 5. BER versus ς with different ξ.

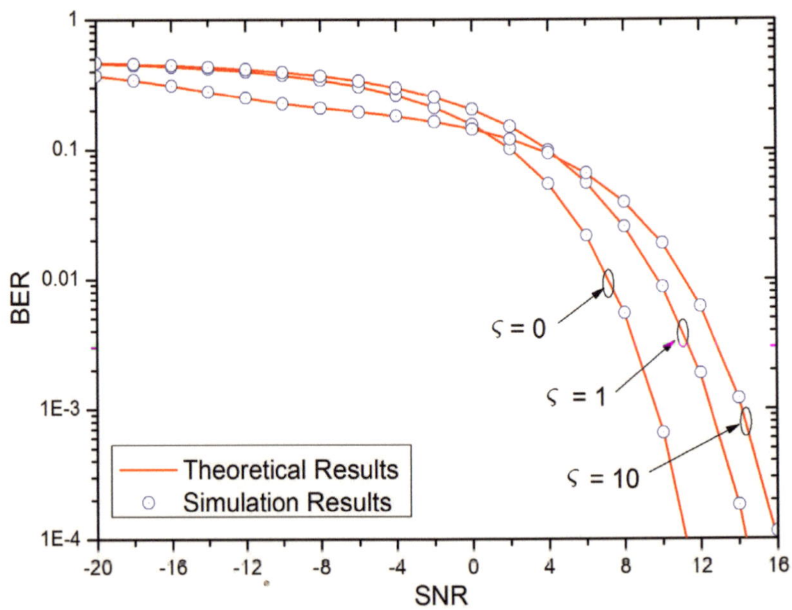

Figure 6. BER versus SNR with different ς.

5.2. Results of BER

Figure 5 shows BER versus ς with different dimming targets ξ. In the simulation, both P and σ are set to be one. It can be seen that the best BER performance is achieved when $\varsigma = 0$, which indicates that the performance for the system with only input-independent noise outperforms that with input-dependent noise. Moreover, with the increase of ς, the BER performance degrades. Furthermore, it can be observed that with the increase of ξ, the value of the BER reduces, which indicates that the system performance improves. In addition, it can be found that the theoretical results show close agreement with the Monte-Carlo simulation results, which verifies the correctness of the derived theoretical expression of the BER.

Figure 6 shows the BER versus the SNR with different ς. It can be observed that the value of the BER decreases with the increase of the SNR. This is because large SNR will generate a small BER, and thus it will result in good performance. Moreover, at low SNR region, the curve with $\varsigma = 10$ achieves the best BER performance. At high SNR region, the curve with $\varsigma = 0$ achieves the best BER performance. Once again, the gap between the theoretical results and the simulation results is so small enough to be ignored, which verifies the accuracy of the derived theoretical expression of the BER.

6. Conclusions

This chapter investigates the performance of the VLC with input-dependent noise. The theoretical expression of the mutual information is derived, which is not in a closed form. Moreover, the closed-form expression of the lower bound on the mutual information is obtained. Furthermore, by employing the OOK, the theoretical expression of the BER for the VLC is derived. Numerical results show that the derived theoretical expressions in this chapter are quite accurate to evaluate the system performance without time-intensive simulations.

Acknowledgements

This work is supported by National Natural Science Foundation of China (61571115, 61372106, 61223001, 61322112, 61401399, and 61531166004), Research Project (2014CB360507), the open research fund of National Mobile Communications Research Laboratory, Southeast University (2017D06), and NUPTSF (NY216009).

Author details

Jin-Yuan Wang[1,2]*, Jun-Bo Wang[2] and Yongjin Wang[1]

*Address all correspondence to: jywang@njupt.edu.cn

1 Peter Grünberg Research Centre, Nanjing University of Posts and Telecommunications, China

2 National Mobile Communications Research Laboratory, Southeast University, China

References

[1] A. Jovicic, J. Li, T. Richardson. Visible light communication: opportunities, challenges and the path to market. IEEE Communications Magazine. 2013;**51**(12):26–32. doi:10.1109/MCOM.2013.6685754

[2] J. Armstrong, Y. A. Sekercioglu, A. Neild. Visible light positioning: a roadmap for international standardization. IEEE Communications Magazine. 2013;**51**(12):68–73. doi:10.1109/MCOM.2013.6685759

[3] L. Zeng, D. C. O'Brien, H. L. Minh, G. E. Faulkner, K. Lee, D. Jung, Y. J. Oh, T. Won. High data rate multiple input multiple output (MIMO) optical wireless communications using white led lighting. IEEE Journal on Selected Areas in Communications, 2009;**27**(9):1654–1662. doi:10.1109/JSAC.2009.091215

[4] K.-I. Ahn, J. K. Kwon. Capacity analysis of M-PAM inverse source coding in visible light communications. Journal of Lightwave Technology. 2012;**30**(10):1399–1404. doi:10.1109/JLT.2012.2185780

[5] J.-B. Wang, Q.-S. Hu, J. Wang, M. Chen, J.-Y. Wang. Tight bound on channel capacity for dimmable visible light communications. IEEE/OSA Journal of Lightwave Technology. 2013;**31**(23):3771–3779. doi:10.1109/JLT.2013.2286088

[6] R. Jiang, Z. Wang, Q. Wang, L. Dai. A tight upper bound on channel capacity for visible light communications. IEEE Communications Letters. 2016;**20**(1):97–100. doi:10.1109/LCOMM.2015.2497694

[7] J.-B. Wang, Q.-S. Hu, J. Wang, M. Chen, Y.-H. Huang, J.-Y. Wang. Capacity analysis for dimmable visible light communications. In: IEEE International Conference on Communications, June, 2014, Sydney. IEEE; 2014, pp. 3331–3335. doi:10.1109/ICC.2014.6883835

[8] J.-Y. Wang, J. Dai, R. Guan, L. Jia, Y. Wang, M. Chen. On the channel capacity and receiver deployment optimization for multi-input multi-output visible light communications. Optics Express. 2016;**24**(12):13060–13074. doi:10.1364/OE.24.013060

[9] A. Chaaban, Z. Rezki, M.-S. Alouini. Fundamental limits of parallel optical wireless channels: capacity results and outage formulation. IEEE Transactions on Communications. 2016;**65**(1):296–311. doi:10.1109/TCOMM.2016.2621743

[10] A. Chaaban, Z. Rezki, M.-S. Alouini. Low-SNR capacity of parallel IM-DD optical wireless channels. IEEE Communications Letters. LCOMM.2016.2633339

[11] J. Grubor, S. Randel, K.-D. Langer, J. Walewski. Broadband information broadcasting using LED-based interior lighting. IEEE/OSA Journal of Lightwave Technology. 2008;**26**(24):3883–3892. doi:10.1109/JLT.2008.928525

[12] L. Hanzo, H. Haas, S. Imre, D. O'Brien, M. Rupp, L. Gyongyosi. Wireless myths, realities, and futures: from 3G/4G to optical and quantum wireless. Proceedings of the IEEE. 2012;**100**(Special Centennial Issue):1853–1888. doi:10.1109/JPROC.2012.2189788

[13] S. M. Moser. Capacity results of an optical intensity channel with input-dependent Gaussian noise. IEEE Transactions on Information Theory. 2012;**58**(1):207–223. doi:10.1109/TIT.2011.2169541

[14] A. Lapidoth, S. M. Moser, M. A. Wigger. On the capacity of free-space optical intensity channels. IEEE Transactions on Information Theory. 2009;**55**(10):4449–4461. doi:10.1109/TIT.2009.2027522

[15] T. Komine, M. Nakagawa. Fundamental analysis for visible-light communication system using LED lights. IEEE Transactions on Consumer Electronics. 2004;**50**(1):100–107. doi:10.1109/TCE.2004.1277847

[16] J.-Y. Wang, Z. Yang, Y. Wang, M. Chen. On the performance of spatial modulation-based optical wireless communications. IEEE Photonics Technology Letters. 200916;**28**(19):2094–2097. doi:10.1109/LPT.2016.2585502

[17] J. M. Kahn, J. R. Barry. Wireless infrared communications. Proceedings of the IEEE. 1997;**85**(2):265–298. doi:10.1109/5.554222

[18] J.-B. Wang, Q.-S. Hu, J. Wang, Y.-H. Huang, J.-Y. Wang. Channel capacity for dimmable visible light communications. In: IEEE Global Communications Conference, December 2013, Atlanta. IEEE; 2013, pp. 2424–2429. doi:10.1109/GLOCOM.2013.6831437

[19] J.-Y. Wang, J.-B. Wang, M. Chen, J. Wang. Capacity bounds for dimmable visible light communications using PIN photodiodes with input-dependent Gaussian noise. In: IEEE Global Communications Conference; 8–12 December 2014; Austin, TX. IEEE; 2014, pp. 2107–2112. doi:10.1109/GLOCOM.2014.7037112

Chapter 9

Receiver Performance Improvement in Radio over Fiber Network Transmission

Lisawati Sainawi and Khadijah Ismail

Additional information is available at the end of the chapter

http://dx.doi.org/10.5772/intechopen.68583

Abstract

Nowadays, wireless demand is raised rapidly since the number of mobile-cellular telephone and broadband internet are springing up. Thus, many base stations (BS) and a lot of money are needed to satisfy the demand. Radio over fiber (RoF) is the solution to this problem since it can operate a lot of BSs that connected to central station (CS) by using optical fiber, as well as offering cost-effective solution. Due to some factors, received signals of RoF tend to be in a low quality. Those factors will lead to some problems such as high value of bit error rate (BER), low value of Q factor, and the receiver could not operate in high data rate network. Wavelength division multiplexing (WDM) network can be one of the solutions of those problems where different signals are transmitted through a single-mode fiber. Bit error rate must be decreased to a certain value, and the Q factor should be increased. The design of RoF will be simulated by using Optisystem software. The performance of RoF's receiver is measured and analyzed based on the obtained BER, value of Q factor and height of opening of eye diagram.

Keywords: radio over fiber, performance, wavelength division multiplexing, receiver, bit error rate, Q factor, eye diagram, optical fiber, eye pattern

1. Introduction

Any form of telecommunication system that having light as the transmission medium is known as optical communication. Optical communication system consists of a transmitter, channel and a receiver. The transmitter will **encode** a message into an optical signal, the channel will carry the signal to its destination, and the receiver will reproduce the message from the received optical signal.

Radio over fiber (RoF) is a technology where RF signal modulates light and then transmitting it over an optical fiber link. Both wireless network and optical are supported by this technology. It is essential for communication system to have high capacity and subcarrier frequency since wireless signal sometimes intend to loss channel at the time of data transmission [1].

RoF is very convenient system since it is less costing and power consumption. This is because RoF lets the electrical signal modulates the optical source and after that the optical signal will travel along the optical fiber to the remote station. When the RF signal is modulated straight to the optical link, the power consumption drops while the antenna side has high frequency radio carriers. The reduction of cost in RoF technology can be explained in two ways. The first one is central station (CS), which provides resources that can be shared by variety of base stations (BS), and secondly, BS only executes simple function. Furthermore, the BS is in a small size and less cost consuming.

Basically, in this technology, central station (CS) is connected with many base stations (BS) by using optical fiber. BS only functions as a converter of optical signal into a wireless signal and vice versa, while at CS, all process involving modulation, demodulation, coding and routing are executed [2]. By using high linear optic link, RoF system distributes the RF signal between CS and BSs. **Figure 1** shows the basic construction of radio over fiber network transmission.

At the transmitter side, data or information from internet or other CS are fed onto modem in the CS during downlink process. Optical signal from the optical source is modulated by the RF signal. After that, the modulated signal will pass through optical fiber toward the BSs. As soon as the signal reaches the BS, photodetector (PD) will function as a detector to detect the modulated optical signal. The PD will also recover the signal before the signal is transmitted through antenna of the BS toward the mobile host. With the similar concept, reverse process is executed for uplink process between mobile host and CS. When the BS receives the signals, optical signal from the LD will be modulated to amplified and transmitted straight toward CS.

In order to determine either the receiver of the RoF in a good quality or not, we can measure and analyze the bit error rate (BER), Q factor value and the eye opening of the resulting result. Bit error rate is the number of received bits of a data stream over a communication channel that has been altered due to noise, interference and distortion orbit synchronization errors. International Telecommunication Union (ITU) has stated that the minimum value of BER of RoF must be below than 10^{-9}. Basically, the value of BER is depending on the measurement time and factors lead to the error. The value of Q depends on the value of BER. There are several ways to determine either our obtained q factor is suitable for our BER value or not. The easiest way is to determine it from BER and Q factor graph. **Figure 2** shows the graph that indicates the relationship between the value of BER and Q factor.

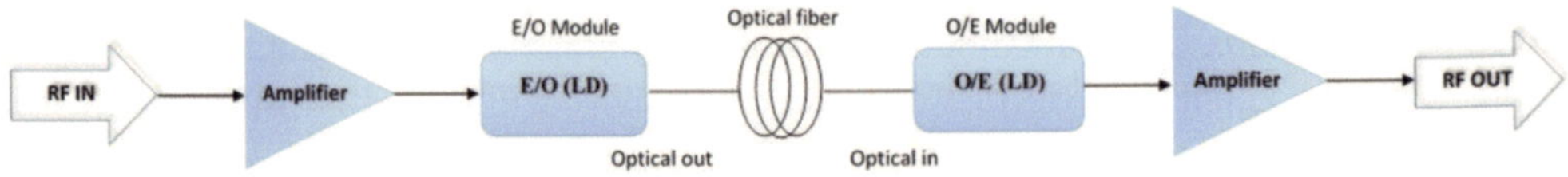

Figure 1. Basic construction of radio over fiber.

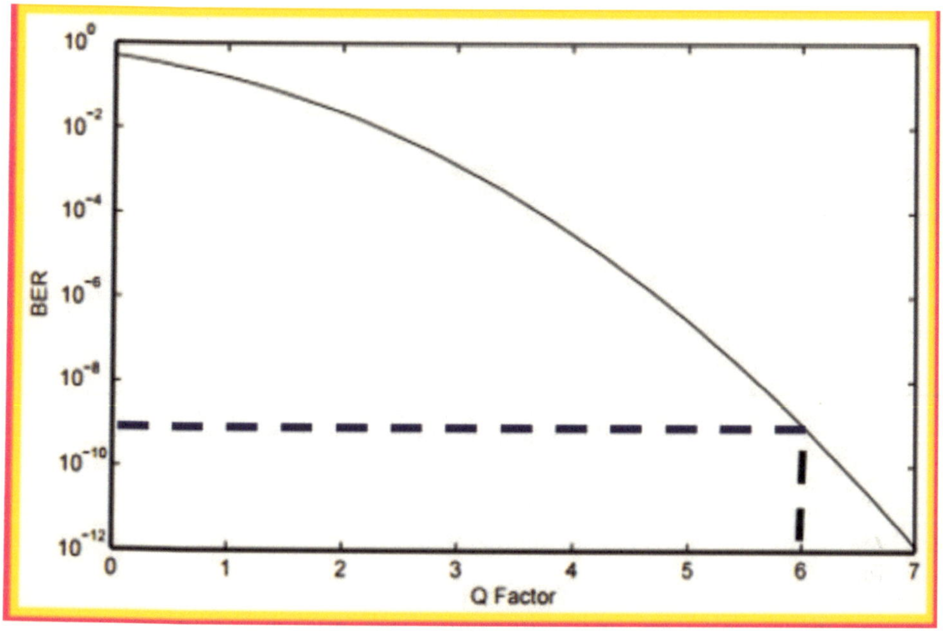

Figure 2. BER versus Q factor.

From the graph, we can see that the value of Q factor increases when the BER decreases. From the graph also, we could see that the value of Q factor for 10^{-9} of BER is approximately 6. Thus, RoF system should have a value of Q factor more than 6 in order to obtain a very good performance of its receiver.

The receiver performance of RoF can also be measured by analyzing the eye diagram of the result after the design has been simulated. Eye diagram analyzer shows multiple traces of a modulated signal to produce eye diagram. This eye diagram is an oscilloscope display where the eye pattern diagram corresponds to minimal signal distortion due to intersymbol interference (ISI) and noise appear in the system [3]. The measurements of eye pattern are made based on time domain. The effect of waveform distortion will appear on the screen of regular BER test equipment. There are many information that could be obtained from eye pattern display. Time interval over which signal that has been reach at the receiver could be sampled without any error causes by ISI could be defined by the width of the eye opening. Noise margin can be determined by looking at the height of eye opening at specified sampling time.

2. Overview of radio over fiber

In 1990, RoF is firstly demonstrated for mobile telephone services or cordless. In this technology, highly linear optic fiber links are used to connect CS and BSs so that RF signal distribution between CS and BSs can be done. Processes involving modulation, demodulation, coding and routing are executed mostly at CS. That means, the only processes occur in BSs only converting optical signal to electrical signal or vice versa. A lot of studies and research have been done just to investigate the limitation and generate new idea to increase the performance of RoF technologies.

Basic RoF consists of all hardware that enable to foist RF signal on an optical carrier at the transmitter side. It also needs fiber optic link to distribute the signal from CS to BSs. At the receiver side, RoF required all the hardware needed to recover the signal from the optical carrier.

2.1. Optical transmission link

2.1.1. Optical fiber

Optical fiber is a platform or a medium to carry information in the form of light from a point to another point. Functioning as a waveguide that will allow the propagation of light, a fiber is a thin filament of glass. One of the advantages of fiber is it provides a path for light with some losses due to the concept of total internal reflection.

In communication system, there are three types of optical fiber, which are step-index multimode, step-index single mode and graded index. Step-index multimode fiber is measured from cladding to the core to cladding as it serves an index of refraction profile that steps from low to high to low. For step index single mode, only one path is allowed for the light to travel in the fiber. Graded index provides large core diameter and higher bandwidth of single-mode fiber.

Optical fiber provides two regions that have low attenuation. First region is at approximately 1300 nm, which has attenuation less than 0.5 dB/km and bandwidth 25 THz. Second region is at approximately 1550 nm, which has attenuation less than 0.2 dB/km and also has bandwidth 25 THz [4]. This combination of two regions will make the total bandwidth as much as 50 THz. Due to the both low attenuation regions, signal loss in data transmission will be very small. Thus, we do not need a lot of amplifiers and repeaters.

2.1.2. Attenuation, dispersion and nonlinearities in fiber

In a fiber optic, attenuation can affect the signal power during the propagation of the signal through distances. Attenuation is a must aspect need to be considered in order to determine the longest distance that signal can go for a given sensitivity of the receiver and the power of the transmitter.

Widening of pulse duration when it propagates through a fiber is known as dispersion. When one pulse starts widening, the pulse is interfering with another pulse besides it. Thus, intersymbol interference (ISI) can happen. Because of this phenomenon, the pulse spacing and the maximum transmission will be limited. There are several types of dispersions and one of them is intermodal dispersion. Intermodal dispersion can happen when several modes of similar signal travels at different velocities through the fiber. Single-mode fiber would not have this kind of dispersion.

Chromatic dispersion is another type of dispersion. Most system will have this type of dispersion since there is no laser able to create a signal that having a single wavelength. Wavelength will be functioning as an index of refraction in dispersive medium. Certain wavelength will travel more fast than other wavelength if the signal that being transmitted has more than one wavelength [5]. Waveguide dispersion might be happen when the propagation of not similar wavelengths depends on characteristics of the waveguide such as indices and shape of fiber core and cladding. Chromatic dispersion is almost 0 in single-mode fiber at 1330 nm. About 1330 nm is also a low attenuation region, and thus, fibers with 0 dispersion can be achieved by using advance techniques such as dispersion shifting.

Nonlinearities may cause an attenuation, distortion and cross-channel interference. Its effect is able to affect the performance of wavelength division multiplexing (WDM) system. In WDM system, nonlinear effects can affect the spacing between similar wavelength channels, limit the maximum power of any channel and able to limit the maximum bit rate.

2.2. Optical transmitter

2.2.1. Optical sources

The most common of light sources used in fiber optic communications are laser diode and light-emitting diode (LED). The benefit of these devices is both have output power for wide range applications. The power also can be directly modulated where the input current is varied to the devices. The efficiency is also high, and they are compatible with the optical fiber.

The diffrence between LED and laser diode is that the laser diodes gives a coherent output where the optical energy is produced in an optical resonant cavity. **Figure 3** shows a basic structure of a laser.

Two mirrors in the laser will form a space between both of them called cavity, a lasing medium that occupied the cavity and a device for excitation. Lasing medium will receive current by the excitation devices and will produce a photon of light. The photon will reflect off the mirrors at both end of the cavity and will go through the medium again.

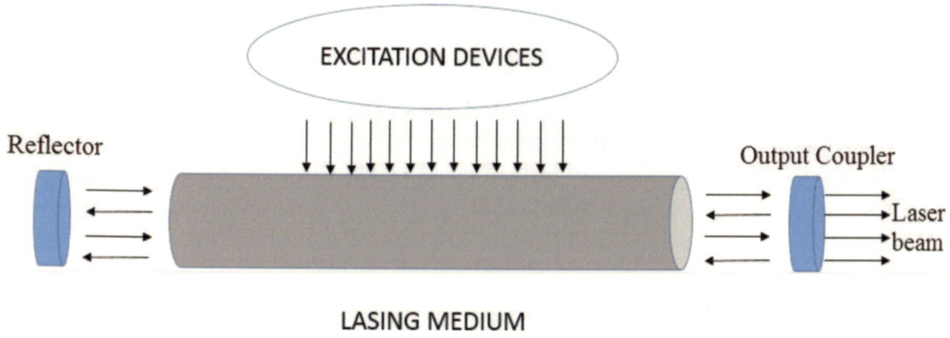

Figure 3. Laser structure.

2.2.2. Optical modulation and line coding

To transmit information across an optical fiber, the data are compulsory to be encoded or modulated onto a laser signal to allow the data being transported through the optical. There is variation of method for modulation. This is including an analog techniques such as frequency modulation (FM), phase modulation (PM) and amplitude modulation (AM). For digital signal, method of modulation includes amplitude shift keying (ASK), phase shift keying (PSK) and frequency shift keying. Among these techniques, binary ASK is the most preferable due to its simplicity. In the systems implementing ASK, the laser is switched on and of to achieved modulation techniques [6]. Current optical communication system also reported on the usage of on-off-Keying (OOK) and DPSK modulations.

To transport digitized information in a communication link, format of transmission the signal must be considered. The signal format is so important since the receiver needs to extract accurately the timing information from the incoming signal. Line coding has several principal functions and one of it is to minimize the errors which causes by noise or any interference effects in the bit stream. The easiest method for encoding data or information is unipolar return to zero (NRZ) code. Logic 1 is represented by a light source that fills the whole bit period, while logic 0 is represented by no pulse transmitted. The process turns the voltage on and off and that's why it is known as (ASK) or on-off keying (OOK). It is essential for NRZ to have a minimum bandwidth and NRZ must be simple to generate and decode.

2.3. Optical receiver

Optical receiver consists of signal processing circuitry, an amplifier and a photodetector [7]. The first thing receiver did when received a signal was converting the optical signal into an electrical signal. After that, the signal will be amplified to an optimum level so that the following process can be done. It is essential to determine and predict the performance of the system based on mathematical models of many receiver stages in order to design a receiver. Also, when designing a receiver, noises and distortions contributed by component in every stage must be considered. Plus, the receiver must have the ability in detecting weak or distorted signals, ability in making a decision on type of received signal and ability to reshaping the distortion signal. This is why it is more complicated compared to process of designing a transmitter. Bit error rate is the most important criteria in measuring RoF system. Other than that are Q factor and the opening of eye diagram.

The height of the eye diagram opening indicated the level of signal distortion. The upper level of the eye represents binary '1' and the bottom level represents binary '0'. Higher eye opening height is desirable, as this indicating that the binary '1' and '0' can be distinguished well. The height of eye opening at a specified time corresponds to the noise margin achieved.

2.3.1. Photodetector

It is compulsory for a receiver to have a device that can interpret the information in the optical signal. Photodetector is a device that can convert the incoming photonic stream into a stream of electrons. When the optical signals reached the receiver, it is in weak and distort condition after undergoes the optical fiber. Thus, photodetector should be sensitive to the emission wavelength range of the optical sources being used. Photodetector also must have an addition of noise to the system, and most importantly, it has to response fast to handle the target data rate. Also, photodetector must be immune to the change of temperature and must be compatible with the physical dimensions of optical fiber. After the photodetector changes the optical signal into an electrical signal, the signal will be amplified and will undergo the threshold device. To determine either the bit is 0 or 1, the presence of light is referred to during the bit duration. It depends on either the electron stream is below or above a certain threshold.

PIN photodiode is the most usable semiconductor photodetector. The structure of this device as shown in **Figure 4** consists of p and n regions. Both regions are separated by intrinsic (i) region [8]. For normal operation, an optimum reverse-bias voltage is supplied across the photodiode to allow the intrinsic region completely depleted of carriers. This will cause the n and p carrier concentrations that become less than impurity concentrations in the region. When a photon flux Φ penetrated into the device, the flux is absorbed.

In PIN photodetector, the light absorption causes the formation of electron-hole pairs. Then, the hole and electron are drifted to the opposite direction, causing the flow of current. More current flows as more light enter the photodetector, which giving rise to the number of electron-hole pairs.

When the energy of an incident photon is higher than or equal to the semiconductor's band-gap energy, the photon excite an electron from valence band to the conduction band by give up its energy. The absorption process will form an electron-hole pairs known as photo carriers. Normally, the photodetector is designed so that those carriers are mainly generated in the depleted intrinsic region. This region has the most absorption of incident light. The large amount of electric field will make the carriers separating between each other and will be collected across the reverse biased junction. Thus, current flow will increase in the external circuit where every carrier pair generated has one electron flowing.

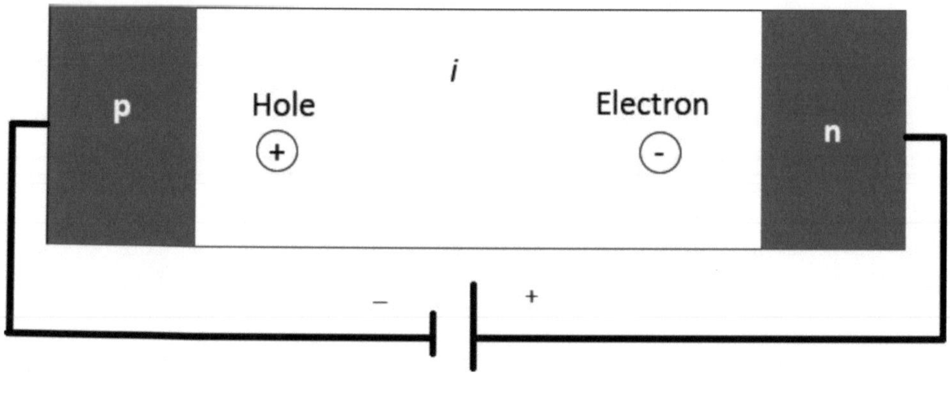

Figure 4. PIN photodiode structure.

2.3.2. Optical amplifier

Optical amplifiers can contribute a lot to long haul or local networks even optical signal still can transmit a long distance without amplifiers. There are several techniques of optical amplification. 1R (regeneration) techniques provide a booster to power up the signal. It does not restore the shape and the timing of signal. It also serves a total data transparency. 2R (regeneration & reshaping) amplification is a technique where the optical signal will be transformed into electrical signal before it directly used to modulate a laser.

3R (regeneration, reshaping, relocking) techniques change the data stream into electronic signal and after that retransmit the signal optically to amplifies the signal [9]. Noise can be eliminated much through the reshaping process where the signal produces back the original pulse shape of each bit. Mostly, reshaping works on digital signal but for some cases it may also works on analog signal but relocking does not work on analog modulated sign.

Basically, there are three main types of optical amplifiers. They are semiconductor optical amplifier (SOA), doped fiber amplifier (DFA) and Raman amplifier. The main function of all amplifiers is to boost the power level of incident light through optical power transfer process. Also, this process can be done through a stimulated emission. Using the concept of laser diode, the mechanism of SOA and DFA is creating the population inversion which is needed for simulated emission. Optical amplifier does not have the ability to generate coherent optical output since it does not have optical feedback which is compulsory for lasing. Thus, optical amplifier can only increase the signal levels.

The amplifier gained an energy from a pump, which is an external source as shown in **Figure 5**. Technically, the external sources are supplying an energy to the electrons in the active medium. This will cause the electrons increased and served a population inversion. The excited electrons will drop to lower level because the electrons are triggered by the incoming photon signal through the process of simulated emission. The output of the signal will be amplified since one incoming photon stimulates a cascaded effect in which way equal energy of photons is emitted by a lot of excited electrons when they hit the ground state.

Advantage of SOA is that this amplifier can be implemented if both signal processing and switching functions were call in optical networks. The drawbacks of SOA is it has rapid gain response that will cause the gain at specific wavelength fluctuate with signal rate for speed up

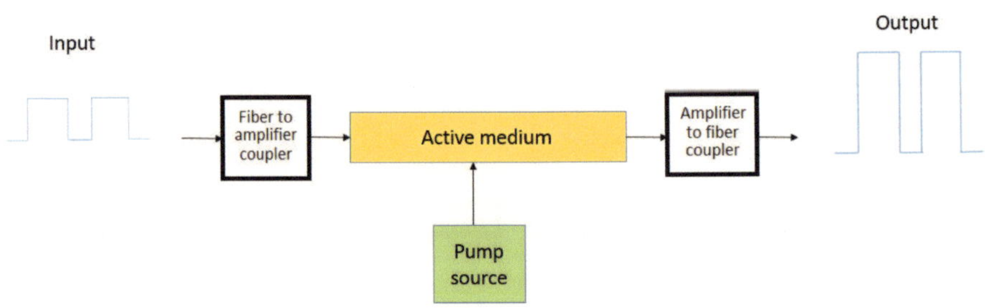

Figure 5. Signal amplification by an optical amplifier.

to some Gb/s. The overall gain also can be affected, and this will cause the signal gain of other wavelengths fluctuate. Thus, cross-talk effect can occur.

For DFA, the length of fiber will be doped with an element that able to amplify the light [10]. The most common element is erbium. The main advantage for this type of amplifier is it can pump the devices at some different wavelengths. DFA also has small coupling loss and low dependence of gain on light polarization. Plus, DFA is immune from interference effects since the gain responses are constant for signal modulations greater than few kilohertz.

Raman amplifier has a transfer of optical power form high power pump wavelength to light wave signals at longer wavelengths. Raman amplification does not need process of population inversion. It is based on stimulated Raman scattering (SRS) which is a nonlinear effect. This effect normally occurs in fibers at high optical powers. Raman gain mechanism was achieved through a discrete amplifier. This type of amplifier can be used in any wavelength band since the gain is based on SRS where the induced transfer of optical power from shorter pump wavelengths to longer signal wavelengths.

3. Overview of wavelength division multiplexing (WDM)

Various methods can be executed to gain full duplex transmission of RoF system and wavelength division multiplexing (WDM) is one of it. WDM has characteristic where the discrete wavelength will form an orthogonal set of carriers that can be separated, routed and switched without interfering with each other [11]. With WDM, many wavelengths are able to be transmitted over a large distance and the downlink and uplink data can be transmitted at the same time through single-mode fiber.

The design of WDM involves some requirements such as selecting various optical sources that has narrow spectral emissions bands. The most straightforward and simplest method can be done by choosing a series of individual lasers that emits at its own specific wavelength. The process of selecting optical sources requires small number of wavelength channel but can be cumbersome for links carrying many wavelengths. In order to implement WDM network, a combination of many passive and active component is need. The difference between passive and active components is passive components and is limited in their ability since passive components did not require an external control to operate, whereas active component can be controlled and has a wide network ability.

WDM contributes simplification to the network. With this method, different wavelengths are allocated toward individually BSs. Thus, the network will be simple and the service upgrades can be done easier. WDM multiplexes optical signals that came from multiple sources, and the signal will be amplified before it is transmitted through optical fiber to under goes demultiplexer so that it can be addressed to each BS as shown in **Figure 6**.

There are two types of WDM architecture: coarse wavelength division multiplexing (CWDM) [12] and dense wavelength division multiplexing (DWDM) [13]. Typically, CWDM systems provide eight wavelengths which separated by 20 nm, from 1470 to 1610 nm. To increase the

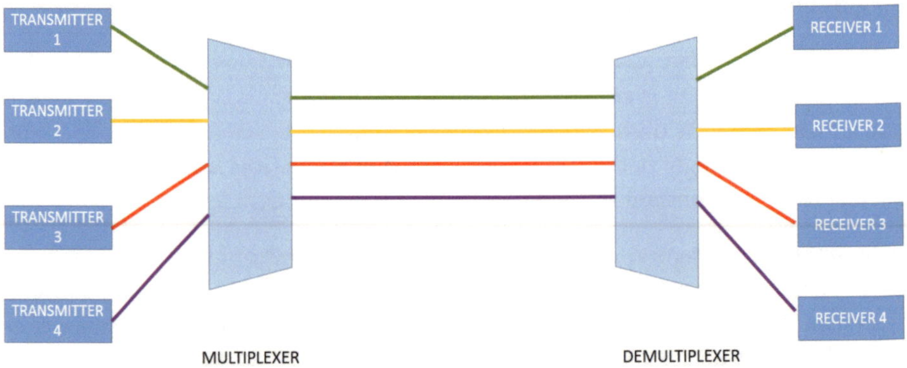

Figure 6. Wavelength division multiplexing.

channel of CWDM to 16, number of wavelength could be increased by using 1310 nm window. One of the advantages of CWDM is that the cost of the optics is 1/3 of the cost of same DWDM optic. This makes that the CDWM is preferable than DWDM. CWDM is able to match the basic capabilities of DWDM but with lower cost and capacity. Typically, CWDM is used for short-range communications. In addition, CWDM equipment is more compact and cost-effective if compared to DWDM designs.

DWDM network systems provide up to 96 wavelengths, which normally has less than 0.4 nm spacing. DWDM is used for long-haul transmission where wavelengths are compacted tightly together. Erbium-doped fiber amplifiers (EDFA) can help the system of DWDM to work over thousands of kilometers. CWDM is not implemented in long haul transmission, where the distance reaches up to thousands of kilometers, causing simpler overall sytem components requirement. This leads to lesser cost implementation, despite of having some limita-tions especially in propagation distance.

4. Methodology

The methodology used in designing RoF system consists of familiarization with Optisystem software, designing system, generate component and simulation object, running the simulation and analyze the data. Optisystem software is used to design, construct and simulate the RoF topology.

Optisystem software is a comprehensive software that provides a platform to plan, test, and simulate optical links in the transmission layer of modern optical networks. Optisystem is also an optical communication system simulation package for the design, testing, and optimization of virtually any type of optical link in the physical layer of a broad spectrum of optical networks, from analog video broadcasting systems to intercontinental backbones.

A system level simulator based on the realistic modeling of fiber-optic communication systems, Optisystem, possesses a powerful simulation environment and a truly hierarchical definition of components and systems. Its capabilities can be easily expanded with the addition of user components and seamless interfaces to a range of widely used tools. Optisystem is compatible with Optiwave's Optiamplifier and OptiBPM design tools. Optisystem serves a wide range of applications, from CATV/WDM network design and SONET/SDH ring design to map design and transmitter, channel, amplifier, and receiver design. Optisystem contains a MATLAB component that enables the user to call MATLAB within its environment to incorporate new components or models into the software. Optisystem uses the MATLAB.dll files to evaluate the MATLAB script in the component to perform the calculations.

Designing RoF system on Optisystem software includes generation of component, simulation object and makes connection between all of them. It is very important to design a very good topology so that the maximum performance of RoF's receiver can be achieved.

5. Simulation

5.1. System design

To send two signals, two continuous wave lasers are used in order to resonate 193.1 and 193.2 THz, respectively, as shown in **Figures 7** and **8**, respectively. To obtain an electrical network that simple and low in speed, a pulse generator is needed. A pseudo-random bit (PRBS) generator is used to operate a Gaussian pulse generator so that baseband signal can be generated. 1 Gbps was set as the bit rate. The PRBS generator indicates the random source of data. To simulate the signal, 1 Gbps per channel was set as the data rate for the simulation.

As the carrier frequency is 2.7 GHz for transmitter 1 and 1.7 GHz for transmitter 2, an amplitude modulator is operated by the 1 Gbps baseband signal. After that, the amplitude modulator will operate the LiNb Mach-Zehnder modulator (MZM)'s port 1. The main function of the amplitude modulator is to translate the baseband signal onto the RF clock. At a frequency of 49.25 MHz for both transmitter, the SCM carrier generator will mix with the RF signal. A 90° hybrid coupler is applied to the signal where it is utilized to separate that the input signal becomes 2 outputs that having 90° phase shift difference between each other. The separated signals are sent to the two arm of M-Z Modulator. At the both transmitter sides, to allow the signal being transmitted along the fiber, the signals change from digital to analog form.

Wavelength digital multiplexer (WDM) is used to multiplex all signals from the transmitter. The bandwidth of the multiplexer was set to 10 GHz. EDFA amplifier with power 10 dBm is used to boost up the power of the optical signal. Then, through an optical fiber, the signals are transmitted. This process can be seen in **Figure 9**.

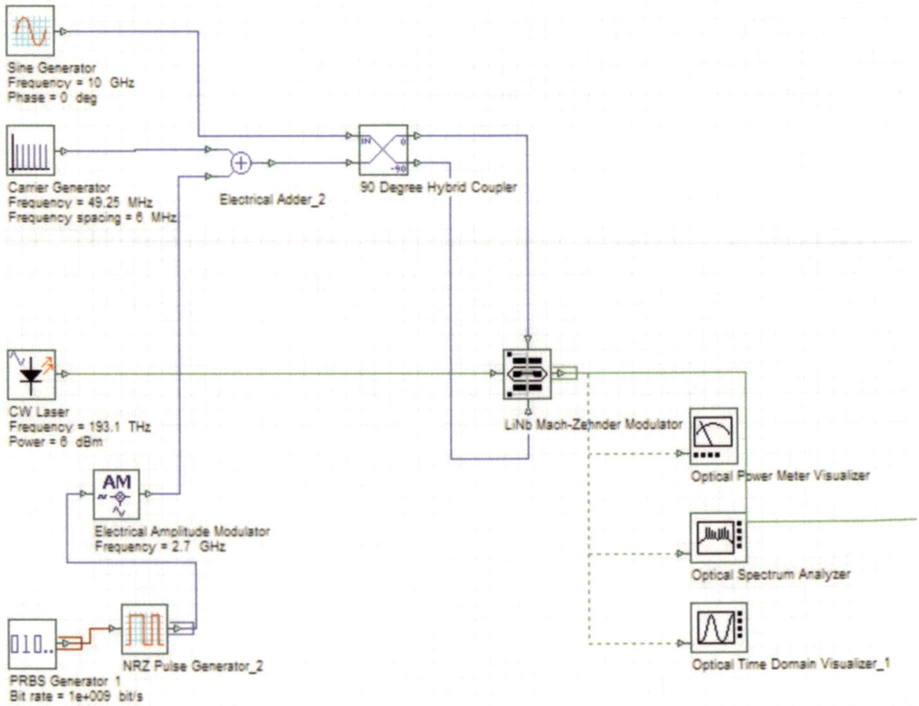

Figure 7. Transmitter 1.

As soon as the signal reaches the receiver side, demultiplexer is used to demultiplex the signals before the signals being distributed to its own receivers. 16 GHz is chosen as the bandwidth of the demultiplexer. At the receiver 1 as shown in **Figure 10**, the signal is once again amplified by using EDFA amplifier with power of 20 dBm. After that, photodetector is used to convert the optical signal into electrical form. Then, bandpass Bessel filter received the signal and filtered the signal. The received signal is amplified by using electrical amplifier with 15 dB gain before demodulated by using demodulator. 3R regenerator technique is applied to the signal before the BER analyzer analyses the received.

Same concept is applied at the second receiver. The EDFA amplifier is used to boost the power of the optical signal after the signal is demultiplexed. Photodetector will transform the signal into electrical form. The converted signal will be amplified by electrical amplifier, and the signal will be passed to the Bessel filter. After the signal has been filtered, demodulator is used to demodulate the signal before the 3R regenerator amplification is applied to the signal. Lastly, BER analyzer will analyze the signal. The configuration of the second receiver can be seen in **Figure 11**.

http://dx.doi.org/10.5772/intechopen.68583

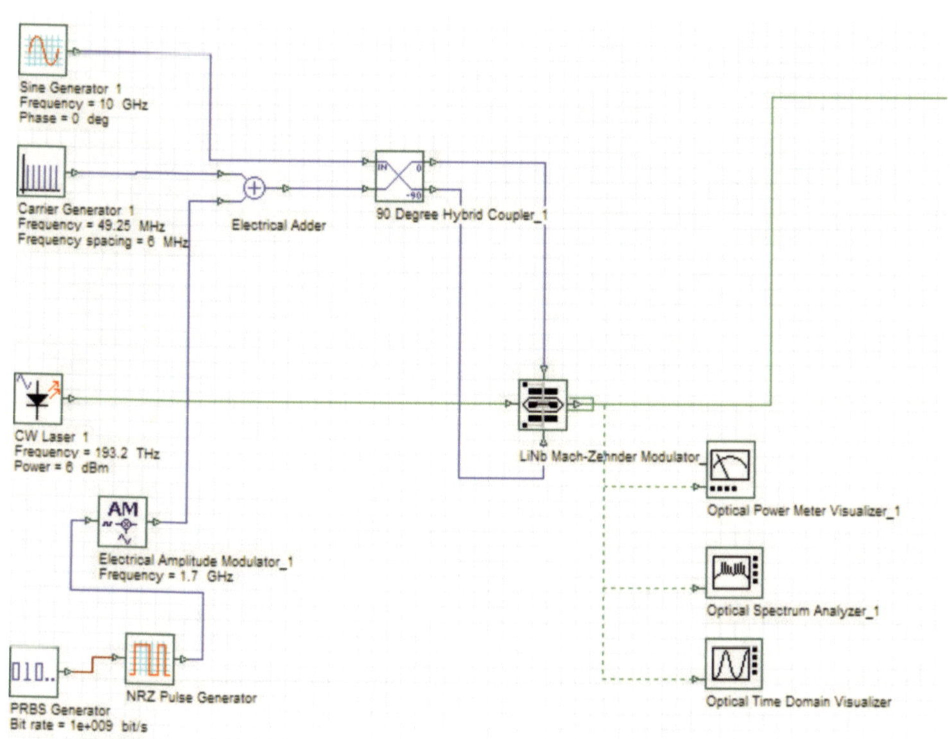

Figure 8. Transmitter 2.

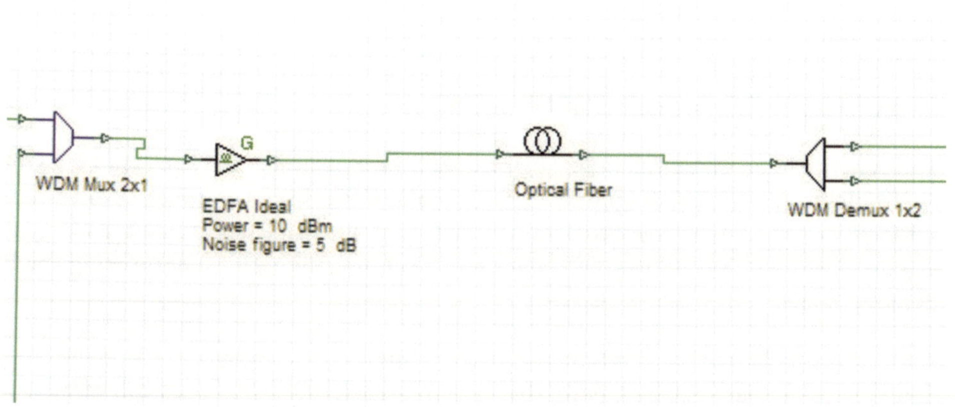

Figure 9. Transmission of data through optical fiber.

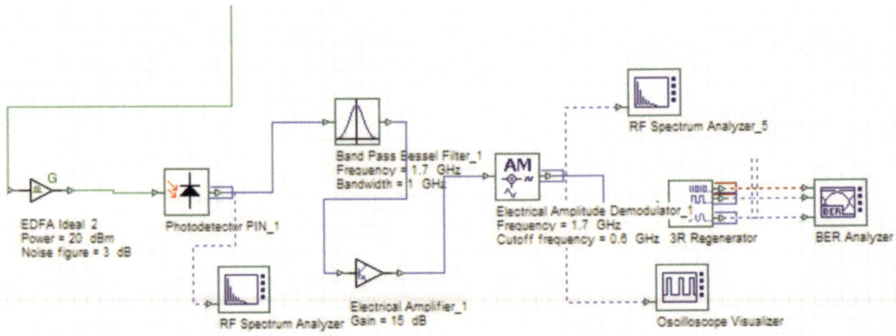

Figure 10. Receiver 1.

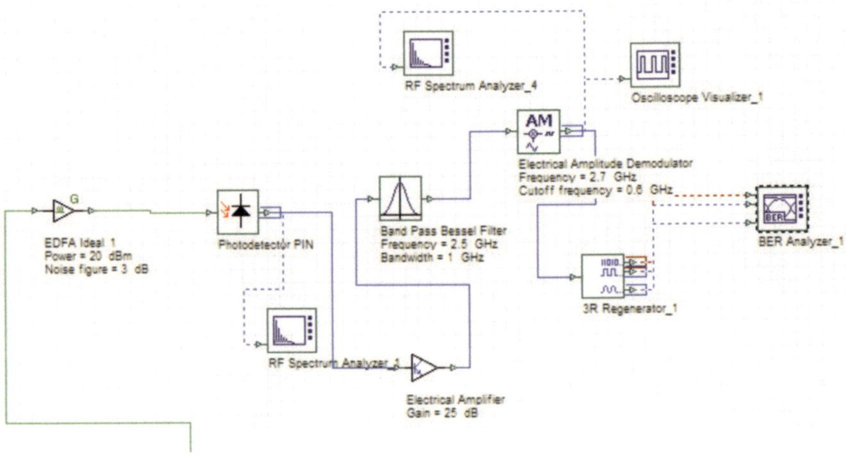

Figure 11. Receiver 2.

5.2. Single parameter optimization (SPO)

SPO can benefit a lot in the simulation. It helps to optimize parameters so we could set a target for the simulation's result. With optimization tools, the software can optimize the fiber length of the EDFA so that a maximum gain could be obtained. It can also calculate the attenuation or the gain in order to get a desired Q factor and minimize the BER by optimizing the fiber length of the system.

After all components had been connected and SPO had been inserted to the topology, the simulation is run to see the result. After the run button is clicked, it will calculate all the calculation of the system. To measure the BER and the Q factor, just simply double click at the BER analyzer component. It will give an eye diagram showing eye opening, BER value and Q factor value. The example of the result is shown as in **Figure 12**.

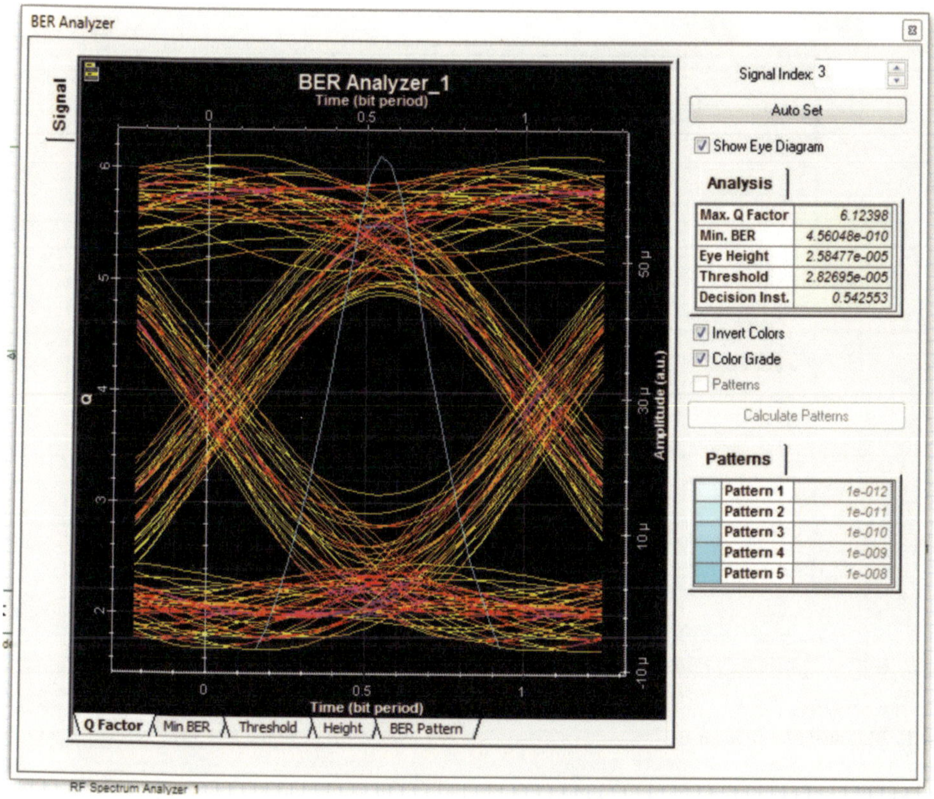

Figure 12. Example of BER analyzer's result.

As shown in the example above, the result tells us that the BER is 4.560^{-10} and the Q factor is 6.124. There is an eye opening in the result of the example which is 2.585^{-5}. The example is a quite good result since the BER is less than 10^{-9} and the Q factor is more than 6. If all the receivers receive an output just like the example above, all objectives had been achieved.

6. Results

After the simulation is done, the performance of the receiver is analyzed by referring the BER analyzer and the eye diagram at the two receivers. All the results can be seen in **Figures 13–16**.

For receiver 1, the BER analyzer shows that the BER for the received signal is 3.54×10^{-20}. The Q factor for this receiver is 9.13. This shows that this receiver has a very good performance since the number of BER is above 10^{-9} and the Q factor is more than 6.

As mentioned before, the height of the eye diagram defines the immunity of the received signal to the noise. The eye diagram for receiver 1 shows an eye opening of 0.0004.

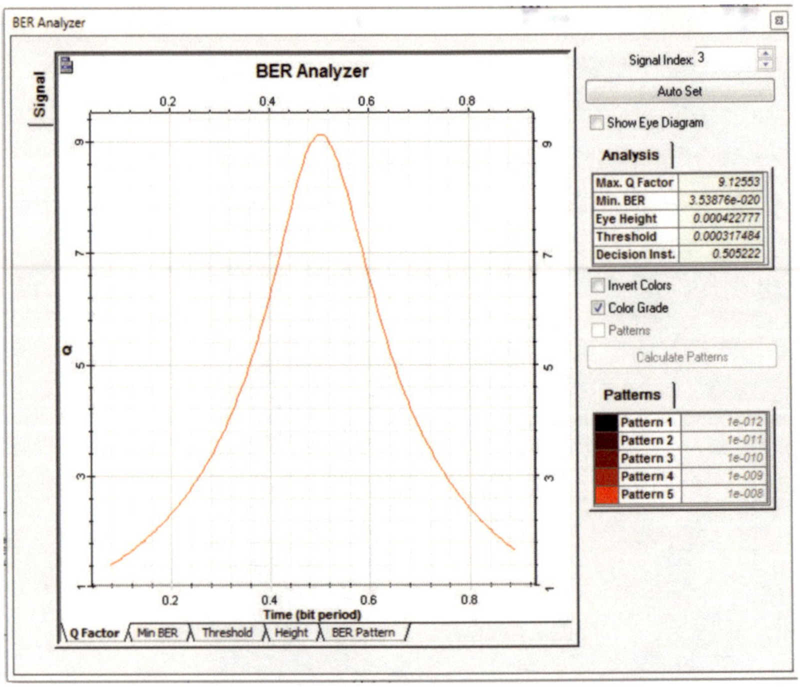

Figure 13. BER result for receiver 1.

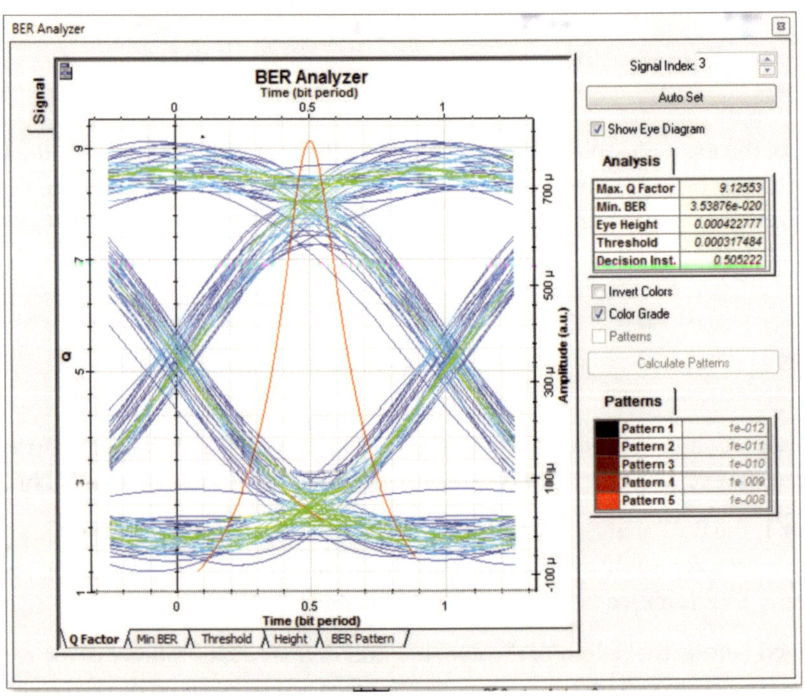

Figure 14. Eye diagram for receiver 1.

http://dx.doi.org/10.5772/intechopen.68583

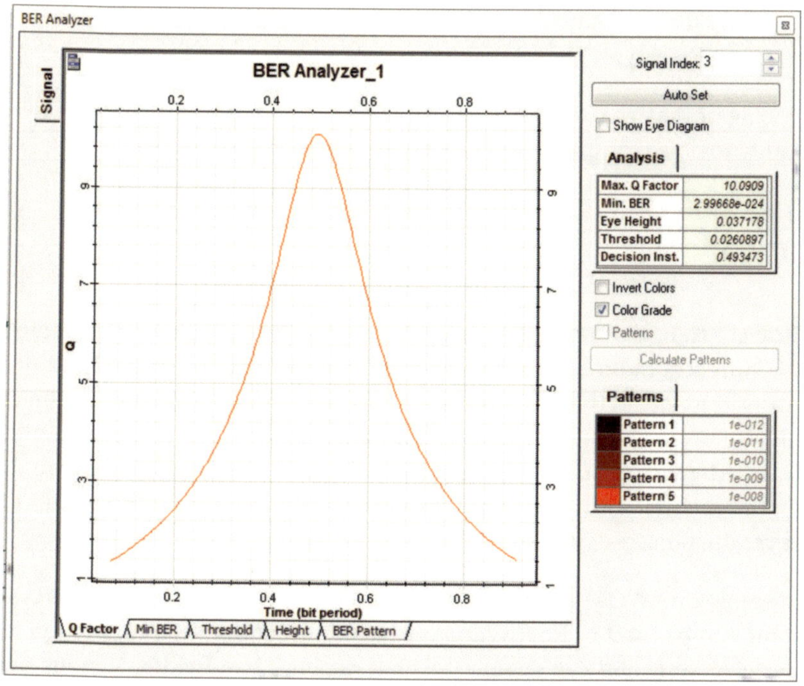

Figure 15. BER result for receiver 2.

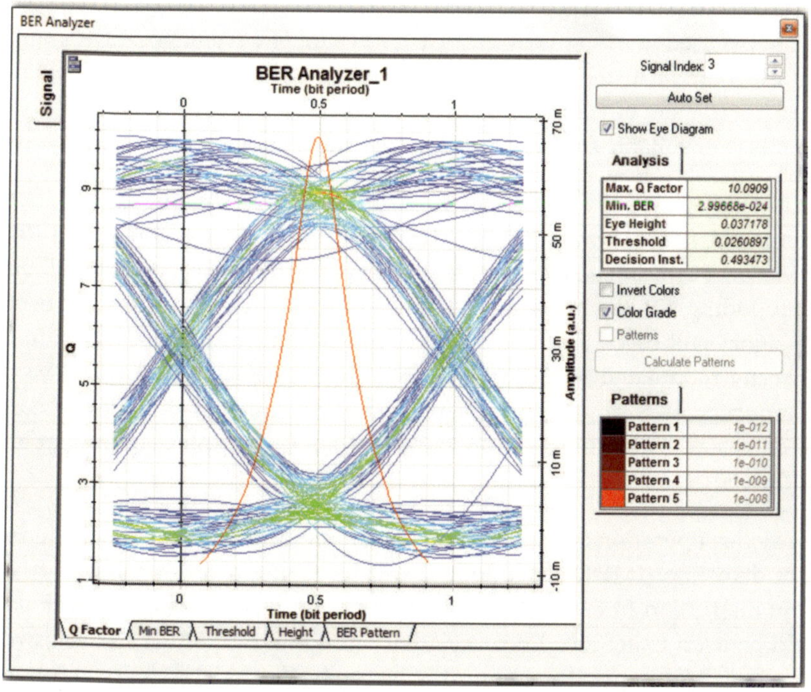

Figure 16. Eye diagram for receiver 2.

For receiver 2, the received signal has BER of 3×10^{-24}. The Q factor for the received signal is 10.09. Thus, the BER analyser shows that this receiver also received a good quality of signal.

The eye opening of the eye diagram is 0.0371. It shows that received signal in receiver 2 has more immunity to noise compared to the received signal in receiver 1.

7. Conclusion

Basically, RoF is a transmission system of analog signal since it distributes a radio waveform from CS to BS direct at radio carrier frequency. A lot of way can be studied and developed to increase the performance of the RoF. This is why RoF is an ideal technology for the future generation. RoF also provides low attenuation and broad bandwidth. It is also immune to electromagnetic interference (EMI), which is a radio-frequency interference where disturbance generated by an external sources that affects an electric circuit by electromagnetic induction, coupling or conduction.

RoF system employing WDM technique is simulated using Optisystem 12.0. WDM is among the best method to be used in RoF. WDM enables transmission of multiple signals through a single fiber over large and can exploit the fiber network bandwidth. Various transmission formats can be supported by various optical channels. Plus, WDM enables the capacity of an optical fiber increase compared to simple point to point link that only enable to carry only single wavelength.

EDFA amplifier is used in this system because it has a lot of advantages. EDFA amplifier provides low noise figure. It also has an independent polarization. Thirdly, the dynamic range is large. The power transfer efficiency of EDFA amplifier is quite large. It also has relatively flat gain. Thus, it is very suitable for long-distance communication.

RoF is a technology that really cost-effective because in this technology, all of the expensive components can be shared by several BS. The installation and maintenance of RoF are easy since the concept of centralized configuration enables all sensitive components being located in safer surrounding. To modulate an optical sources, an electrical signal is used, and the modulated signal will be carried out through optical fiber to the receiver side. When the signal is directly modulated to the optical fiber, power consumption could be decreased, while the frequency of radio carriers is very high at the antenna side. Power consumption also can be decreased by having a simple radio station. All complex equipment and devices are kept at the CS.

RoF also provides operational benefits in terms of operational flexibility. Intensity modulation direct detection (IMDD) technique could be made in order to operate as a linear system. Low dispersion fiber (SMF) with combination with modulated RoF sub carriers (SCM) could be used to achieve linear system. The same radio over fiber network can be used to distribute multi-operator and multi-service traffic which results in huge economic savings.

Author details

Lisawati Sainawi and Khadijah Ismail*

*Address all correspondence to: khadijah_etc@hotmail.com

National Defence University of Malaysia, Kuala Lumpur, Malaysia

References

[1] Prakasam S, Karthikeyan R. OFDM signal improvement using radio over fiber for wireless system. IJCNWC. 2013 June;**3**:287–291

[2] Kim HB and D. Der Ingenieurwissenschaften. Radio Over Fiber Based Network Architecture

[3] Johny J, Shashidharan S, Rof II, and Sing DEU. Design and simulation of a radio over fiber system and its performance analysis; 2017. pp. 655–658

[4] Yunazar F. Modeling and performance analysis of the OFDM scheme for radio over fiber system [thesis]. Malaysia: Universiti Teknologi Malaysia; 2009. p. 74

[5] Iftekhar SF. Radio Over Fiber Transmission by Sub Carrier Multiplexing; August, 2009

[6] Reddy V. Simulation and analysis of radio over fiber (RoF) systems using frequency up-conversion technique. 2016;**133**(12):36–43

[7] Azadeh M. Fiber Optics Engineering. United States: Springer; 2009. p. 374

[8] Keiser G. Optical Fiber Communications. 4th ed. New York: McGraw Hill; 2011. p. 654

[9] Das B, Abdullah MFL, Shah M, Shahida N, and Bukhsh Q. All regeneration for optical communication network using optical 3R regeneration and phase sensitive amplifier. 2015;**8**(8):87–94

[10] Singh YN, Gupta HM, Jain VK. Optical amplifier in broadcast optical network: A survey [thesis]. India: Indian Institute of Technology Kanpur; p. 48

[11] Mohan A and Anisha AP. Performance comparison of radio over fiber system using WDM and OADM with various digital modulation formats. 2015;**4**(2):2013–2016

[12] Basa A, Talukder MZ and Islam MR. Performance analysis and comparison between coarse WDM and dense WDM. 2013;**13**(6)

[13] Robinson S, Jasmine S, and Pavithra R. Investigation on hybrid WDM (DWDM + CWDM) free space optical communication system. 2015 December;**6948**:1187–1192

Section 3

Optical Technology Components

Chapter 10

Evaluation of Parametric and Hybrid Amplifier Applications in WDM Transmission Systems

Vjaceslavs Bobrovs, Sergejs Olonkins, Sandis Spolitis,
Jurgis Porins and Girts Ivanovs

Additional information is available at the end of the chapter

http://dx.doi.org/10.5772/67607

Abstract

Over the past two decades, a rapid expansion of the amount of information to be transferred has been observed. This tendency is explained by the rapid increase of Internet and other service users, as well as with the increasing availability of these services. This rapid growth in the amount of globally transmitted data is also associated with the expansion of the range of services offered, including such resource-consuming services as high-resolution video transmission, videoconferencing, and cloud computing, as well as with increasing popularity of such services. To satisfy this constantly increasing demand for higher network capacity, fiber optical transmission systems have been studied and applied with a growing intensity. Currently, optical transmission systems with wavelength-division multiplexing (WDM) have attracted much attention, as this technology allows using the available optical fiber resources more effectively than alternative technologies.

Keywords: optical amplifiers, parametric amplifiers, hybrid amplifiers, fiber optics, EDFA, WDM

1. Introduction

According to the latest Cisco forecast, the total amount of global IP traffic in 2016 reached 1.1 zettabytes, whereas in 2018 it will reach 1.6 zettabytes. The forecasted increase in the monthly transferrable IP traffic over the period from 2013 to 2018 is shown in **Figure 1a**. Studies performed by Cisco show that in comparison with 2012 the amount of Internet traffic transferred in the peak hours in 2013 increased by 32%, whereas the average daily volume of transferrable Internet traffic increased by 25% [1]. If this tendency remains, then in 2018 the volume of

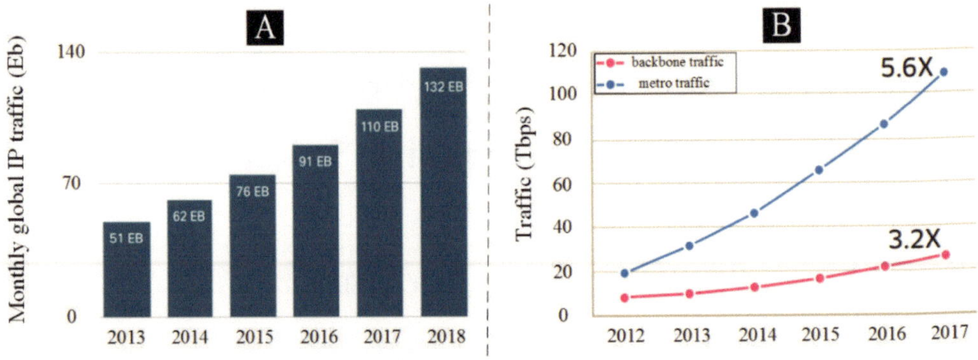

Figure 1. Cisco forecast of the monthly transferrable IP traffic (A) and Bell Labs forecast of the transferrable data amount in backbone and metro networks (B) [1, 2].

transferrable Internet traffic during the peak hours will reach 1 petabit per second, whereas the daily average will reach 311 terabits per second [1, 3]. According to the Bell Labs forecast, results of which are shown in **Figure 1b**, during the period from 2012 to 2017, the increase of traffic in backbone networks will reach 320%, whereas in metro networks, it will reach by 560% [2].

It is possible to increase the wavelength-division multiplexing (WDM) system throughput capacity either by increasing the data transmission speed in channels or the number of channels. The wavelength band that is used for transmission in WDM systems is limited due to the wavelength dependence of optical signal attenuation in optical fibers [4, 5]. In modern transmission systems, the minimum attenuation of single-mode optical fiber is 0.2 dB km^{-1}, and it is observed in the "C" wavelength band, which corresponds to wavelengths from 1530 to 1565 nm. Regardless of the fact that the attenuation value is so low, its impact accumulates with every next kilometer. In long-haul transmission systems, where transmission lines are several hundreds and even thousand kilometers long, the attenuation substantially degrades the quality of the received signal, as the photodetector sensitivity is limited [6–8]. As the number of channels increases, the attenuation caused by the optical signal division also increases, especially in cases where power splitters are used [9]. However, by increasing the speed of data transmission, it becomes necessary to reduce the optical noise produced by optical components (light sources, modulators, amplifiers, receivers, etc.), as higher transmission speed signals have lower noise immunity.

Therefore, solutions are needed for compensating the ever-increasing accumulated signal attenuation in an ever-broader wavelength range. Currently, erbium-doped fiber amplifiers (EDFAs) are most commonly used around the globe for compensation of optical signal attenuation. The amplification bandwidth of EDFAs is strictly limited (for conventional EDFA solutions, it is only 35 nm), which restricts the wavelength range used for the transmission in existing systems [10–12]. It is, thus, necessary to seek for new solutions to amplifying optical signals and for opportunities of expanding the range of amplified wavelengths and increasing the attainable amplification level for the already-existing optical signal amplification solutions. This can be achieved by combining amplifiers of various types. In such a way, it is possible to combine the positive properties and partly compensate the drawbacks of different types of amplifiers.

http://dx.doi.org/10.5772/67607

During recent years, the need to increase transmission capacity of existing optical networks together with requirements for reducing the total cost of construction and maintenance of optical networks has induced increasing interest in all-optical signal processing [13–16]. In contrast to solutions with optical-electrical-optical (O/E/O) signal conversion, which induces the so-called bottlenecks in optical transmission systems, all-optical signal processing is performed in real time, whereas the signal is transmitted through a nonlinear medium [17]. Therefore, all-optical signal processing allows avoiding the part of transmission capacity limitation that is caused by O/E/O signal conversion.

The progress in nonlinear material research has resulted in commercial production of optical fibers and other components with high values of the nonlinear coefficient. Therefore, the optical power, required to initiate fiber nonlinearities, has become lower [15]. Fiber nonlinearity is the main mechanism that is used for all-optical signal processing. Optical amplifiers are the only optical devices capable of rising the power of optical signal high enough to induce manifestation of nonlinear effects during transmission. That is why the usage of optical amplifiers for all-optical signal processing purposes has been intensively studied all over the world during recent years, and various applications of optical amplifiers have been demonstrated [13–16, 18–20].

2. Main principles of optical amplification

Amplification of optical signals is based on the energy transfer from pumping optical radiation or another type of energy to the amplifiable optical signal. This process is implemented differently in various types of optical amplifiers. In general, the amplification process uses the stimulated emission phenomenon in the amplification environment, such as, for instance, semiconductor optical amplifiers or doped fiber optical amplifiers. Furthermore, such nonlinear optical effects such as Raman, Brillouin, and four-wave mixing (FWM) are used to amplify optical signals in cases of Raman, Brillouin, and parametric optical amplifiers, respectively [21].

The mechanism of amplifying optical signals is based on occurrence of stimulated light emission in the gain medium. The light emission phenomenon can be explained using the Rutherford-Bohr atomic model. Bohr has stated that atoms may jump from one energy state to another, by performing what is known as the quantum jumps, corresponding to a change of orbit. This orbit change requires a change in the energy level; therefore, if the atom jumps from the higher energy state to the lower energy state, it will produce a photon. A photon contains energy, which corresponds to the difference between the initial higher energy level and lower occupied level energy, as the overall energy of the process must remain unchanged. This assumption derives from the law of conservation of energy [22]. Thus, photon energy can be determined according to the following equation [23]:

$$E_{photon} = E_2 - E_1 = h\,v_{photon} \tag{1}$$

where E_{photon} is the generated photon energy, E_1 and E_2 are the high and low energy level, h is the Planck constant, and v_{photon} is the generated photon frequency.

Optical amplifiers can be classified according to the nature of the amplification process [23]:

a. Amplifiers, in which amplification is obtained, using linear properties of the material (semiconductor optical amplifiers (SOAs) and amplifiers on rare-earth element-doped fiber basis (xDFA))

b. Amplifiers, for which the principle of operations is based on nonlinear properties of the material (Raman optical amplifiers, Brillouin optical amplifiers, and fiber optical parametric amplifier (FOPA))

A second way of classifying optical amplifiers is according to the medium, in which amplification takes place:

- Amplifiers, in which semiconductor material is used (SOA)
- Amplifiers, which are produced on the basis of optical fibers

The main parameters that are used to characterize optical amplifiers are the level of amplification, the gain bandwidth, the saturation power of the amplifier, the polarization sensitivity of the produced gain, and the amount of signal impairments produced by the amplifier.

The achievable level of amplification is determined as the relation of the output signal power to the power of the same signal in the input of the amplifier. Amplifiers are sometimes also described with amplification efficiency, which describes the amplification as a function of the pumping power. The unit of measurement of efficiency of amplification is dB/mW [24].

The bandwidth of the amplifier produced gain is applied to the wavelength or frequency range, in which the use of the amplifier is effective, namely, where it can ensure an increase in signal power. This value is especially important in WDM transmission systems, as it limits the number of channels in such systems [23].

The saturation point for an optical amplifier is the maximum attainable output power value, namely, when the optical signal power in the amplifier output no longer increases while raising the signal power at the amplifier input. When the input power is increased above the saturation point, all carriers in the gain medium are already in a saturated status, and a higher level of energy transfer to the amplified signal is no longer possible. The saturation power is defined as the output power, at which 3 dB decrease in amplification is observed, in respect to the maximum possible level of amplification [23].

The dominating source of noise in optical signal amplifiers is the amplified spontaneous emission (ASE), which originates in the gain medium [25]. The amount of noise generated by amplifiers depends on various factors. The most important of these are the gain medium material parameters (e.g., the spontaneous lifetime of the energy level), gain spectrum, noise bandwidth, amplifier saturation, and population inversion parameters. The problem of noise generated by an amplifier is most explicit in systems, where it is required to use multiple amplification stages, therefore placing the amplifiers in a cascade, such as backbone optical networks. Each amplifier in such cascades not only amplifies the transmitted

http://dx.doi.org/10.5772/67607

signal but also the noise generated by the amplifier from the previous amplification stage and additionally adds ASE noise of its own [23]. To assess the amount of ASE noise generated by the amplifier, the noise figure (NF) parameter is normally used. This value describes the optical signal-to-noise ratio (OSNR) changes, as the signal passes through the amplifier [23, 26].

In the studies conducted by the authors, using simulation software OptSim, the performance of SOA, EDFA, lumped Raman amplifier (LRA), and the distributed Raman amplifiers (DRA) under equal operating conditions has been compared. The simulation scheme introduced for this purpose is displayed in **Figure 2**. Such a structure of the WDM transmission system simulation model will also be used further in the research, when the operations of an amplifier are analyzed.

The performance of different types of amplifiers has been compared in a 16-channel dense wavelength division multiplexing (DWDM) transmission system with 10 Gbps transmission speed per channel, 50 GHz channel spacing, and non-return-to-zero on-off keying (NRZ-OOK) (on-off keying) modulation format. In each case, also the length of the dispersion-compensating fiber (DCF) has been determined. Optical amplifiers have been used as in-line amplifiers. The comparison of SOA, EDFA, LRA, and DRA performance is available in **Table 1**.

The largest transmission distance has been achieved in a system with the DRA. Here, just like in the case of LRA, the attainable amplification is limited by the impact of fiber nonlinearity on the quality of the amplified signal. A 1150 mW co-propagating pumping radiation is used for DRA pumping. The amplification process occurs in the transmission line section between the DRA pumping source and the receiver block. Thus, the single-mode fiber (SMF) attenuation reduces the signal amplification rate in the direction from the amplifier to the receiver block, which allows achieving much larger amplification than in the case of LRA, and accordingly increases the attainable transmission distance. Irrespective of the

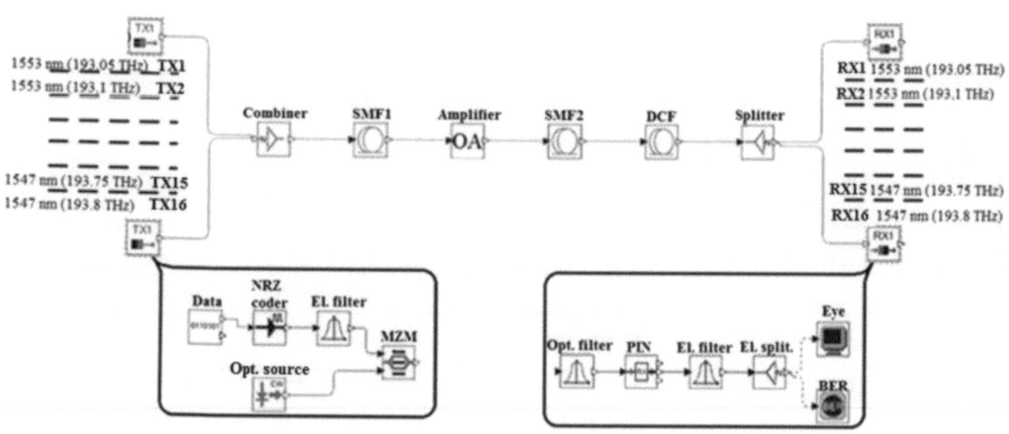

Figure 2. Simulation model of the 16-channel 10 Gbps DWDM transmission system used for comparison of optical amplifier performance.

Amplifier type	-	SOA	EDFA	LRA	DRA
Transmission distance (km)	69	112	135	119	146
DCF length (km)	5	15	20	17	20
Gain in wavelength range from 1546 to 1553 nm (dB)	-	17.4	23.4–25.1	19.9–20	24.9–25
NF in wavelength range from 1546 to 1553 nm (dB)	-	-	4.5–4.6	3–3.1	−8.6
Level of interchannel cross talk in the channel with the highest bit error rate (BER) (dBm)	−55.5	−50	−47.9	−48.3	−49.3

Table 1. Summary of the results obtained in the 16 channel 10 Gbps DWDM transmission system depending on the type of amplifier used (Column 2—without using an amplifier).

fact that the average amplification in the case of DRA is larger just only by 0.7 dB than in the case of the EDFA amplifier, the achieved transmission distance is larger by 11 km than in the system with EDFA. This can be explained by the low amplification efficiency of the Raman amplifiers at low powers of the amplified optical radiation. Thus, the signal, the power of which is much larger than the noise power, will be amplified more effectively than the noise generated by the amplifier. Nevertheless, such characteristic of the amplifier should also be interpreted as a serious drawback of the distributed Raman amplifiers, as the need arises to use powerful pumping lasers (1150 mW strong pumping radiation is necessary to achieve amplification of 25 dB). EDFA pumping source power is equal to 316 mW. EDFA is able to ensure a high level of signal amplification; however, this could be achieved only in a 35 nm wavelength region in the "C" optical band. The typical noise figure of EDFAs is higher than in the case of LRA and DRA. The main deficiency of SOAs is a very high number of produced signal impairments; therefore, this type of amplifiers is rarely used in WDM systems, even though their gain spectrum is much broader in comparison with EDFAs.

Taking into account the excessive number of SOA produced signal impairments, the strong wavelength and unevenness of the EDFA produced gain, and the low amplification effectivity of Raman amplifiers, it is clear that, if Cisco and Bell Labs forecasts are correct, then it will be necessary to find another optical signal amplification solution that could ensure a higher level of amplification over a broader wavelength band and at the same time that would amplify signal impairments as little as possible.

The first possible solution is to combine the aforementioned optical amplifiers into a hybrid optical amplifier, which would allow compensating for the negative properties of various amplifier types, for instance, to expand and equalize the EDFA gain spectrum, or would reduce the SOA-generated noise proportion in the amplifier output.

Another possible solution is the use of fiber optical parametric amplifiers (FOPAs). This type of amplifiers can ensure a high level of amplification over a broad wavelength band, and, if compared to other lumped amplifier types, given an optimized configuration, they produce very small number of signal impairments. Moreover, parametric amplifiers can also be used for all-optical signal processing purposes, for example, for wavelength conversion [27, 28],

http://dx.doi.org/10.5772/67607

dispersion compensation [29], time-division-multiplexed signal demultiplexing [20], and 2R and 3R all-optical signal regeneration (2R—signal power and form regeneration; 3R—signal power, form, and phase regeneration) [30, 31].

3. Hybrid optical signal amplification

This chapter is dedicated to studies of hybrid optical amplifiers, which were obtained by applying the combinations of currently commercially used optical amplifiers (SOA, EDFA, and Raman amplifiers). The possibilities of applying hybrid Raman-EDFA and Raman-SOA solutions in WDM transmission systems for improving the operations of existing lumped in-line amplifiers have been studied and demonstrated. Due to the excessive number of SOA produced signal distortions and the strong wavelength dependency of EDFA produced gain, the implementation of EDFA-SOA hybrid solution has not been considered.

The unevenness of the EDFA gain spectrum and signal distortions caused by ASE noise significantly affect the performance of the whole transmission system, especially in systems with several amplification spans. To demonstrate the impact of the unevenness of EDFA gain spectrum and of the generated signal distortions, a 16-channel 10 Gbps DWDM transmission system simulation model has been introduced with four amplification spans. Equal power of the optical flow has been ensured at each amplifier input.

The obtained results are shown in **Figure 3**. After each amplification span, BER value of the detected signal increases by 2–3 orders (given the same input signal power). Upon comparing the EDFA gain spectra after the first and fourth amplification span, it is found that amplification decreases on average by 11.6 dB, whereas the amplification difference between the channels increases from 1.3 to 4.3 dB. The following conclusions are drawn:

- Every additional EDFA not only generates the amplified spontaneous emission noise but also amplifies the noise produced by the previous amplification spans. This significantly degrades the quality of amplifiable signal.
- The ASE power level after each amplifier is gradually increasing. Accordingly, part of the erbium ion population inversion is used to amplify the noise generated in the previous amplification spans. As a result, the part of the obtained population inversion, which was used for signal amplification, has decreased.

The slope of the gain spectrum increases after each amplification span. Uneven amplification is undesirable in multichannel WDM systems, especially in systems with several cascaded EDFA in-line amplifiers, as it leads to difference between power levels of various channels, which, accordingly, will lead to a signal quality degradation in channels with a lower amplification level.

Summing up all the aforementioned results, it has been concluded that it is necessary to configure the EDFA amplifier in a way to obtain the overall amplification spectrum that is as even as possible in the frequency range used for transmission, as well as to reduce the number of EDFA produced signal distortions.

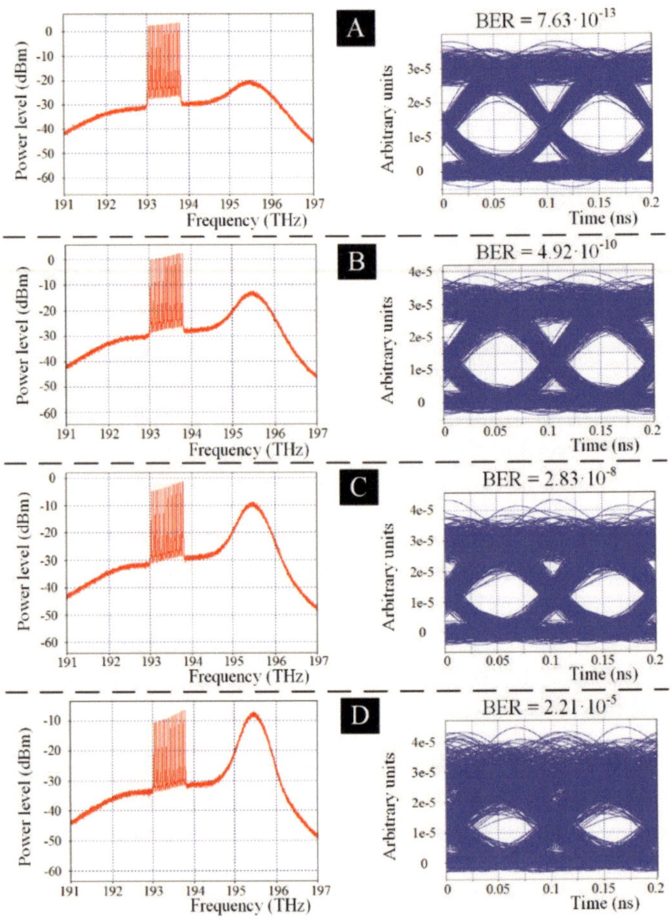

Figure 3. Optical spectra (the power level depending on frequency) at the output of the EDFAs (to the left) and eye diagrams of the signal detected in the ninth channel (to the right) after first (A), second (B), third (C), and fourth (D) stages of amplification.

3.1. Raman-EDFA hybrid amplifier

In the Raman-EDFA optical amplifier combination, most noise is generated by the EDFA amplifier. Therefore, in most cases, the Raman amplifier is used as a preamplifier in such cascades. EDFA amplifiers provide lower noise figures when functioning closer to the saturation point. Therefore, in hybrid amplifiers, EDFA with a relatively short doped fiber should be used (the longer the doped fiber, the higher level of amplification is obtained by the photons generated by spontaneous emissions). For further analysis of the hybrid Raman-EDFA solution, a simulation model is used, which is shown in **Figure 4**.

In the simulation model, the optical flows that are produced by the 16 transmitters are combined and transferred through a 150 km long standard single-mode fiber (SMF1). The signal power level at the SMF1 fiber output in all 16 cannels has reached −37.1 ± 0.1 dBm. The overall optical flow has been amplified by the EDFA in-line amplifier or by the hybrid Raman-EDFA

http://dx.doi.org/10.5772/67607

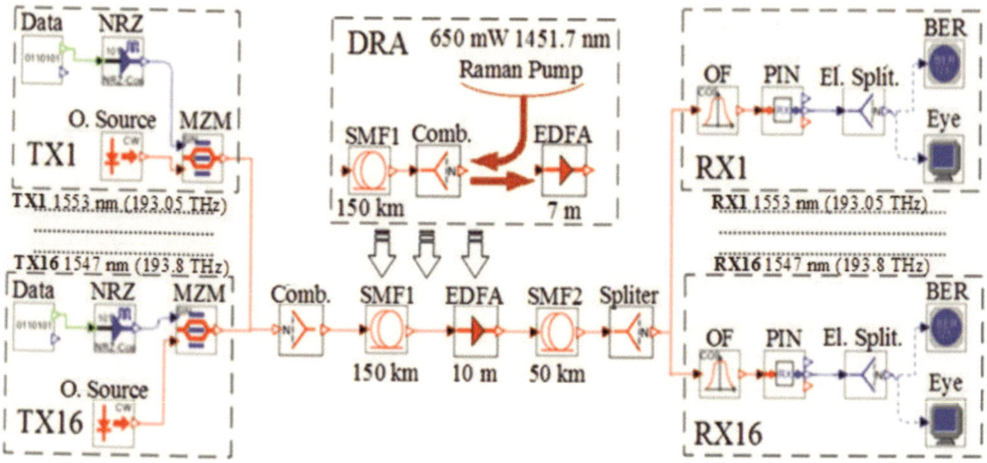

Figure 4. Simulation model of the 16-channel 10 Gbps DWDM transmission system with an EDFA in-line amplifier or with a hybrid Raman-EDFA amplifier.

amplifier (arrows in **Figure 4** show the layout of the hybrid amplifier) and afterward transferred through a 50 km long SMF (SMF2). Dispersion compensation has been performed using a fiber Bragg grating (FBG), and then the optical flow has been divided among 16 receivers, using an optical power splitter.

After comparing the gain spectra produced by the EDFA in-line amplifier and the hybrid Raman-EDFA amplifier (see **Figure 5**), it has been found that implementation of the hybrid solution allows reducing the gain difference among all 16 channels from 1.5 dB (in the case of the EDFA) to 0.1 dB (in the case of the hybrid amplifier).

As can be seen in **Figure 6**, implementation of the hybrid solution has ensured OSNR improvement in all 16 channels from 1.7 up to 2.6 dB, that is, an average increase of ~2 dB. Such OSNR improvement can be explained by the following facts:

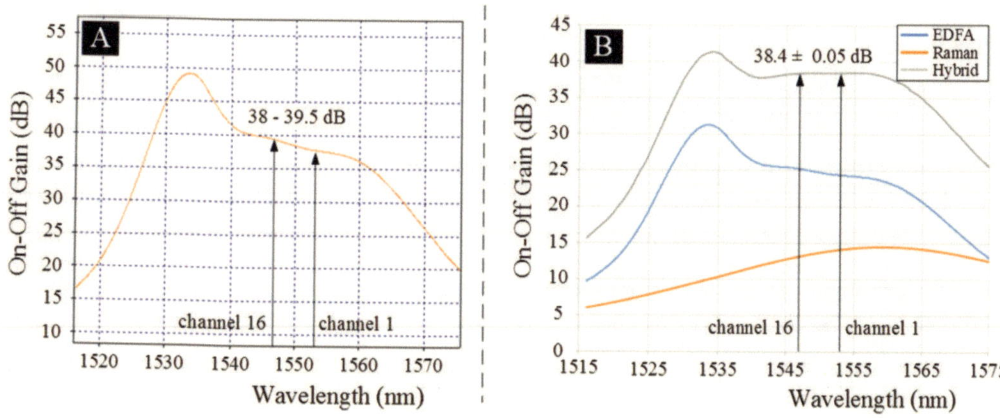

Figure 5. Gain spectra of the EDFA in-line amplifier (A) and of the hybrid Raman-EDFA amplifier (B).

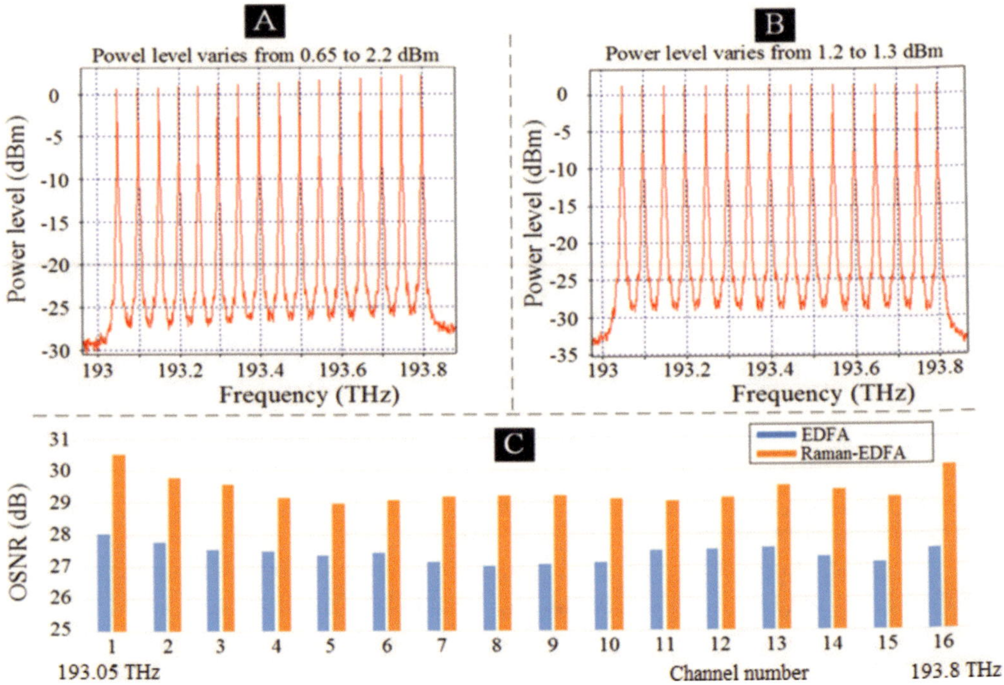

Figure 6. Signal spectra at the output of the EDFA (A) and at the output of the hybrid Raman-EDFA amplifier (B) and OSNR comparison among all 16 channels in the system with the EDFA in-line amplifier and the hybrid Raman-EDFA amplifier (C).

- The usage of the distributed Raman amplifier has raised signal power at the input of the EDFA by 13.1–14.1 dB; therefore, the EDFA functions closer to the saturation point.
- The EDFA fiber length has decreased by 3 meters, which allows reducing the required input signal power for saturation of the EDFA.
- The coherent nature of stimulated Raman scattering (SRS) ensures that in SMF1 optical fiber, the signal is amplified more effectively than the low power optical noise, which allows obtaining negative noise figure values (from −0.4 to −0.6 dB in the wavelength region used for transmission), and accordingly improved OSNR.

In the case of the hybrid amplifier, it has been found that raising the signal power at the input of the EDFA and reducing the length of the erbium-doped fiber (EDF) allow obtaining lower noise figure values by 0.3–0.4 dB for the EDFA.

Upon performing a comparison of operations of the aforementioned EDFA and Raman-EDFA solutions, it can be concluded that the hybrid amplifier can ensure more even amplification over a broader wavelength region and higher OSNR values. However, more powerful lasers are necessary for implementing such solutions, which increases the costs of developing this solution. For the EDFA in-line amplifier, 316 mW of pumping power is required to amplify

http://dx.doi.org/10.5772/67607

the −37.1 dBm input signal by more than 38 dB. In the case of the hybrid solution, the Raman amplifier required 650 mW of pumping power to ensure that gain is high enough and that its slope can compensate the slope of the EDFA with 200 mW pump gain spectrum, but the total pumping power of the hybrid amplifier has reached 850 mW. However, the hybrid solution ensured gain difference below 1 dB over a 23 nm wavelength range (from 1538 to 1561 nm, by 17 nm more than that used for transmission of all 16 channels), which allows significantly increasing the number of channels in WDM transmission systems.

3.2. Raman-SOA hybrid amplifier

The Raman-SOA hybrid solution is configured in a way to reduce the number of signal distortions produced by the semiconductor optical amplifier and also to increase the attainable transmission distance. The introduced simulation model of the transmission system used for studying this amplifier combination is similar to the one used previously (see **Figure 7**). Wavelength grid is chosen based on ITU-T G.694.1 recommendation where the central frequency is 193.1 THz.

The transmission line span length between the transmitter block and SOA is specifically selected to ensure optimum signal power at the input of the semiconductor amplifier. Inserting the distributed Raman amplifier in a cascade before the semiconductor amplifier would increase the signal power in SOA input, which would lead to a more explicit manifestation of nonlinear optical effects in the semiconductor material and would, accordingly, deteriorate the quality of the amplifiable signal. Therefore, it is the semiconductor amplifier that is used as the first in the cascade.

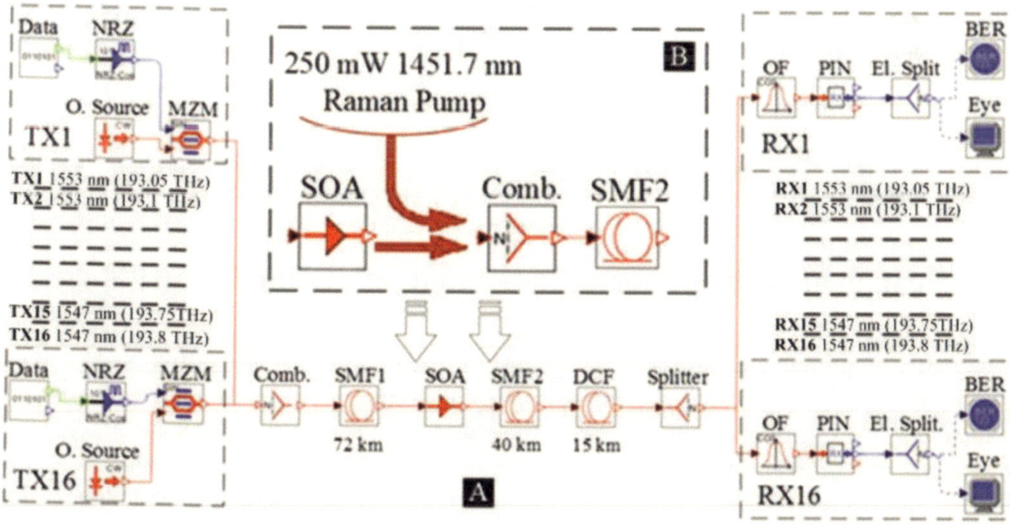

Figure 7. Simulation model of the 16-channel 10 Gbps DWDM transmission system with the SOA in-line amplifier (A) or with a hybrid Raman-SOA amplifier (B).

The implementation of the hybrid Raman-SOA solution allows using such mode of the semiconductor amplifier, in which it produces minimum distortions of the amplified signal, whereas the amplification deficit, which occurs after reducing the pumping current value by 43 mA, is compensated by the DRA with a 250 mW 1451.7 nm co-propagating pump. The implementation of the Raman-SOA hybrid solution allows increasing the attainable transmission distance by 12 km. The gain spectrum of the DRA is shown in **Figure 8a**. Eye diagrams for channels with the highest BER value in a system with the SOA amplifier (ninth channel f = 193.45 THz) and in a system with the Raman-SOA hybrid amplifier (tenth channel$f$ = 193.5 THz) are shown accordingly in **Figure 8b** and **c**. From **Figure 8a**, it can be seen that the DRA produced amplification is large enough to compensate the amplification deficit of 5.3 dB that occurs after reducing the SOA pumping current by 43 mA. SOA amplifier (9th channel) and Raman-SOA hybrid amplifier (10th channel) be selected, because they are the worst channels.

After comparing **Figure 8b** and **c**, it has been found that implementation of the Raman-SOA hybrid solution allows obtaining approximately the same BER level as in the case of SOA in-line amplifier, but at signal power lower by 1.5 times. This shows that, by using SOA together with the distributed Raman amplifier and introducing relevant SOA pumping current adjustments, it is possible to substantially lower the amount of SOA produced noise and, therefore, to improve the quality of the amplified signal.

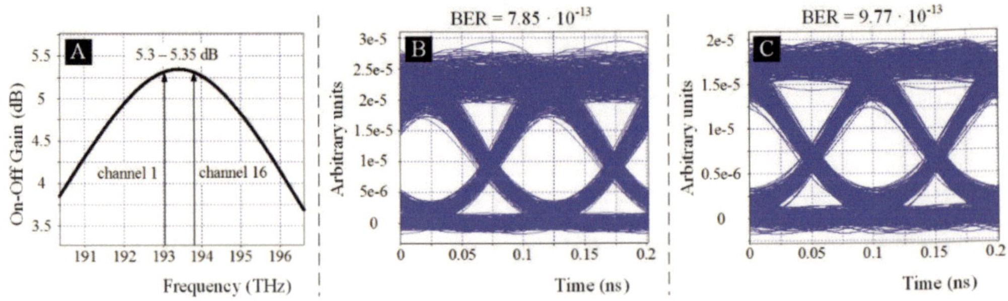

Figure 8. DRA produced gain spectrum (A) and eye diagrams of the ninth channel in the system with the SOA in-line amplifier (B) and the tenth channel in the system with the Raman-SOA hybrid amplifier (c).

4. Evaluation of parametric amplifiers and its application

Parametric amplifiers can be based on degenerate FWM (in the single-pump case) and on nondegenerate FWM (in the dual-pump case). FOPA produced gain will reach its maximum, if the phase-matching condition is met or if the phase-mismatch parameter k is equal to zero. In the case of a single-pump FOPA, irrespective of the broad amplification range, the amplification spectrum is not even. In the experimental transmission system, described before, the gain −3dB bandwidth has reached 2.2 THz (see **Figure 9**). It has been found that for ensuring an optimum operation mode of a single-pump FOPA, it is necessary to maintain a small negative linear phase deviation in respect to the zero-dispersion frequency, which would compensate the nonlinear phase mismatch. That is why the pumping radiation wavelength must be slightly larger than the fiber zero-dispersion wavelength (ZDWL).

http://dx.doi.org/10.5772/67607

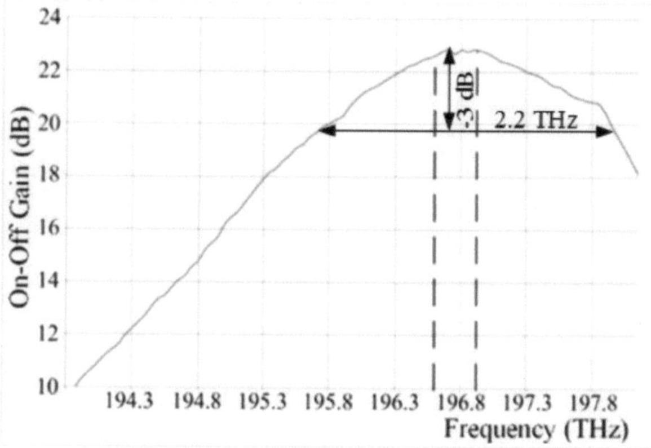

Figure 9. Gain spectrum of the single-pump FOPA with 660 mW 1553.9 nm pumping radiation.

The gain spectrum bandwidth is very dependent on the nonlinearity parameter of the medium and on the pumping radiation power and the length of the gain medium highly nonlinear fibers (HNLFs). Thus, by increasing the fiber length, it is possible to achieve a higher level of amplification, but in this case, the gain spectrum width will be reduced accordingly (the longer the fiber, the larger the accumulated phase mismatch is). Due to this reason, when constructing FOPA amplifiers, it is not recommended to use HNLFs that are longer than 1 km. If they are configured in a way to achieve as wide gain spectrum as possible, it is required to use as short HNLF as possible, but to maintain the achievable amplification level, the pump power must be increased, or a fiber with a higher nonlinearity coefficient must be used. When selecting the pumping radiation parameters, one must keep in mind that, by changing the pumping radiation power, also the nonlinear phase mismatch is changed. Therefore, along with adjusting the pump power, its wavelength also needs to be reconfigured.

The performance of single-pump parametric amplifiers is affected by various factors, which must be taken into account when designing a specific FOPA. It is necessary to selectively choose the pumping radiation parameters to ensure as high amplification efficiency as possible and to avoid occurrence of excessive channel-channel four-wave mixing (CC-FWM) and pump-channel four-wave mixing (PC-FWM) produced interchannel cross talk, which in its turn is produced due to excessive pumping. The SBS threshold increase is also very important; otherwise, the amplification effectiveness will decrease, and the amplified signal will be distorted.

One of the most effective solutions of increasing the SBS threshold is phase modulation of the pumping radiation. However, as can be seen from the results shown in **Figure 10**, if the choice of frequencies modulating the pumping radiation phase is not thoroughly considered, a substantial expansion of spectrum of the idler spectral components will occur. Therefore, in systems, in which idlers are used for all-optical signal processing, it must be ensured that the chosen pumping radiation phase-modulation does not produce excessive spectral broadening of idlers (in the results shown in **Figure 10**, idler spectral broadening has reached 54% at the level of −15 dB from the maximum power spectrum). For initiating the FWM process, it is also necessary to preserve the angular momentum among the four photons involved in the parametric interactions, as the parametric gain has explicit polarization dependence.

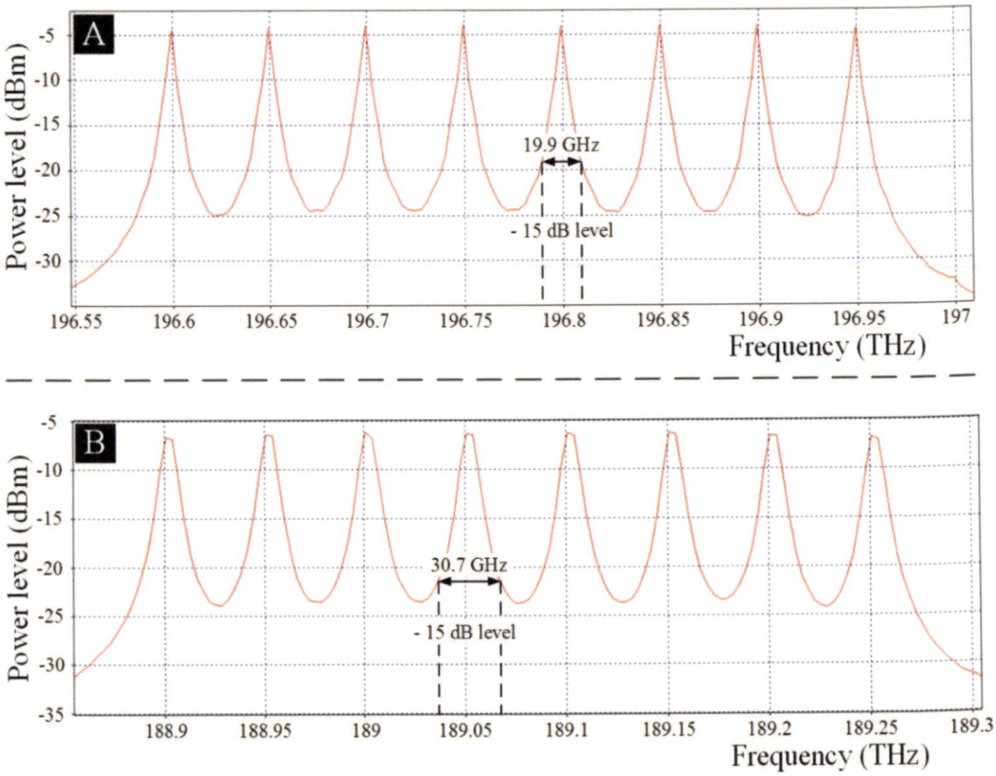

Figure 10. Spectra of the amplified signal (A) and the generated idlers (B) at the output of the single-pump FOPA.

Unlike single-pump FOPAs, dual-pump FOPAs can ensure even amplification over a very broad wavelength band. To achieve even amplification in a broad wavelength band, it is necessary to ensure that the wavelengths of the pumps are placed symmetrically in respect to the gain medium ZDWL, whereas the frequency distance between ZDWL and pumping radiations must be large enough (depending on the specific FOPA configuration), to avoid the impact of PC-FWM-generated components on the quality of the amplified signal.

Since dual-pump FOPAs both degenerate and nondegenerate FWM, for one input signal, it is possible to obtain at least five idlers (see **Figure 11**, where ω_1 and ω_2 are the pumping source frequencies, but ω_3 is the signal), which are directly related to the amplifiable signal frequency. This leads to amplification spectrum depressions at frequencies near the frequencies of the pumps. It has been concluded from the results obtained in this study that at 0.5 mW pump power it is recommended that the amplified signal frequency is at a distance of at least 1.2 THz from pump frequencies.

Just like in the case of single-pump FOPAs, dual-pump FOPAs also require the usage of one of the methods for mitigation of the negative impact of SBS. However, in the dual-pump case, it is important to note that by manipulating with the phase of both pumping radiations it can

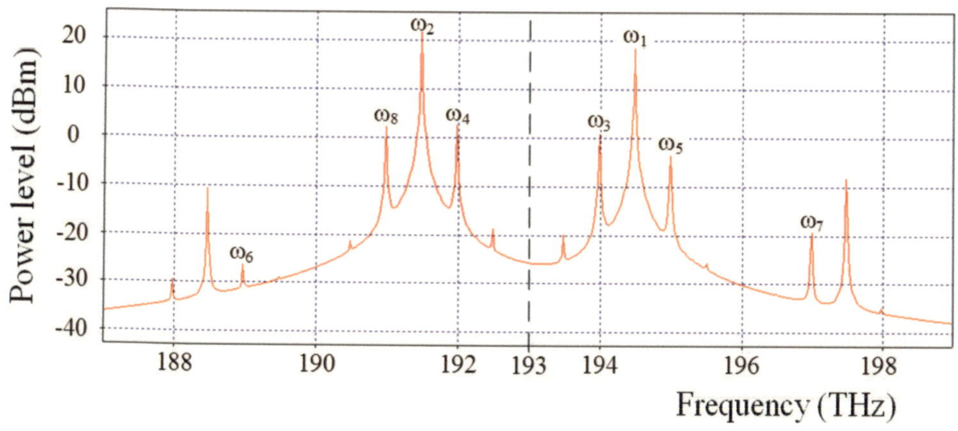

Figure 11. Optical spectrum at the output of a dual-pump FOPA with 200 mW 191.5 THz pumps and 500 meter long HNLF.

be achieved that the relevant idler will not experience spectral broadening. The amplification efficiency in the case of dual-pump FOPAs is highly dependent on the SRS triggered energy transfer between the pumps. It is not possible to fully avoid this effect. To reduce its impact, normally higher power is used for the pump with the higher frequency than for the other pumping radiation, thus achieving that the average power difference between the pumps is minimum over the entire gain medium.

In traditional WDM transmission system architecture, one optical source is required to produce a single-channel carrier. It is not the most cost-effective solution, as, by increasing the number of transmission channels, the number of required light sources increases accordingly. Due to this reason, an increasing number of studies are conducted to find such transmission system architecture, which would be able to ensure a higher number of signal carriers using fewer optical sources [32–34]. FOPAs during the process of parametric amplification generate idler spectral components, which, in essence, are phase-conjugated copies of the amplified signal. These idlers could be used not only for wavelength conversion or 2R and 3R signal regeneration but also for increasing the number of carriers on the transmitter side of a WDM transmission system.

Therefore, a model of a dual-pump FOPA has been introduced for doubling the number of existing carriers in a WDM transmission system. For this reason, a simulation model of a 32-channel DWDM transmission system has been created with 10 Gbps transmission speed per channel, 100 GHz channel spacing, and NRZ-OOK modulation format. This system simulation model is displayed in **Figure 12**. The authors chose SMF length of 20 km, because it is the typical line length of optical access networks. EDFA preamplifier is used here for insertion loss compensation of transmission line and other transmission elements.

The main feature in the simulation model, which is presented in **Figure 12**, is that the FOPA is placed before the transmitter block or at the 32-channel modulator inputs. The optical

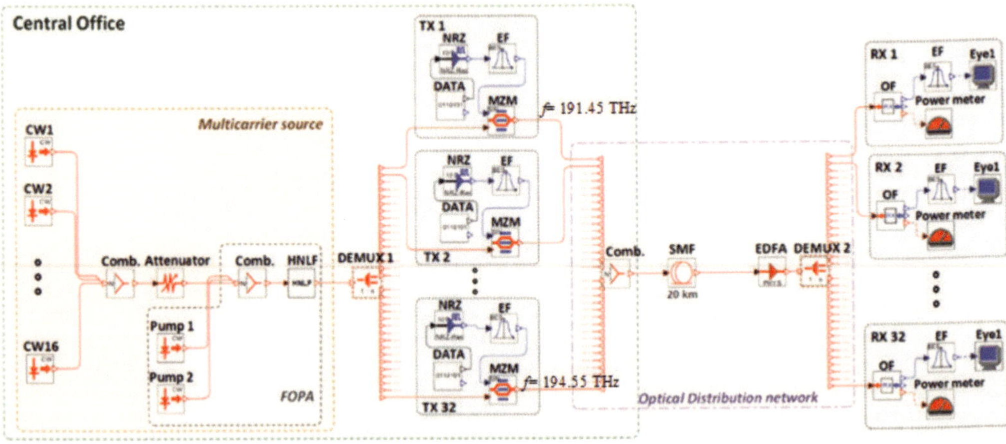

Figure 12. Simulation model of the 32 channel 10 Gbps WDM transmission system with the provided multicarrier source solution, which is based on wavelength conversion using a dual-pump FOPA.

multicarrier source consists of continuous radiation lasers (CW1–CW16), an optical attenuator, two powerful pumping sources, two optical splitters, and a 500 m long HNLF. One of the main goals of this experiment is to obtain 32 carriers with even frequency distribution (equal channel spacing), which can be achieved by using idlers ω_4. Taking into account that the distribution of idlers ω_4 and the initial light source frequencies are symmetrical in respect to the gain medium ZDWL, it has been decided to place the carrier with the lowest frequency higher by 50 GHz than the HNLF ZDWL (193 THz). Therefore, the frequencies of the 16 initial carriers are distributed in a range from 193.05 to 194.55 THz with 100 GHz channel spacing (see **Figure 13a**). The optical flow sent through the parametric amplifier is not modulated and basically represents a continuous radiation set. At the output of the HNLF, a combination is obtained consisting of 16 initial carriers, 16 idlers ω_4 (generated as a result of parametric processes), 2 pumps, and other third-order spectral components (see **Figure 13b** and **c**). The pump power for both pumps is set to 400 mW each (26 dBm), and 190 and 196 THz frequencies are temporarily chosen.

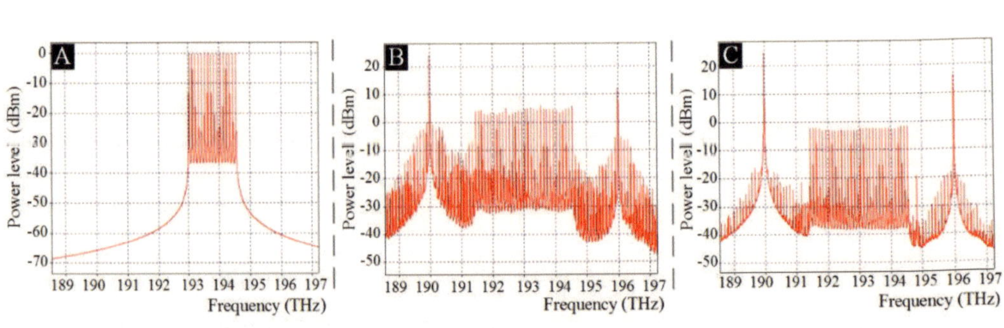

Figure 13. Optical spectrum at the input of the FOPA when initial carrier power is set to 0 dBm (A) and optical spectra at the output of the HNLF when the power of the initial carriers is set to 0 dBm (B) and to −10 dBm (C).

It has been found that when given an excessively high level of input signal power, CC-FWM processes will trigger explicit interchannel cross talk (see **Figure 13b**). Due to this reason, when the idlers obtained as a result of parametric processes are used for increasing the number of carriers, it is necessary to limit the power of the carriers at the amplifier input. Based on the obtained results, the power level of the initial carriers at the input of the amplifier is reduced to −10 dBm. The alignment of idlers in respect to the central frequencies of the throughput band of demultiplexer filters is achieved by changing the frequency of the first pumping radiation (frequency obtained in simulations −196.01 THz). With the aforementioned amplifier configuration, the maximum power level difference among all the 32 channels has been reduced to 1.9 dB.

To assess the performance of the proposed system architecture solution, BER value dependence on the received signal power in the channel with the poorest signal quality (the highest BER) is obtained. These results are compared to the corresponding results obtained in a system with traditional architecture (with 32 laser sources, which function in a continuous radiation mode). The obtained results are shown in **Figure 14**. It has been found that power penalty of 1.8 dB exists between the system with the proposed multicarrier source and the conventional 32-channel solution. It is important to note that part of the obtained power penalty is directly related to the large amount of ASE produced by the EDFA used as a preamplifier for ensuring the necessary signal power at the input of the received block.

There are at least two alternatives to the proposed system architecture, which can produce more than one carrier per optical source: spectrum-sliced systems [33, 35] and systems, which are based on FWM use for producing third-order spectral components (without the initial carriers) [34, 36]. Nevertheless, the use of idlers ω_4 produced by FOPA for doubling the number of carriers in WDM systems ensures the best carrier signal stability and, therefore, the highest quality of the transmittable signal.

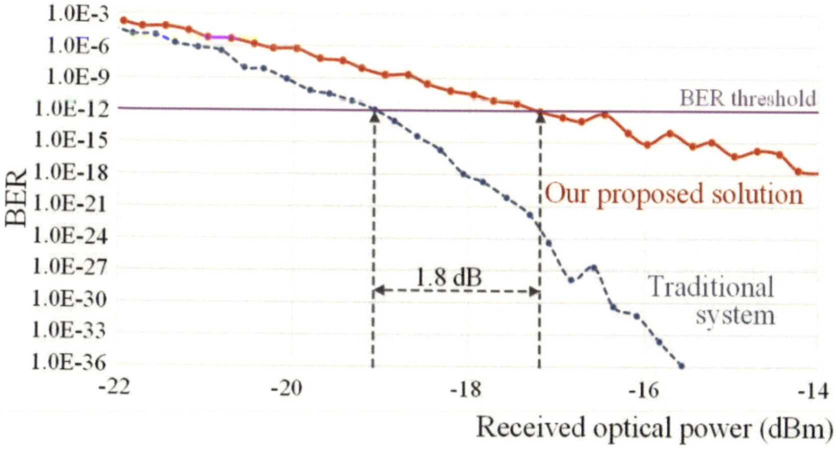

Figure 14. BER value dependence on the power of the detected signal for the 12th channel (f =192.55 THz; λ = 1557 nm) in the system with the proposed multicarrier source solution (solid line) and in the system with traditional architecture (dashed line).

It has been found that for dual-pump FOPAs the maximum amplification efficiency is achieved when both pumps are linearly polarized with the same state of polarization (SOP) and their SOP corresponds to the SOP of the amplified signal. However, when the SOP of both linearly polarized pumps is orthogonal to the SOP of the amplified signal, amplification decreases to its minimum value, and in a broad frequency region, it is equal to zero. The results obtained in this paper have shown that the same situation is observed also in the case of single-pump FOPAs.

This property of parametric amplifiers can be used for emphasizing one state of polarization from a combination of two orthogonally polarized optical components. The key problem, which is observed when the FOPA is used for emphasizing a specific state of polarization, is ensuring the conservation of the relative positioning of signal and pump SOP throughout the entire length of HNLF. This problem occurs due to the following reasons:

- Due to the effect of fiber birefringence, SOP of optical radiation changes along the fiber, and, as result, random SOP rotation is observed. It is very difficult to compensate such a random SOP change, as the rotation rate is affected by various factors, such as temperature, the frequency of the transmitted radiation, internal and external mechanical loads, etc. It is possible to avoid rotation of SOP of the pumps and the amplified signal by using polarization-maintaining HNLF as the gain medium.
- When the amplified signal and the pumps are propagating in the gain medium, additionally to fiber birefringence, their states of polarization are also affected by self-phase modulation (SPM) and cross-phase modulation (XPM) nonlinear effects. Therefore, when configuring the parametric amplifier, it is first necessary to avoid excessing pumping; otherwise, it can lead to a more explicit occurrence of SPM and XPM, which decreases the efficiency of the FWM process in the gain medium.

It is not possible to completely avoid changes in relative positioning of the SOP of the signal and the SOP of the pump. To demonstrate this, a simulation model is introduced, where a single-pump FOPA (500 mW, 1533.9 nm) amplifies a signal with −31 dBm total optical power at the input of the HNLF. At first, both the signal and the pump are linearly co-polarized. During the simulation, the SOP of the pump is rotated in respect to the SOP of the amplified signal, and the power of the signal is observed at the output of the FOPA. It has been found that by using polarization-maintaining fibers, under the influence of SPM and XPM, a change in the relative positioning of the signal SOP and the pump SOP is observed. As a result of this change, even when the signal is orthogonally polarized in respect to the SOP of the pump at the input of the HNLF, the signal obtains 1.5–1.6 dB gain. When the SOP of the amplified signal is co-polarized with the SOP of the pump, the obtained amplification reaches 18.3 dB, which is by 16.7 dB higher than in case of orthogonal relative positioning of the SOP at the input of the HNLF.

As it has already been previously mentioned, the polarization dependence of the parametric gain can be used for emphasizing radiation with a specific SOP from the flow of orthogonally polarized optical components, which in its turn can be used for emphasizing polarization-multiplexed signals and 2PolSK to NRZ-OOK modulation format conversion.

http://dx.doi.org/10.5772/67607

For conversion of 2PolSK signal to NRZ-OOK modulation format, cases of single-channel and multichannel systems are considered. To avoid changes in relative positioning of the states of polarization between pumping radiations, single-pump parametric amplifiers are used in both cases. In both cases, the FOPA is placed at the receiver (or receiver block) input.

At first, a single-channel transmission system simulation model is introduced with 2PolSK modulation format, 150 km long optical fiber, a FOPA preamplifier (which simultaneously performs modulation format and wavelength conversion functionality), and two receivers for detecting the converted NRZ-OOK signal at signal and idler frequencies. The introduced simulation model is displayed in **Figure 15**.

In case of a single-channel system, the primary task is to assess the new modulation format conversion solution created within the scope of this paper, by obtaining a power penalty introduced specifically by the process conversion of the modulation format. Based on the obtained results, 535 mW, 1554.1 nm pumping radiation is chosen, the phase of which is modulated with the following frequency tones: 180 MHz, 420 MHz, 1.087 GHz, and 2.133 GHz. Such configuration ensures 14.8 dB gain for the logical "1" component of 2PolSK signal, which is sufficient for ensuring BER value below the 10^{-12} mark for the obtained NRZ-OOK signal.

Based on the obtained results, it has been concluded that the idler requires lower pump power to ensure BER values below the 10^{-12} mark, even though the gain for idler is lower by 0.8 dB than the signal (see **Figure 16**). Therefore, in the case of a single-channel system, it is recommended to process the idler spectral component as the informative signal. These results can be explained by the fact that the signal at its initial frequency contains the orthogonally polarized logical "0" component, which for the obtained NRZ-OOK signal is interpreted as noise.

It has been found that the power of the obtained NRZ-OOK signal that is necessary to ensure BER value below 10^{-12} is −23.65 dBm, whereas the necessary idler power is −23.8 dBm (see **Figure 17**). In a standard single-channel system with NRZ-OOK modulation format, the signal power required to ensure BER value below the 10^{-12} threshold has reached −24 dBm. Thus, there is a power penalty of 0.4 dB between the NRZ-OOK signal from the standard

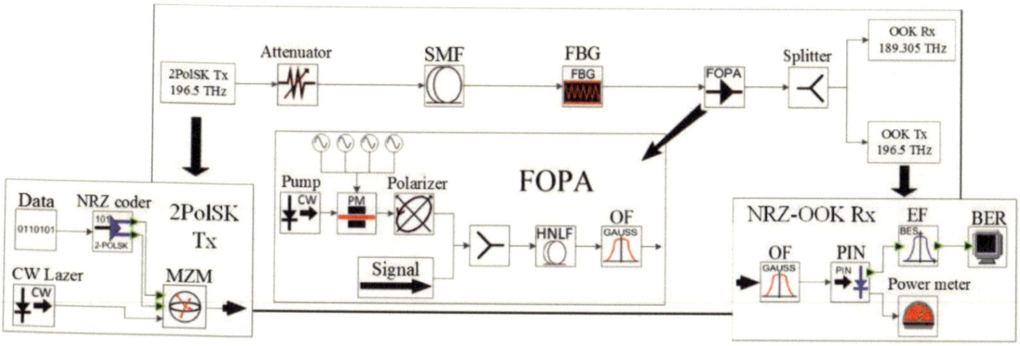

Figure 15. Simulation model of the single-channel transmission system, where a single-pump FOPA with linearly polarized pumping radiation is used for 2PolSK to NRZ-OOK modulation format conversion.

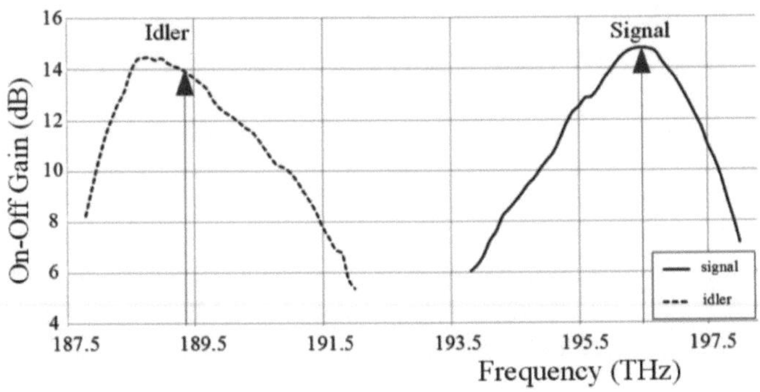

Figure 16. Gain spectrum produced by the single-pump FOPA with linearly polarized pumping radiation at signal frequencies (solid line) and at idler frequencies (dotted line).

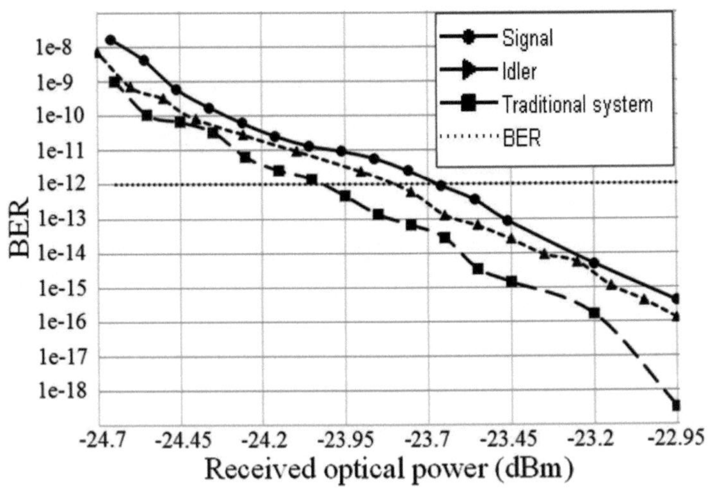

Figure 17. BER value dependence on the detected signal power in the standard single-channel transmission system (dashed line) and in the system with modulation format conversion at the initial signal frequency (solid line) and at idler frequency (dotted line).

single-channel system and the converted signal. It is important to note that the power penalty between the NRZ-OOK signal from the standard single-channel system and the generated idler is lower by 0.1 dB (only 0.2 dB). These results are explained by the fact that the idler produced during the FWM process does not contain the logical "0" component of the initial 2PolSK signal, which in this case is interpreted as noise for the converted NRZ signal. The obtained power penalty values are also attributable to the relative intensity noise, which is transferred from the pumping radiation to the amplifiable signal, as well as to the phase SOP mismatch between the pump and the amplified signal that occurs due to SPM and XPM.

In the case of the multichannel system, the goal is to assess the performance of the developed modulation format conversion solution in the presence of interchannel cross talk. When converting 2PolSK signal to NRZ-OOK modulation format, using FOPA with linearly polarized pumping radiation, one must pay special attention to the control of the level of interchannel cross talk produced by the CC-FWM interactions because if the SOP pumping radiation and SOP signal coincidence, the FWM process takes place with its maximum efficiency, including also production of the CC-FWM interchannel cross talk.

To assess the performance of the proposed solution in the presence of interchannel cross talk, a 16-channel 10 Gbps DWDM transmission system is introduced with 2PolSK initial modulation format and 100 GHz channel spacing (see **Figure 18**). In this system, the access network is divided into two branches, eight channels in each. The first branch consists of eight channels, occupying frequency range from 194.5 THz (1541 nm in wavelength) to 195.2 THz (1536 nm), whereas the second branch is occupying frequency range from 196.2 THz (1528 nm) to 196.9 THz (1523 nm). Only those results are included in this paper, which are obtained in the second access network branch, where the signal is divided among eight receivers using an optical splitter with 10.5 dB attenuation.

Based on the obtained results, in the second access network branch, 790 mW 1554.15 nm pumping radiation is used, the phase of which is modulated with the same frequency tones as in the case of a single-channel system: 180 MHz, 420 MHz, 1.087 GHz, and 2.133 GHz. It has been found that to ensure BER values below the 10^{-12} threshold, all eight idlers require pump power that is by 35 mW higher than in the case of the signals at their initial frequency. The obtained gain for the idler spectral components is lower by at least 2.2 dB, whereas the gain spectrum slope near its maximum is higher than in the initial signal frequency band (see **Figure 19**). The obtained level of amplification in the initial signal frequency band changes in the range from 30.3 to 30.9 dB among all eight channels; thus, the amplification difference between the channels reaches 0.6 dB. Between the idler spectral components, such difference has reached 2.6 dB (from 26.1 to 28.7 dB), but the biggest amplification difference between the signal at its initial frequency and the corresponding idler has reached 4.2 dB. This explains the need for pump power exceeding 35 mW to ensure BER values below the 10^{-12} mark in the idler frequency band.

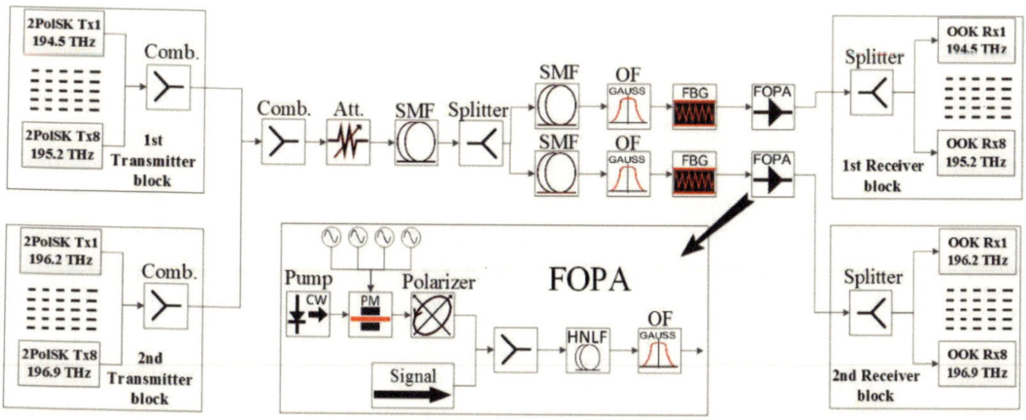

Figure 18. Simulation model of the 16-channel WDM transmission system with two access network branches, where in each branch the FOPA preamplifier is used for 2PolSK to NRZ-OOK (on-off keying) modulation format conversion.

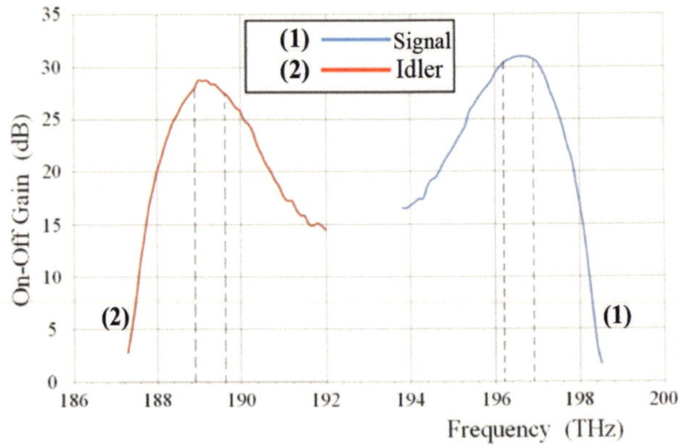

Figure 19. Gain spectrum ensured by the FOPA with 790 mW 1554.15 nm linearly polarized pumping radiation at initial signal frequencies (1) and idler frequencies (2).

To assess the performance of the proposed solution, the results obtained in the second branch of the access network are compared to the standard eight-channel DWDM system without signal amplification. The detected signal power required to ensure a certain BER value in the fifth channel of the second access network branch both at the initial signal and idler frequencies is compared with the same results obtained in the standard eight-channel DWDM system. As seen in **Figure 20**, in a system with modulation format conversion, to ensure BER values below the 10^{-12} mark, the required signal power is at least −23.5 dBm. In the standard eight-channel system, the corresponding required power level is −23.9 dBm; therefore, in this case, the power penalty between the signal with the converted modulation format and the signal from the standard eight-channel solution is 0.4 dB. It must be particularly emphasized that, contrary to the single-channel system, in the multichannel system, the idler BER values are higher than those of signals at their initial frequencies—when receiving the idler corresponding to the fifth channel, at least −23.15 dBm is required to ensure a BER value below the 10^{-12} threshold, which is more by 0.3 dB than receiving the signal of the fifth channel at its initial frequency.

It has been found that in case of the idler, larger amplitude fluctuations are observed, which densify the logical "0" and "1" component levels of the eye diagrams. The cause behind generating noise of such range is CC-FWM produced interchannel cross talk, which is produced as a result of parametric amplification and creates third-order spectral components, the frequencies of which correspond to the frequencies of the amplified signals. As mentioned previously, the gain difference between the idlers is much larger (by 2 dB) than between the signals at their initial frequencies. Therefore, more explicit manifestation of CC-FWM processes is observed, which also produces additional interchannel cross talk. Moreover, the cross talk caused by CC-FWM is not only transferred from signals at their initial frequencies to the idlers but is also generated between the idlers.

It has also been found that the pumping radiation-phase modulation leads to spectral expansion of the idlers by approximately 40%, which, accordingly, results in additional interchannel

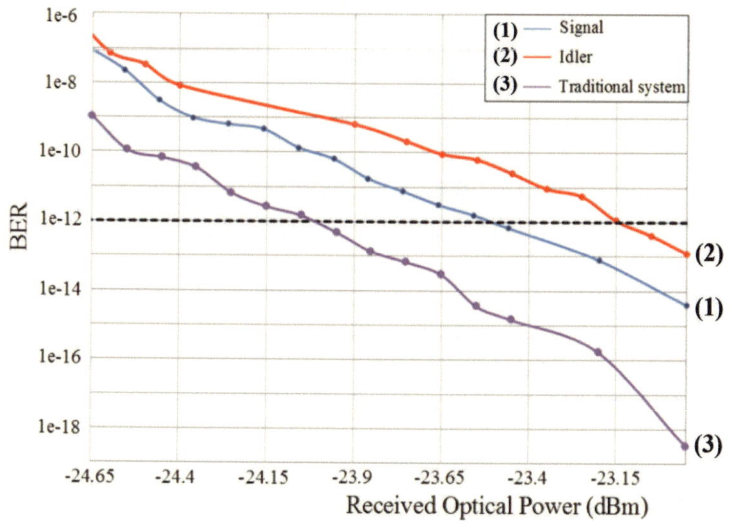

Figure 20. BER value dependence on the detected signal power in the standard eight-channel transmission system (3) and in the fifth channel on the second branch of the multichannel system with modulation format conversion at the initial signal frequency (1) and idler frequency (2).

cross talk. Interchannel cross talk caused by the CC-FWM interactions and idler spectral broadening is the main reason why, in the case of idlers, the power penalty in relation to the standard system is larger by 0.4 dB than for signals at their initial frequencies.

The second studied application of parametric gain polarization dependence is the emphasizing of a signal with a specific SOP from a combination of two polarization-multiplexed NRZ-OOK signals. For this purpose, a two-channel 10 Gbps transmission system with NRZ-OOK modulation format and polarization multiplexing is introduced (see **Figure 21**). Both signals, the SOP of which are mutually orthogonally allocated, are transmitted using the same frequency −196.5 THz.

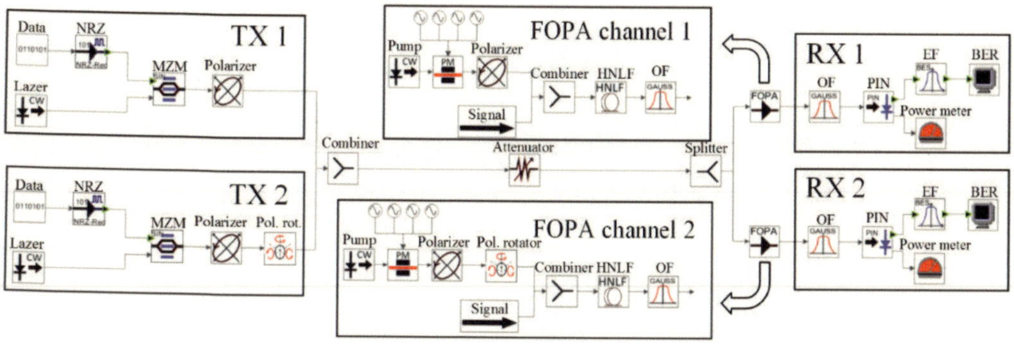

Figure 21. Simulation model of the two-channel optical transmission system with FOPA for division and amplification of polarization-multiplexed signals.

Based on the obtained results, a decision has been made to use 530 mW 1553.9 nm pumping radiation in the case of both FOPA, because this is the lowest pump power that can ensure BER values below the 10^{-12} threshold in both the first and the second channel. Previously described results have shown that phase modulation of the pump can cause spectral broadening of idlers. In a situation, when the probability that both orthogonally polarized optical radiations are observed simultaneously in the logical "1" level is high, the mutual deviations of the SOP of optical components have a bigger impact on the quality of the amplified signal than in the case when such simultaneous transmission of logical "1" is not performed (e.g., in the case of 2PolSK signal). Therefore, it is important to minimize the phase mismatch between the pumping radiation and the amplified signals, which can also cause a change of the relative positioning of SOP between the signal and the pump. Bearing in mind this fact and having observed the FOPA produced gain spectrum and in the OSNR at the output of the amplifier, the following frequency tones have been selected for pumping-radiation phase modulation: 0.13 GHz, 0.42 GHz, 1.087 GHz, and 1.94 GHz. The obtained signal gain in the first channel has reached 20 dB, whereas in the second channel, it has reached 20.1 dB. Idler component gain maximum is lower by 0.7 dB (see **Figure 22**).

To assess the performance of the proposed solution, BER value dependence on the received signal power is obtained, and these results are compared with the same results obtained in a standard single-channel transmission system with NRZ-OOK modulation format without optical signal amplification. It can be concluded from the results shown in **Figure 23** that there is a power penalty of 0.8 dB between the signal detected in the first channel and the signal from the standard single-channel NRZ-OOK system. The power penalty for the idler spectral component has reached 0.5 dB.

Unlike the system with modulation format conversion, in this case, a situation has been observed, when, given the same frequency, it is possible that the logical "1" of orthogonally polarized components is observed simultaneously in both channels. Thus, the effect of orthogonally polarized radiation on the divided signal quality is higher than in a system with modulation format conversion. This is the fact, which mainly explains a larger power penalty value than

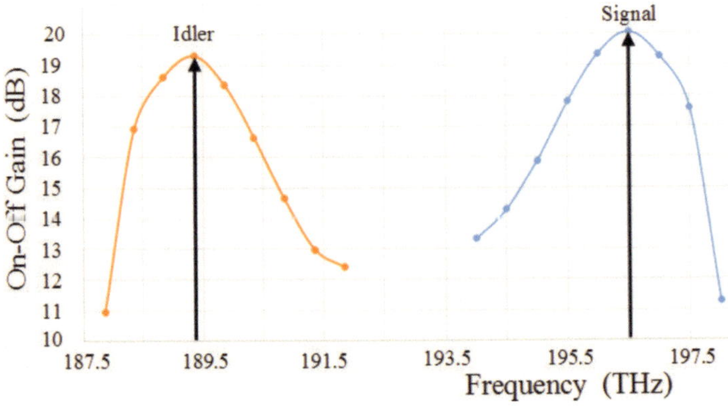

Figure 22. FOPA produced on-off gain for the first channel at idler frequencies (left side) and at the initial signal frequencies (right side).

http://dx.doi.org/10.5772/67607

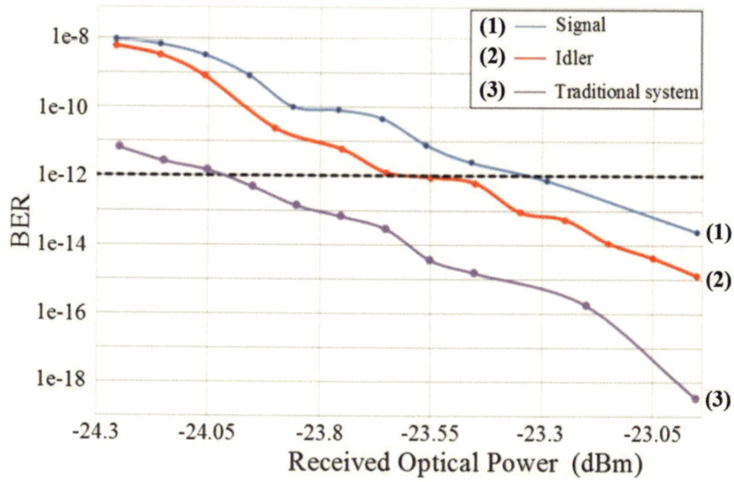

Figure 23. BER value dependence on the power of the detected signal in the standard single-channel system (3) and in the first channel of the system with the chosen FOPA configuration of the signal at its initial frequency (1) and of the idler (2).

in a single-channel system with modulation format conversion. A lower power penalty value in the case of idler component is explained by the fact that the orthogonally polarized radiation of the second channel is not included in the parametric amplification and idler generation process, and, therefore, it is not reflected in the idler itself. Irrespective of the fact that the amplification of orthogonally polarized (second channel) radiation is much lower (1–2 dB), its power level is still sufficiently high to affect the BER value of the divided signal, which in our case produces an additional power penalty of 0.3 dB. The idler use has allowed achieving a lower power penalty in respect to the standard single-channel system with the NRZ-OOK modulation format, whereas to achieve a BER value below the 10^{-12} threshold, it is necessary to use pump power that is larger by 10 mW than that when receiving the initial signal with 196.5 THz frequency.

Upon comparing the amplification spectra in a single-channel system with modulation format conversion and in a system with polarization-multiplexed signal division, it has been concluded that the amplification obtained in the latter case is larger by 5.2 dB (14.8 and 20 dB, respectively), irrespective of the fact that the pumping radiation power differs only by 5 mW. This is explained by the following two factors:

- In a system with signal division from orthogonally polarized signal combination, the signal power level at the input of the HNLF fiber is lower by 3.4 dB (−44.1 dBm). Therefore, the amplifier requires a lower pumping radiation power for ensuring a specific amplification level.
- Secondly, to minimize the idler spectral broadening in a system with polarization-multiplexed signal division, the frequency tones used for pumping radiation-phase modulation are reconfigured, and the achieved amplification difference clearly shows that with the given configuration SBS mitigation is more effective than in a system with modulation format conversion.

Upon summing up all the information presented in this chapter, it can be concluded that polarization dependence of the parametric amplification can be used for emphasizing optical radiation with a specific state of polarization from a flow of two orthogonally polarized optical radiations. FOPA with linearly polarized pumping radiation can be used both for 2PolSK signal conversion into NRZ-OOK modulation format and for signal emphasizing from a flow of two polarization-multiplexed optical signals. In both cases, such FOPA configurations have been found, which ensure that the BER values of the processed signal are below the 10^{-12} mark, and at the same time, none of the proposed solutions cause power penalty that exceeds 1.8 dB in comparison with the relevant standard solutions.

5. Conclusions

The implementation of the hybrid Raman-EDFA amplifier has allowed not only equalizing the gain spectrum but also increasing OSNR in all channels by 1.7–2.6 dB. The usage of the distributed Raman amplifier in cascade as a preamplifier has allowed the EDFA to operate closer to the saturation point, and, therefore, the EDFA noise figure decreased by 0.3–0.4 dB. The OSNR increase is also related to the fact that due to the coherent nature of Raman scattering, DRA amplifies the signal more effectively than the noise, which allows obtaining negative noise figure values (from −0.4 to −0.6 dB).

The implementation of the hybrid Raman-SOA amplifier has enabled the use of such SOA configuration, at which SOA produced the lowest amount of amplified signal distortions. As a result, by using the Raman-SOA hybrid amplifier, it is possible to obtain approximately the same BER level as in the case when the SOA is used as a single in-line amplifier for a signal, which is weaker by 1.5 times. This has allowed increasing the attainable transmission distance by 12 km or by 11%.

While changing the power of FOPA pumping radiation, the nonlinear phase mismatch of the parametric process also changes. Therefore, while configuring the pump power, the wavelength must also be changed accordingly.

Modulation of the FOPA pumping radiation phase, which has been used for increasing the SBS threshold, has caused spectral expansion of idlers by 54%. Therefore, the frequency tones used in systems with wavelength conversion for pumping radiation-phase modulation must be selected in a way that ensures that the spectral expansion of idlers remains as low as possible.

By manipulating with the parameters of dual-pump FOPA, it is possible to achieve an increase in the number of carrier signals from 16 to 32, simultaneously ensuring equal channel spacing of 100 GHz and maximum difference in power levels of 1.9 dB among all channels. It has been found that, in case when each carrier signal power at the input of the FOPA is equal to 0 dBm, the CC-FWM interactions produce considerable interchannel cross talk. Due to this reason, when using the idlers to double the number of carriers, it is necessary to control the power of carrier signals in the amplifier input.

It has been found that even when using polarization-maintaining HNLF fibers, due to the influence of self-phase modulation and cross-phase modulation, a change in interposition of SOP of the signal and FOPA pumping radiation has been observed, as they are transmitted through HNLF. As a result, the signal, the SOP of which is orthogonal in respect to the SOP of the pumping radiation at HNLF input, has been amplified by 1.5–1.6 dB.

In a single-channel system with 2PolSK signal conversion to NRZ-OOK modulation format, power penalty of 0.4 dB has been observed between the NRZ-OOK signal from the standard single-channel system and the converted signal, whereas in the case of the idler spectral component, the power penalty is lower by 0.2 dB. These results can be explained by the fact that the idler produced by the parametric FWM process does not contain the logical "0" component radiation from initial 2PolSK signal, which for the converted NRZ signal is interpreted as noise.

In the multichannel system with modulation format conversion, a more explicit manifestation of CC-FWM processes has been observed among the idlers rather than among channels at initial frequencies. This can be explained by the fact that pump power exceeding 35 mW is required to obtain BER values below the 10^{-12} threshold in all eight channels in the idler frequency band than for signals at their initial frequencies. As a result of such amplification difference, more explicit CC-FWM manifestation has been observed, which accordingly has led to additional interchannel cross talk. Additionally, cross talk generated by CC-FWM has not only been transferred to the idlers from the signals at their initial frequencies but also generated among the idlers.

Unlike the system with modulation format conversion, in the system with signal emphasizing from the flow of two orthogonally polarized signals, a situation is observed, when at the same frequency it is possible that the logical "1" components are observed in both channels simultaneously. Therefore, the influence of orthogonally polarized radiation on the quality of the emphasized signal is larger by 0.4 dB (in the case of idlers, by 0.3 dB) than in a system with modulation format conversion.

Acknowledgements

This work has been supported by the National Research Program in Latvia within the project Nr. 10-4/VPP-4/11.

Author details

Vjaceslavs Bobrovs*, Sergejs Olonkins, Sandis Spolitis, Jurgis Porins and Girts Ivanovs

*Address all correspondence to: vjaceslavs.bobrovs@rtu.lv

Riga Technical University, Institute of Telecommunications, Latvia

References

[1] Cisco Systems Inc., "The Zettabyte Era: Trends and Analysis," white paper, June 2014, pp. 1-24.

[2] BELL LABS Metro Network Traffic Growth, An Architecture Impact Study Strategic White Paper, 2013.

[3] Cisco Systems Inc., "Cisco VNI: Forecast and Methodology, 2013-2018," white paper, June 2014, pp. 1-14.

[4] Agrawal G.P., "Fiber Optics Communications Systems", John Wiley & Sons, USA, 2002, 561 p.

[5] Thyagarajan K. and Ghatak A., "Fiber Optic Essentials", John Wiley & Sons Inc., Canada, 2007, 259 p.

[6] Kikuchi K., "Ultra-long-haul optical transmission characteristics of wavelength-division multiplexed dual-polarization 16-quadrature-amplitude-modulation signals," Electronics Letters, Vol. 46, No. 6, March 18, 2010, pp. 433-434, DOI: 10.1049/el.2010.3533.

[7] Downie J.D. "High-capacity long-haul transmission using ultra-low loss optical fiber," 17th Opto-Electronics and Communications Conference (OECC), July 2-6 2012, pp. 172-173, DOI: 10.1109/OECC.2012.6276426.

[8] Šalik P., Čertik F. and Roka R., "Duobinary modulation format in optical communication systems," Advances in Signal Processing, Vol. 3, No. 1, 2015, pp. 1-7.

[9] 3MTM technical specifications, PLC Optical Splitters," USA, 2008, pp. 1-2.

[10] Baney D.M. and Stimple J., "WDM EDFA gain characterization with a reduced set of saturating channels," IEEE Photonics Technology Letters, Vol. 8, No. 12, Dec. 1996, pp. 1615-1617. DOI: 10.1109/68.544695.

[11] Sharma M. and Sharma V.R., "Gain flattening of EDFA in C-band using RFA for WDM application," 2nd International Conference on Signal Processing and Integrated Networks (SPIN 2015), 19-20 Feb. 2015, pp. 346-351, DOI: 10.1109/SPIN.2015.7095422.

[12] Becker P.C., Plsson N.A. and Simpson J.R., "Erbium-Doped Fiber Amplifiers", Academic Press, USA, 1999, 481 p.

[13] Houbavlis T., Zoiros K.E., Kalyvas M., Theophilopoulos G., Bintjas C., Yiannopoulos K., Pleros N., Vlachos K., Avramopoulos H., Schares L., Occhi L., Guekos G., Taylor J.R., Hansmann S. and Miller W., "All-optical signal processing and applications within the esprit project DO_ALL," Journal of Lightwave Technology, Vol. 23, No. 2, Feb. 2005, pp. 781-801, DOI: 10.1109/JLT.2004.838854.

[14] Oxenlowe L.K., Pu M., Ding Y., Hu H., Da Ros F., Vukovic D., Jensen A.S., Ji H., Galili M., Peucheret C. and Yvind K., "All-optical signal processing using silicon devices," European Conference on Optical Communication (ECOC 2014), 21-25 Sept. 2014, pp. 1-3, DOI: 10.1109/ECOC.2014.6964241.

[15] Willner A.E., Khaleghi S., Chitgarha M. R. and Yilmaz O. F. "All-optical signal processing", Journal of Lightwave Technology, Vol. 32, No. 4, February 15th, 2014, DOI: 10.1109/JLT.2013.2287219.

[16] Chu C.-H., Lin S.-L., Chan S.-C. and Hwang S.-K. "All-optical modulation format conversion using nonlinear dynamics of semiconductor lasers," IEEE Journal of Quantum Electronics, Vol. 48, No. 11, Nov. 2012, pp. 1389-1396, DOI: 10.1109/JQE.2012.2212877.

[17] Ozolins O., Parts R., Bobrovs V., "Impact of cascaded MRRs on all-optical clock recovery from 40 Gbit/s RZ-OOK signal," 9th International Symposium on Communication Systems, Networks & Digital Signal Processing (CSNDSP 2014), 23-25 July 2014, pp. 541-545, DOI: 10.1109/CSNDSP.2014.6923888.

[18] Hisano D., Maruta A. and Kitayama K.-I., "Timing detector using cross gain modulation in semiconductor optical amplifier for adaptive all-optical signal processing," 18th Microoptics Conference (MOC 2013), 27-30 Oct. 2013, pp. 1-2.

[19] Hu H., Jopson R.M., Gnauck A.H., Dinu M., Chandrasekhar S., Xie C. and Randel S. "Parametric amplification, wavelength conversion, and phase conjugation of a 2.048-Tbit/s WDM PDM 16-QAM signal," Journal of Lightwave Technology, Vol. 33, No. 7, April 1, 2015, pp. 1286-1291, DOI: 10.1109/JLT.2014.2370038.

[20] Hansryd J. and Andrekson P.A., "O-TDM demultiplexer with 40-dB gain based on a fiber optical parametric amplifier," IEEE Photonics Technology Letters, Vol. 13, No. 7, July 2001, pp. 732-734, DOI. 10.1109/68.930430.

[21] Shama H.B., Gulati T. and Rawat B., "Evaluation of optical amplifiers," International Journal of Engineering Research and Applications, Vol. 2, No. 1, 2012, pp. 663-667.

[22] Csele M., "Fundamentals of Light Sources and Lasers," John Wiley & Sons Inc., Canada, 2004, 343 p.

[23] Thyagarajan K. and Ghatak A., "Fiber Optics Essentials," John Wiley & Sons Inc., Canada, 2007, 259 p, DOI: 10.1002/9780470152560.

[24] Forestieri E. "Optical Communication Theory and Techniques," Springer Science + Business Media Inc., USA, 2005, 216 p, DOI: 10.1007/b100765.

[25] Kaminow I.P., Li T. and Willner A. E., "Optical Fiber Telecommunications: Components and Subsystems," Academic Press, USA, 2008, 945 p.

[26] Agrawal G.P., "Applications of Nonlinear Fiber Optics," Academic Press, USA, 2001, 473 p.

[27] Karasek M., Honzatko P., Vojtech J., Radil J., "Multi-wavelength conversion at 10 Gb/s and 40 Gb/s based on 2 pumps FOPA," 13th International Conference on Transparent Optical Networks (ICTON 2011), 2011, pp. 1-4, DOI: 10.1109/ICTON.2011.5970951.

[28] Temprana E., Ataie V., Peric A., Alic N., Radic S., "Wavelength conversion of QPSK signals in single-pump FOPA with 20 dB conversion efficiency," Optical Fiber Communications Conference and Exhibition (OFC 2014), 9-13, Mar. 2014, pp.1-3. DOI: 10.1364/OFC.2014.Th1H.2.

[29] Ruo-Ding L., Kumar P., Kath W.L., "Dispersion compensation with phase-sensitive optical amplifiers," IEEE Journal of Lightwave Technology, Vol.12, No. 12, 1994, pp. 541-549, DOI: 10.1109/50.285338.

[30] Wang J., Ji H., Hu H., Mulvad H.C.H., Galili M., Palushani E., Jeppesen P., Yu J.L., Oxenlowe L.K., "All-optical 2R regeneration of a 160-Gbit/s RZOOK serial data signal using a FOPA," IEEE Photonics Conference (IPC 2012), 23-27 Sept. 2012, pp. 108-109, DOI: 10.1109/IPCon.2012.6358512.

[31] Wang J., Yu J., Meng T., Miao W., Sun B., Wang W., Yang E., "Simultaneous 3R regeneration of 4*40-Gbit/s WDM signals in a single fiber," IEEE Photonics Journal, Vol. 4, No. 5, pp. 1816-1822, 2012, DOI: 10.1109/JPHOT.2012.2215955.

[32] Lee H.H., Cho S.H., Lee J.H., Lee S.S., "Excess intensity noise suppressed 100-GHz spectrum-sliced WDM-PON with pre-spectrum-sliced seed light source," 36th European Conference Exhibition on Optical Communication (ECOC 2010), 19-23 Sept. 2010, pp.1-3, DOI: 10.1109/ECOC.2010.5621291.

[33] Spolītis S., Olonkins S., Poriņš J., "Realization of dense bidirectional spectrum sliced WDM-PON access system," 9th International Symposium on Communications Systems, Networks and Digital Signal Processing (CSNDSP 2014), Conference Proceedings, 2014, pp.1-6, DOI: 10.1109/CSNDSP.2014.6923890.

[34] Ivanovs Ģ., Ļašuks I., Ščemeļevs A., "A hybrid TDM/WDM-PON system with Fwm-generated source of multiwavelength optical signals," Latvian Journal of Physics and Technical Sciences, Vol. 47, No. 5, 2010, pp.3-14, DOI: 10.2478/v10047-009-0020-3.

[35] Mathlouthi W., Vacondio F., Rusch L.A., "High-bit-rate dense SS-WDM PON using SOA-based noise reduction with a novel balanced detection," Journal of Lightwave Technology, Vol. 27, No. 22, Nov. 15 2009, pp.5045-5055, DOI: 10.1109/JLT.2009.2026062.

[36] Fok M.P., Shu C., "Bandwidth enhanced multi-wavelength source from an SBS-assisted fiber ring laser using four-wave mixing in a highly nonlinear bismuth oxide fiber," Conference on Lasers and Electro-Optics / Quantum, DOI: 10.1109/CLEO.2006.4628216.

Chapter 11

Next-Generation Transport Networks Leveraging Universal Traffic Switching and Flexible Optical Transponders

Bodhisattwa Gangopadhyay, João Pedro and Stefan Spälter

Additional information is available at the end of the chapter

http://dx.doi.org/10.5772/intechopen.68953

Abstract

Recent developments in communication technology contributed to the growth of network traffic exponentially. Cost per bit has to necessarily suffer an inverse trend, posing several challenges to network operators. Optical transport networks are no exception to this. On one hand, they have to keep up with the expectations of data speed, volume, and growth at the agreed quality-of-service (QoS), while on the other hand, a steep downward trend of the cost per bit is a matter of concern. Thus, the proper selection of network architecture, technology, resiliency schemes, and traffic handling contributes to the total cost of ownership (TCO). In this context, this chapter looks into the network architectures, including the optical transport network (OTN) switch (both traditional and universal), resiliency schemes (protection and restoration), flexible-rate line interfaces, and an overall strategy of handover in between metro and core networks. A design framework is also described and used to support the case studies reported in this chapter.

Keywords: optical transport network, flexible line interfaces, protection and restoration, network planning

1. Introduction

The exponential growth of consumer demands and machine-to-machine network traffic coupled with the downward trend in revenue per bit transported is challenging network operators to adopt a strategy which tackles a twofold problem. The dual nature of the problem, on one hand lies in selecting a network architecture/technology which can efficiently transport traffic originating from multiple sources, be it time division multiplexing (TDM) or Packet, and on the other hand to make use of an increasingly flexible and heterogeneous optical layer, where the characteristics of the optical light paths to be set up (e.g., modulation format, spectral width) are customized to the specific path properties.

Today, typical transmission networks are a layered combination of dense wavelength division multiplexing (DWDM) equipment (the lowest layer above the optical fiber layer), a subwavelength aggregation and grooming layer and an internet protocol (IP) layer. These layers form server/client relationships and are independent of each other. From a technological point of view, the functions of each layer are very different. Higher layer equipment is typically more expensive per bit transported because it needs to do more processing, so the use of the layers must be carefully balanced to deliver cost-optimized networks. The introduction of coherent 100G (100 Gigabits per second) optical transport was a key catalyst which offered massive performance gains over incumbent technologies to exploit more capacity from a single fiber. Telco operators achieved considerable gains as the cost per bit started going down with the introduction of 100G. But this was not the end, as they eyed newer improved scalable architectures to strengthen network operations and total cost of ownership (TCO). These next-generation network architectures aimed at efficiently grooming and aggregating sublambda data streams resulting in cost-optimized well-packed 100G wavelengths which would allow telco operators to survive within challenging capital expenditure (CAPEX) and operational expenditure (OPEX) cost targets in the near future. However, due to their relatively compact and short reach topologies, an abundance of optical fiber and the requirement to interconnect network elements at 10G rates or less, metro networks still mostly relied on a direct-detect 10G optimized optical transport infrastructure, though this situation is bound to change dramatically [1].

Importantly, the abundant deployment of 100G coherent systems in core networks has been attained at the expense of relatively costly line interfaces, performing electrical-optical (EO) and optical-electrical (OE) conversions. In the meanwhile, optical transport network (OTN) emerged as a key building block to complement the capacity gains unleashed by 100G and coherent optics. Efficiency, predictability, and reliability of the transport world and the agility, programmability of the packet world were blended into the ITU G.709 standard (OTN protocol) [2] and thus became an automatic choice. The demanding service level agreements (SLAs) of private E-LANs, E-Lines, and other packet traffic along with wavelength services from the near future could be met by the many features that OTN was offering. Moreover, traffic processing on IP packet level (layer 3) is much more expensive (per GByte) than switching the same amount of traffic in optical channel data unit (ODU) containers by OTN switches. Therefore, OTN provides a more cost-effective platform for subwavelength services to be multiplexed not only at their source node, but also at selected intermediate nodes and, as a result, reduce the amount of expensive (WDM) line interfaces used without having to resort to expensive router equipment to perform this task [3].

Nevertheless, a continuous steep inverse trend between the data volume and cost per bit being carried and the explosive growth of the traffic between data centers has supported the development of higher-order modulation formats, namely, 8-quadrature amplitude modulation

(8-QAM) and 16-QAM, which provide 50% (150G) and 100% (200G) more capacity than standard quadrature phase shift keying (QPSK) albeit at the expense of reducing transparent reach to around half and one-third, respectively. This catered to the need of extremely high equipment density and maximizing optical transport capacity per fiber and per transceiver. However, there was another need growing up from the continuously shifting traffic pattern. And one answer to all these needs mentioned above was the flexible-rate interface modules, which grant software-switchable modulation (QPSK, 8-QAM, and 16-QAM supported in the same device), flexible channel spectral width, and flexible frequency tunability to provide the ideal balance between performance, capacity, and reliability across the most challenging networks [4].

A key aspect of transport networks is their capability to withstand failure scenarios, given the very large amount of traffic they carry. This requirement is usually met via protection and restoration techniques [5]. Protection is a static mechanism to protect against failures, where the resources for both the primary and the backup paths are reserved prior to the data communication. Restoration is a dynamic mechanism where the backup path is not set up until the failure occurs. Survivability using these techniques is usually provided to handle single/multiple link or node failures in the network with each scheme, claiming a different stake in terms of network resources and recovery time. Moreover, protection/restoration can be supported at different layers of the OTN and the selection of which layer to utilize, either the ODU or the optical channel (OCh) layer, also involves similar trade-offs [6].

This chapter is organized as follows. Section 2 overviews the current role of OTN switching in transport networks and introduces the concept of the universal OTN switch, highlighting the motivation behind it and the key benefits of adopting it. In order to support the case studies presented in the remaining of the chapter, a routing and grooming framework is detailed in Section 3. Section 4 addresses the relevance of mechanisms for failure survivability in transport networks, presenting a case study comparing the cost-effectiveness of supporting restoration at different layers of the transport network. Moreover, the benefits of combining universal traffic switching and flexible-rate line interfaces are investigated in Section 5 and are quantitatively assessed via a case study using reference transport networks. Section 6 elaborates on the prospects of further cost savings in metropolitan and core networks as a result of adopting coherent-detection technology in both network segments. Finally, Section 7 presents the concluding remarks.

2. The role of OTN switching in transport networks

Technological advancements have seen different node architectures being proposed, each having their pros and cons. Transponders and muxponders still provide the simplest approach to getting traffic on and off a 100G transmission interface, by multiplexing one or more client interface to a single high-speed line interface. On one hand, this approach incentivizes simplicity and relatively low CAPEX, while on the other hand, penalizes the operator because of its limited ability to groom traffic, inability to perform remote configuration, and being inefficient to combine add/drop traffic with pass-through traffic from other line interfaces.

2.1. First-generation OTN switch

The present day traffic represents a blend of packet traffic and extensively installed legacy TDM traffic. To address this varied traffic pattern and mix, the modern day network architectures requires an OTN switch to cater service-agnostic switching where the multiple client service types can be mapped into ODU frames and the same can be switched at the ODU level. This will not only allow subwavelength services to be aggregated at their source and destination nodes but also allow them to be groomed at intermediated nodes and thus finally contribute to a reduction in the number of expensive WDM line interfaces in use. **Figure 1** depicts the transponder/muxponder and OTN switch architectures. The later architecture introduces a digital OTN switch, which is separated from the WDM box using short-reach optics or integrated with the WDM box to reduce footprint and power consumption [7]. According to Infonetics, the majority of service providers (86%) are choosing OTN switching as the technology best suited to fill 100G optical channels, because it enables efficient aggregation of diverse services and protocols over a single optical link. Noteworthy, it is being embraced and deployed by network providers throughout Asia, Europe, and North America [8].

Another key feature enabled by OTN switching is fast shared protection and restoration schemes with fine granularity, which cannot be achieved with a transponder or muxponder solution. Other than contributing to a reduction of CAPEX, OTN switching does include several other benefits. Among them, high scalability, fast end-to-end service provisioning, multiple traffic type support, subwavelength switching, router port offloading, and client service level mapping are a few to name [7]. But then, certain specific traffic patterns and network topology can result in the switchless architecture to have a CAPEX advantage over

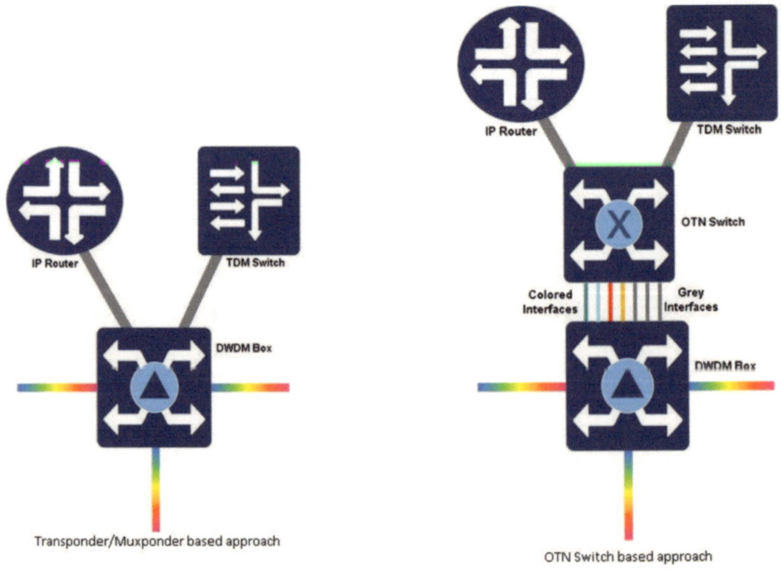

Figure 1. Transponder/muxponder-based approach vs. OTN-based approach.

http://dx.doi.org/10.5772/intechopen.68953

the switched architecture. The same is witnessed when client services data rates match with the WDM channel data rates or even when the traffic type is not a mix and only comprises of packet data, and in these cases, the overall node architecture can be simplified to a switchless one with simplified equipment spanning L3-L0. Another perspective also challenges network operators while including OTN switch and that is contributed by the additional power and space consumption increasing the OPEX.

The traditional or first-generation OTN switch focuses only on switching ODUs and involves dedicated cards for circuit-based switching and dedicated cards for packet-switching, as illustrated in **Figure 2**. Consequently, switching is restricted within the same switch-type domain (packet/circuit). Furthermore, within the same switching domain, interworking between different technologies, such as the OTN and the synchronous digital hierarchy (SDH)/synchronous optical network (SONET), may be also blocked. Essentially, traditional OTN switches fail to deliver "universal client port" functionalities, which would allow to support any mix of client interfaces in a single switching domain, further improving the ability to efficiently utilize the transport network resources.

2.2. The universal OTN switch

Conversely to the traditional or first-generation OTN switch, next-generation universal OTN switch as shown in **Figure 3** can aggregate different protocols on the client side and enable transparent multiplexing of packet and TDM traffic, allowing a single device to be used in multiple applications efficiently.

The universal OTN switch is backed by a universal transport platform which is capable of switching traffic flows based on any L1-L2.5 protocol and on every port, including multiple protocols on the same port simultaneously. Hence, it can offer network operators the best of both worlds by dynamically controlling every flow on every port as a circuit or packet and providing the most efficient, future-proof solution for virtually all applications [9].

These next-generation OTN switches employ universal cards to handle TDM, OTN, carrier Ethernet (CE), and multiprotocol label switching-transport protocol (MPLS-TP) traffic and

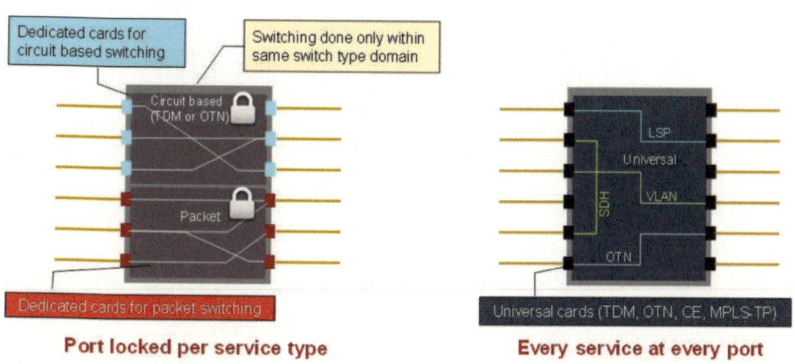

Figure 2. Traditional vs. universal OTN switch.

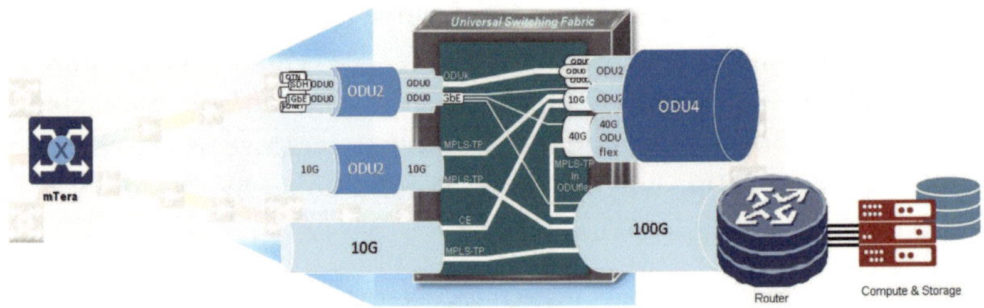

Figure 3. Universal OTN switch.

provide grooming efficiency, granularity, and service classification of a packet switch along with scalability, operational efficiency, and performance of an OTN switch, as illustrated in **Figure 3**. Furthermore, the universal OTN switch can also assist in reducing router ports by distributing services from aggregated router hand-offs with virtual local area network (VLAN) to ODU mapping, as shown in **Figure 4**.

Naturally, several other benefits, common to first-generation OTN switches, are also present and include fast end-to-end service provisioning, rapid restoration, high scalability, sub-wavelength level switching, and easy support of new/multiple traffic types. Likewise, if the service data rates are the same as the data rates of the WDM wavelength channels or when only packet traffic is present, then the universal switch might not have a CAPEX advantage over traditional OTN switch or even the conventional transponder/muxponder with OTN encapsulation.

Depending on the location in the network and the traffic matrix, switching at the packet or STS-1/VC-4 level can have efficiency benefits over switching OTN at the ODU0 (1.25G) level or above, as long as one of two conditions are met. Either there must be a significant number of client interfaces below 1G or there must be the potential for large statistical gains from

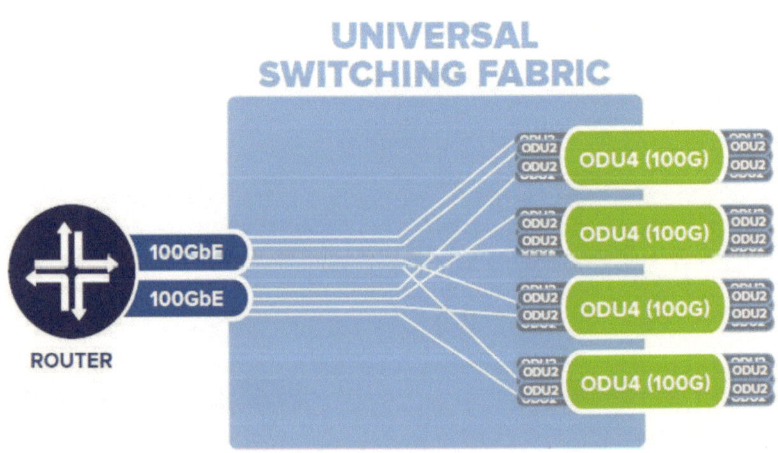

Figure 4. Universal switching fabric [1].

multiplexing uncorrelated bursty traffic flows. These conditions are more likely to occur in a metro network versus the long haul network where statistical gains have often already been achieved at the IP/MPLS layer and client interface speeds are likely to be above 1G.

Importantly, universal switching platforms enable OTN, SONET/SDH, and packet switched traffic to share the same high-speed interface, with SONET/SDH mapped to ODU2, packet traffic mapped to ODU flex, and the remaining capacity available for OTN switched traffic, thus making the most efficient use of each high-speed interface and the optical spectrum it consumes. In addition, universal switching provides investment protection against changes in traffic patterns and client types. With universal fabrics, and the ability to define in software interfaces and virtual interfaces for OTN, CE (Bridging, VLAN cross-connect), or MPLS-TP/ virtual private LAN service (VPLS), and without impacting the capacity of the switch, operators can easily evolve from first-generation to universal OTN switching.

3. Routing and grooming framework

The planning of a transport network requires appropriate dimensioning algorithms to guarantee that all traffic demands can be successfully supported and at the expense of minimum CAPEX, therefore giving the network operator the best positioning to run a profitable business. Moreover, the multilayer nature of transport networks exploiting both OTN switches and reconfigurable optical add/drop multiplexers (ROADMs) and the additional requirements entailed by protection and restoration mechanisms have to be taken into account by the network design framework.

The multilayer routing and grooming framework developed and implemented to support the different network scenarios analyzed in the remaining of this chapter can accommodate the different constraints imposed by the specific node architectures and available line rates and consists of the following main steps:

Step 1. For each node pair *sd*, compute a set of *K* routing options (*K* shortest paths for unprotected demands, *K* shortest disjoint cycles for demands with protection/restoration) over the network graph *G*(*V*, *L*), where *V* denotes the set of nodes and *L* denotes the set of links, and store them in set Π.

Step 2. Order all traffic demands from the same planning period according to a given criteria (e.g., largest first, longest first).

Step 3. For the next ordered traffic demand t_d perform the following steps:

a. Set $k = 1$. Create an auxiliary graph, G(V', L'), where each network node $v \in V$ belonging to routing solution $\pi_k \in \Pi$ is mapped as node $v' \in V'$ and where each existing light path overlapping with π_k and with available capacity to support t_d is mapped as link $l' \in L'$. For unconnected nodes in V', determine the feasibility of a new overlapping light path with data rates of 100G and 200G. If feasible, map links $l_{100}' \in L'$ and $l_{200}' \in L'$.

b. Set the cost of all links in L', such that their cost verifies $c(l') \ll c(l_{200}') < c(l_{100}')$, that is, reusing an existing light path is always preferred over creating new light paths and creating a new 200G light path is slightly more economic than creating a 100G light path (as to give preference to more spectral efficient 200G light paths in case of a tie).

c. Route over the auxiliary graph to determine the least cost path/cycle and let the routing and grooming solution cost be $c(\pi_k^*)$. Increment k. If $k = K$, go to **(Step 3d)**. Otherwise, repeat the steps from **(Step 3a)**.

d. Return the routing and grooming solution for t_d, over all K solutions, with smallest cost. If none is found, the traffic demand is blocked.

Step 4. If all traffic demands have been considered, end the algorithm. Otherwise, repeat from **(Step 3)** for the next traffic demand.

Importantly, the framework is ready to support flexible-rate line interfaces, namely, capable of operating at 100G (QPSK) and 200G (16-QAM), and it can easily incorporate more line rates (e.g., 150G and 300G via 8-QAM and 64-QAM modulation formats, respectively) or can be executed for a single line rate by disabling all others. Moreover, in the case of utilizing shared restoration mechanisms to recover from failure scenarios, the fact that network resources can be shared by the backup paths of ODUs/OChs whose working paths are link- or node-disjoint also needs to be modeled. In order to control the degree of resource sharing, the number of ODUs/OChs that can share the same resource is limited by a maximum resource sharing value S.

4. Protection/restoration in transport networks

Due to several failure issues, one of the key aspects to be considered while planning and operating telecommunication networks is resiliency. With the evolution toward 5G, everyday network operators and equipment vendors struggle to come up with innovative network resiliency mechanisms which are robust, faster, and cost-effective. Transport networks are no alien to this behavior and the resiliency schemes for the same dates back to the time when SDH/SONET-based networks were at helm. Currently, OTN is the predominant transport technology. Given the multiple layers defined within the OTN, resilience mechanisms acting at the electrical and optical domains are available, the former being enforced at the ODU layer via ODU switching network elements, whereas the latter are supported at the OCh layer through ROADMs [6].

As referred, in addition to the layer at which failure survivability mechanisms are enforced, these mechanisms can be further classified as protection or restoration. Usually, protection relies on dedicated backup resources determined in advanced, reserved, and preconfigured for a particular service and which can be quickly triggered to replace the working resources. On the other hand, restoration does not require dedicated resources, i.e., backup resources can be shared, and the backup path and its resources are only assigned upon recovery from a failure. Moreover, restoration is typically triggered by the control plane, i.e., via the generalized multiprotocol label switching (GMPLS). GMPLS-based restoration can by dynamic, designated as dynamic source rerouting (DSR) and where backup resources are determined once a failure is detected, or preplanned, named preplanned shared restoration (PSR), in which case the backup resources are known in advance of the failure. The main advantage of DSR is that it can grant (best effort) survivability against a larger number of failure scenarios, namely, when multiple failures take place. On the other hand, PSR provides faster recovery of the failed service/demand since it avoids the time required to compute the backup path and associated resources [6].

The economics and qualitative behavior of PSR in transport networks are evaluated in the following, with particular emphasis on comparing the use of such scheme at the ODU and OCh layers and identifying in which network and traffic scenario is each one of them expected to be a more suitable option for a transport network operator.

Based on the resiliency schemes and the quality-of-service (QoS) they cater to, the working and protection/restoration paths are either link disjoint or node disjoint (the later forces the links to be disjoint though). Resources on the protection/restoration path (backup path) can be shared among multiple demands contrary to the same being dedicated to a single demand and this is enabled by the shared restoration scheme as illustrated in **Figure 5**. The sharing of resources is a novelty when compared to 1:1 dedicated protection schemes, which increase CAPEX due to a higher number of line interfaces required. Solid red and yellow highlights the working path of both the demands, respectively, while both these demands are using the same backup path. This is only possible when the working paths of both the demands are link disjoint and both the demands simultaneously stay unaffected by a single link failure and thus the same resource can be used for restoring either of the demands.

When resource sharing is enforced, savings are expected with respect to the amount of required additional resources, when compared to dedicated schemes (e.g., 1 + 1 protection), although restoration schemes usually provide slower recovery to failures. Furthermore, a

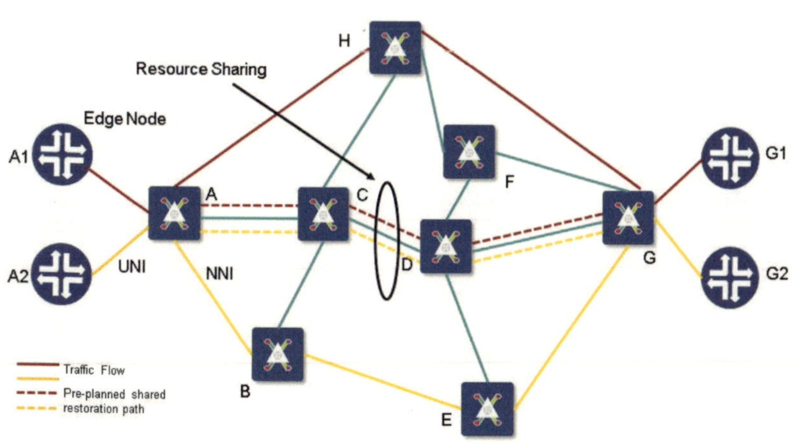

Figure 5. Illustration of resource sharing for shared restoration schemes.

qualitative and quantitative comparison of both schemes (ODU and OCh based restoration) can be performed, addressing key aspects such as restoration time, network element complexity, switching granularity, planning complexity, and line interface count. **Table 1** intends to summarize the qualitative comparison of both schemes.

In order to gain further insight on how PSR at the ODU and OCh level compare in terms of their resource requirements, above all in number of line interfaces, a planning case study is presented in this section. The study is realized over the 31-node backbone network covering Italy that is already used in previous studies [10] and illustrated in **Figure 6**. It is assumed that each network link supports up to 96 wavelength channels and each wavelength is operated at 100 Gb/s (i.e., carries one ODU4). A very simplified performance model is used to determine the transmission reach between regeneration sites: it consists of a maximum transmission reach of 1500 km and a reach penalty of 60 km per node traversed. With respect to the traffic pattern, 30% of the node pairs were randomly selected to exchange traffic demands. Each client traffic demand is mapped into an appropriate ODU k, with $k = \{0, 1, 2, 3\}$, which correspond to data rates of {1.25, 2.5, 10, 40} Gb/s, respectively. The traffic load offered to the network is evenly distributed over the different ODU rates, meaning that each ODU k accounts for around 25% of all offered traffic load [6]. As to understand the impact of the network traffic load on the effectiveness of both restoration schemes, the total traffic load is varied from 2 up to 20 Tb/s, with load increments of 2 Tb/s.

A comprehensive comparison requires not only to compute the resource requirements of enforcing ODU or OCh restoration for every traffic demand, but also to benchmark these results against the resources needed when the traffic demands are either unprotected or require 1 + 1 protection. The routing and grooming framework of Section 3 is used to plan the network. An upper bound on the number of demands sharing a common restoration resource is set to $S = 10$. Importantly, in the case of PSR at the OCh layer, the traffic is first groomed into ODU 4s and afterward the resulting ODU 4 are routed considering also the protection/restoration mechanism being enforced.

	Restoration layer	
	ODU	**OCh**
Restoration time	• Faster: few hundreds of milliseconds	• Slower: hundreds of milliseconds up to seconds
ROADM complexity	• Low: can use simple ROADMS	• High: colorless and directionless
ODU switching	• Mandatory	• Not required
Switching granularity	• Finer: starting from 1.25 Gb/s	• Coarser: e.g., 40 or 100 Gb/s
Planning complexity	• Multilayer: with intermediate grooming	• Single-layer: without intermediate grooming
Line IF count	• Expected to be higher	• Expected to be lower

Table 1. Qualitative comparison of ODU and OCh restoration.

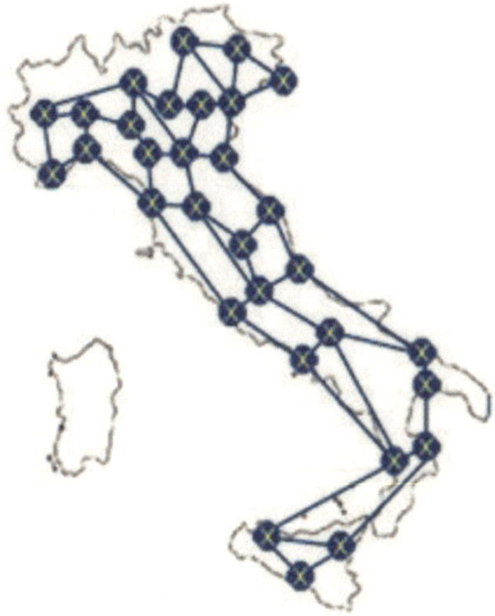

Figure 6. Network topology – Telecom Italia National Backbone 31 Node network.

For both ODU and OCh protection/restoration, disjointness is applied at the link level and the number of candidate routing paths/cycles K is set to 5 working and backup paths. **Figures 7** and **8** present the line interface count and wavelength channel utilization as a function of the offered traffic load when using resilience mechanisms acting at the ODU layer. The wavelength channel utilization is defined as the fraction of wavelengths being used overall network links.

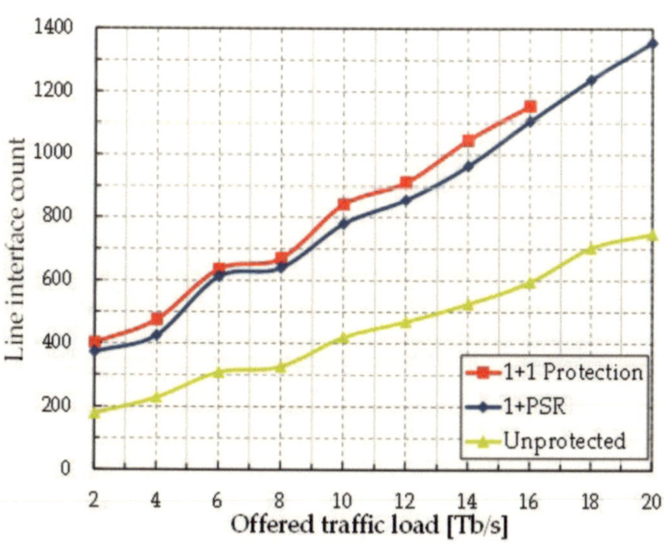

Figure 7. Line interface count for ODU restoration and protection.

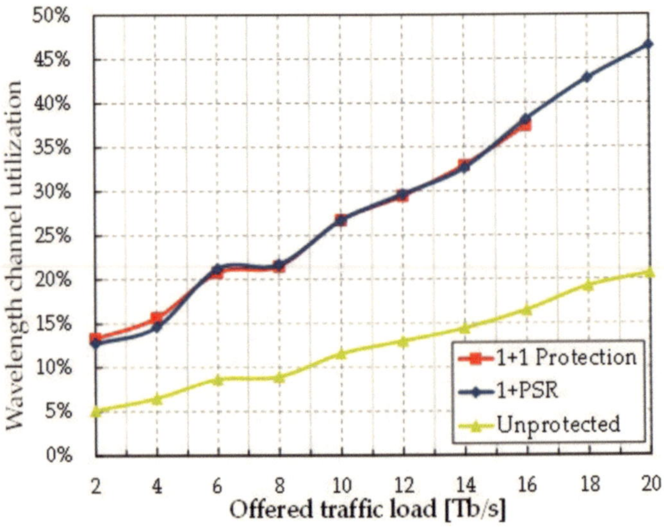

Figure 8. Wavelength channel utilization for ODU restoration and protection.

The equivalent pair of plots when supporting resilience mechanisms that operate in the OCh layer is depicted in **Figures 9** and **10**.

The main outcome of this comparison is that OCh PSR is a more cost-effective scheme than ODU PSR in a network of this size and supporting the traffic pattern defined as input [6]. In addition, it is also clear that PSR enables the network to withstand a single link failure with less resource overprovisioning than that of 1 + 1 protection.

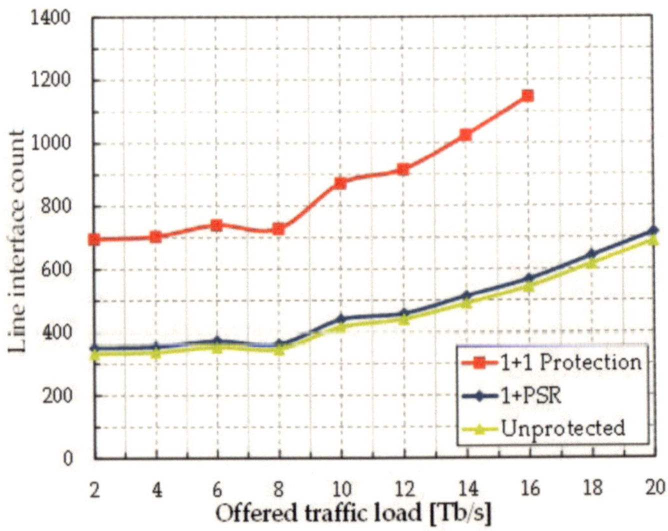

Figure 9. Line interface count for OCh restoration and protection.

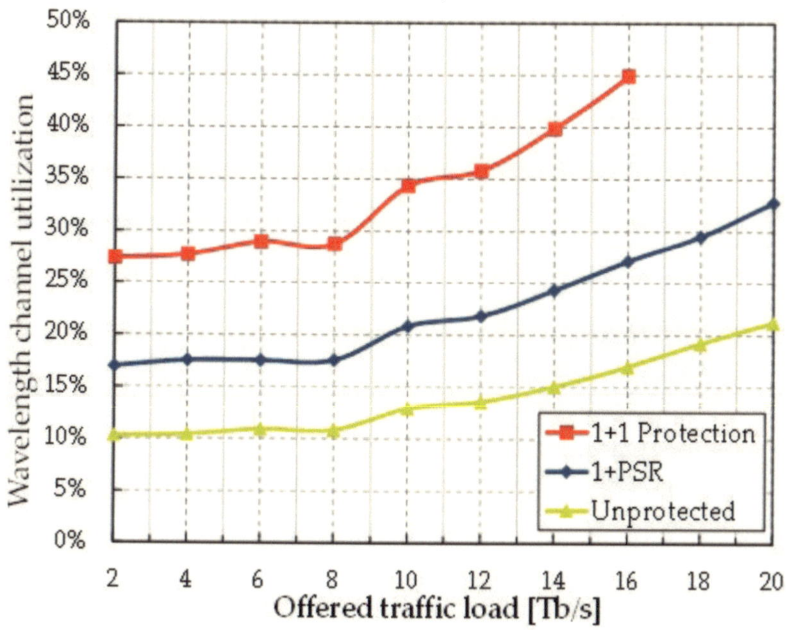

Figure 10. Wavelength channel utilization for OCh restoration and protection.

5. Combining universal OTN switching with flexible-rate DWDM networks

The wide deployment of 100G in core networks began with QPSK modulation format, where binary electrical signals are converted to a format with four constellation points, which is transmitted over two orthogonal polarizations. The applied coherent detection and advanced digital signal processing technologies enable detecting arbitrary multilevel schemes, which can be used to transmit more bits per time slot (e.g., 16-QAM, 64-QAM).

To cater network operators with the capability to address continuously shifting traffic patterns and increased capacity demands, flexible-rate interfaces are being introduced to deliver optimal transmission reach, performance, and agility. **Figure 11** exemplifies the demand reach distribution in a variety of long-haul and ultra-long-haul networks along with the expected cumulative distribution of the utilization of BPSK, QPSK, 8-QAM, and 16-QAM modulation formats for a fixed symbol rate providing data rates of 50, 100, 150, and 200 Gb/s, respectively.

There are several benefits obtained from the flexible-rate interfaces, but to highlight a few will be cost-optimized coverage from intracity data center interconnection (DCI) to ultra-long-haul demands, single sparing blade for all modulation schemes, and restoration of higher reach modulation schemes using higher reach modulation schemes. A joint design considering flexible-rate and universal OTN switching technology can unfold crucial benefits that go beyond the separate claims of improved capacity on the line side (flexible-rate) and more efficient traffic aggregation (universal OTN switching).

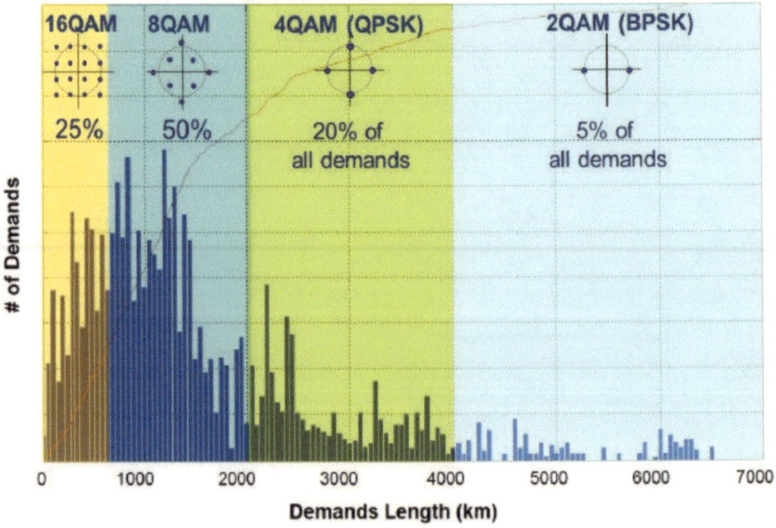

Figure 11. Example of demand length distribution and expected cumulative distribution of the modulation format utilization in core networks with Raman amplification [11].

Recent work shows evidence that if both features are combined in an optimal way, a considerable reduction of up to 30% could be achieved in terms of the number of deployed light paths, with proportional savings in a number of line interfaces and spectrum used [12]. In the scope of this work, two long distance optical transport networks, as depicted in **Figure 12** and inspired in networks operated by Telecom Italia/TIM, were considered. One is a more recent Italian national backbone, which has 44 nodes and 71 fiber links and has already been used in other studies [13, 14].

It meets the needs of circuits at a national level, mainly for IP router interconnection and for connectivity of big clients. With its shortest paths under 2200 km and backup paths (disjoint

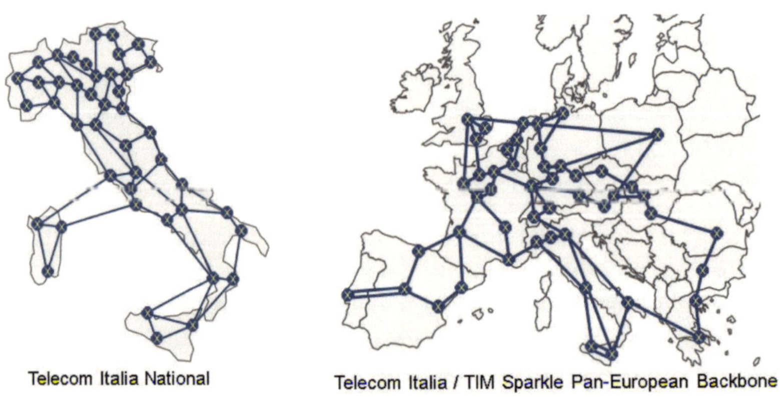

Figure 12. Network topologies – Telecom Italia National Backbone 49 Node network and Sparkle.

from the shortest one) under 2600 km, it is between a regional and long-haul network. The other network is TIM Sparkle Pan-European backbone [15], a geographically expanding network that currently covers Central, Southern, and Eastern Europe with 49 nodes and 72 fiber links. It classifies as an ultra-long-haul network (shortest paths under 5500 km and backup paths under 7000 km). With respect to the client rates to be serviced by the networks during their entire lifecycle, the following data rates are considered: 10G, 40G, and 100G for the Italian National Backbone and 1.25G, 2.5G, 10G for the European Backbone. Traffic comprises both Internet Packet traffic and SONET/SDH TDM traffic. Two traffic periods are considered wherein the later phase, traffic was extrapolated to be at the end of 4 years of the current period, with 25% growth for each year. Total traffic for the two periods under analysis and the partitioning of bandwidth among different client rates for the Italian national backbone and European backbone network is as described in Ref. [14]. Using the multilayer optimization algorithm described in Section 3, five routing options were calculated for each node pair ($K = 5$) and the algorithm was run considering all four combinations given by traditional versus universal OTN switching and using only 100G (QPSK) line rates versus optimizing the line rate between 100G (QPSK) and 200G (16-QAM) according to the properties of the routing path [16].

The required number of light paths, which impacts the number of required expensive line interfaces as well as the amount of spectrum used, is shown in **Figure 13**.

As expected, for TIM national backbone network, the benefit of supporting higher data rates with the same line interface is more effective in reducing the number of light paths when compared to the Sparkle network, mostly because the former topology benefits from average shorter routing paths. Noticeably, universal switching always grants savings when compared to its traditional counterpart due to the packet-level aggregation and universal nature of the switch. Moreover, the combined effect of using both universal switching and flexible-rate leads to the highest savings in light path count and, consequently, the most cost-effective solution. Although not shown here due to the lack of space, savings in router port count are also achieved, which further contributes to decrease TCO when leveraging universal OTN switching and flexible-rate line interfaces. Thus, a joint network design considering both flexible-rate and universal switching technology can unfold crucial benefits that go beyond the separate claims of improved capacity on the line side (flexible-rate) and more efficient traffic aggregation (universal switching).

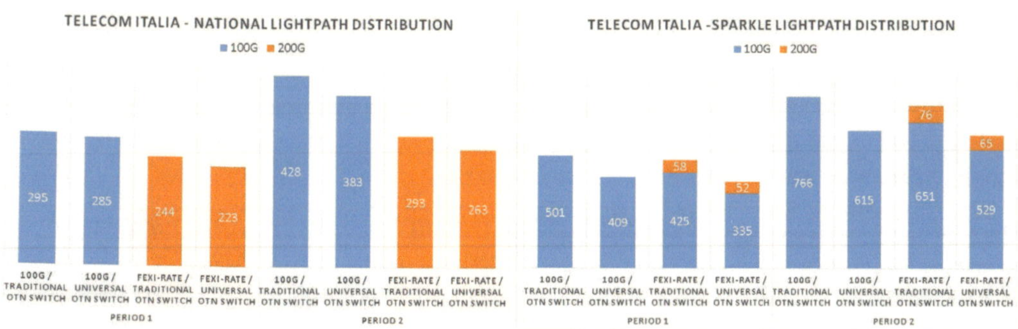

Figure 13. Number of light paths as function of the network and node architecture.

6. Transparent handover between metro and core segments of next-generation transport networks

Transport networks can hierarchically be broadly classified into access, metro, and core domain. End-user connectivity is catered to by the access network and a few dozen kilometers could be covered based on the specific technology requirements and the density of user population. In between, the core and the access part, lies the metro network which on one hand aggregates traffic coming from the access networks and on the other hand transports intrametro traffic (e.g., intrametro data center interconnection) and covers often up to a few hundred kilometers. Following the level of traffic aggregation, metro networks could further be segmented into metro aggregation and core. At the end, covering a bigger geographical area (e.g., state or country) is what is called the core network interconnecting the metro networks and typically spans over several hundreds to thousands of kilometers.

In the metro transport networks, the predominant technology and topology are in the form of SDH/SONET rings. But while these SDH/SONET rings had intrinsic scalability limitations, there was always an increasing need for capacity. And this gave rise to the introduction of coarse wavelength division multiplexing (CWDM) and DWDM in these networks catering to wavelength switching support at intermediate nodes thus reducing optical-electrical-optical (OEO) conversions which was initially a cost burden to the operator. In present days, the metro network is mostly infested with 2.5G or 10G wavelengths using direct-detection modulation formats and as always the increasing demand of data will be catered to using higher data rate channels (e.g., 40G and 100G) in the coming days. Compared to these, core network has a different behavior and is severely dependent on DWDM and ROADMs exhibiting asymmetrical meshed topology. The wavelength channels have a good share of 10G, 40G, and 100G. Moreover, QPSK is now the dominant modulation format for the higher data rates of 40G and 100G with coherent-detection replacing direct-detection. The utilization of higher-order modulation formats, such as 16-QAM, enables to increase channel capacity. But, the poor reach performance of these higher modulation formats penalizes a widespread deployment of the same. To mitigate this negative effect, adoption of super channels is being looked upon as an alternative to core networks where multiple carriers are grouped together to realize these higher bit-rates (e.g., 2 × 100G QPSK to create a 200G super channel on a 100GHz spectrum) [17]. Another important aspect to look upon is the optical add-drop multiplexers (OADMs) being used in core and metro networks. Metro networks are comparatively simpler with lesser nodal degrees and thus simpler OADMs of the likes of broadcast-and-select or fixed add/drop ROADMs or to the extent of nonreconfigurable fixed OADMs (ROADMs) are in use today. On the other hand, core networks use much higher capacity and also wavelength channels and thus require expensive OADMs.

The traffic flow between the above described metro and core network at the boundary node is depicted in **Figure 14** and the same is represented by metro-to-core (M2C) in the next part of this chapter.

To be more detailed, let us presume one data channel running from one metro network to another metro network via a core network and is designated as metro-to-core-to-metro

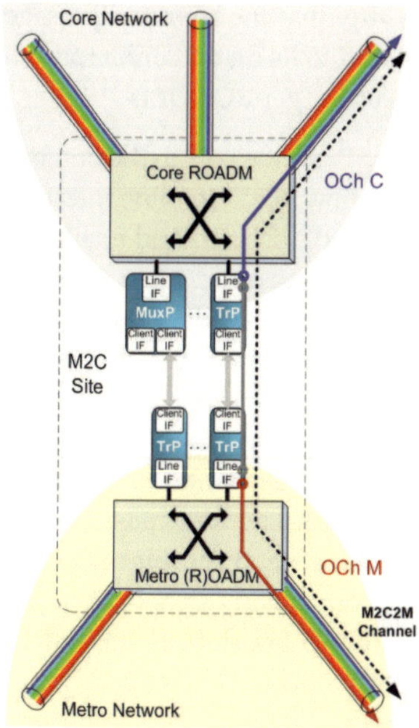

(a) Traditional all-opaque handover

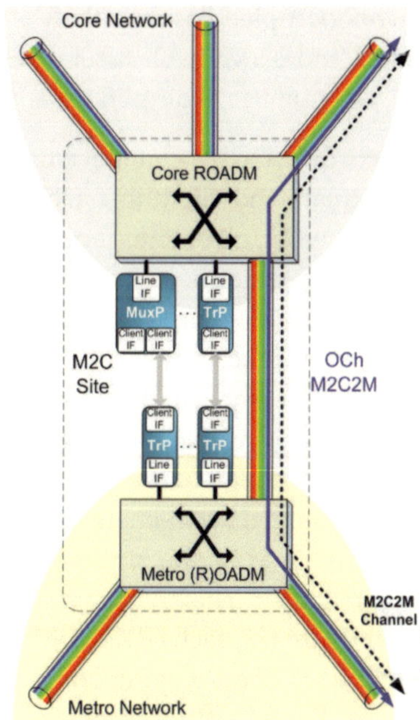

(b) Enhanced transparent handover

Figure 14. Architectures for interconnecting metro and core networks (TrP – transponder, MuxP – muxponder). (a) Traditional all-opaque handover; (b) enhanced transparent handover.

(M2C2M) channel. The current mode of operation forces this M2C2M channel to ride on an optical channel in the metro domain (OCh M) till the metro (R)OADMs. At this (R)OADM node, the same OCh M gets terminated at a line interface only to be transferred to a transponder or muxponder which maps the same into the optical channel (OChC) ferrying inside the core network. At the boundary node in between the core and metro node, a similar process makes sure that the channel is handed over from the core to the metro network. It is to be noted that, these handover sites could be a single site at the same physical location (housing the (R)OADMs) or could be physically separated sites within a comparatively shorter distance hosting each (R)OADM looking toward the core and metro network.

This architecture represents an all-opaque architecture depicting a clear enough demarcation between the metro and core network segments where the hardware (client interfaces) performs the handover. As it already undertakes the OEO operation, these sites are used for grooming subrate data signals into higher data-rate pipes/channels. For example, the M2C2M channel with a data rate of 2.5G can be multiplexed/groomed into 10G and 100G OCh M and OCh C, respectively. Therefore, the M2C2M channel can be multiplexed/groomed in OCh M along with other subrate data channels of equal or smaller capacity starting at the same node and destined to another metro network, while in the core network the M2C2M channel can be multiplexed/groomed with other data channels that start and end in the same metro networks pair. This process directly influences the fill ratio of the core network light paths and thus contributing to the reduction of CAPEX as a result of lesser number of expensive line interfaces and also the network occupied bandwidth. It is noteworthy that the bandwidth usage in the core network is much more a valuable asset when compared to its metro counterpart due to longer fiber links, higher number of amplifiers, and advanced, larger ROADMs.

In this approach, the M2C2M channel crosses one line interface pair in the first metro network where they originate and client interface pair in the first M2C site, line interface pair in the following core network, client interface pair in the second M2C site, and finally a line interface pair in the terminating metro network, thus spending in total 6 and 4 line and client interfaces, respectively. But still these are being shared with other smaller data rate M2C2M channels. But this is not the entire scenario because soon we start encountering higher data rate M2C2M channels (e.g., 10G or even higher) which cannot be multiplexed/groomed into the already existing OCh C due to nonterminating OChs in the same metro networks or capacity non-availability in the existing channels. Such use-cases foresee savings by the transparent handling of M2C2M channel between metro and core networks, which is possible if an additional fiber link exists in the M2C site in between the (R)OADMs and this higher data rate M2C2M channel is mapped on it. This transparent M2C interconnection can only be deployed by augmenting nodal degree of each (R)OADMs and additional booster/pre-amplifier deployment for each direction while employing one extra port of splitter/combiner and wavelength selective switches (WSSs) [18, 19]. Adding an optical switch between add/drop ports subset and using them directly for the optical bypass will be a good alternative [20].

In order to gain insight on expected design, implementation, and operational differences between a metro core network with traditional all-opaque M2C handover versus the same network with enhanced transparent handover, a subset of these differences is highlighted in **Table 2**, to highlight the different aspects to be assessed with care before opting for either of

	All-opaque M2C handover	Enhanced transparent M2C handover
Interface requirements	• Separate (local) optimization of line IFs required in the metro and core networks • Intensive usage of client IFs at M2C sites	• Joint (global) optimization of line IFs used at metro and core networks • Savings in client IF count at M2C sites
(R)OADMs at M2C sites	• All traffic handover done via the add/drop ports of both (R)OADMs	• Both (R)OADMs have to be augmented with one additional degree
Spectral efficiency	• Independent channel format selection at the metro and core networks	• Format used in OCh M2C2M impacts spectral efficiency of both networks
	• Best suited (cost-wise, performance-wise) format can be selected in each network: e.g., direct-detect 10 Gb/s in metro, more spectral efficient but expensive 100 Gb/s in core	• Additional spectral inefficiencies associated to coexistence of direct detect and coherent formats in the core network: extra spacing between neighboring channels [20]
Planning complexity	• Lower: due to sequential planning of metro networks first and core network afterward	• Higher: requires integrated planning of metro and core networks

http://dx.doi.org/10.5772/intechopen.68953

Resource provisioning	• Separate provisioning of resources at each network segment	• Requires end-to-end provisioning of the OCh M2C2M
	• Easy support for deploying different vendors hardware (e.g., line interfaces, common equipment) and software (e.g., network management systems) in the metro and core networks	• Complex support for deploying different vendors equipment: interworking between optical and network management systems, limited interoperability between control plane instances, use of alien wavelength concept, shortening transparent reach

Table 2. Qualitative comparison of all-opaque metro-to-core handover with enhanced transparent handover.

these strategies for a M2C traffic handover. Further, the analysis targets to capture technology landscape influenced differences in today's metro and core networks [21].

7. Conclusions

The overriding purpose of this chapter is to depict the relative importance of the universal OTN switch and flexible-rate line interfaces in the context of different network scenarios and traffic conditions. To accomplish the same, results from different network studies were presented. It is evident that, when the traffic pattern involves multiple subrates coming from varied data sources (TDM, Packet), universal OTN switch yields the maximum savings from CAPEX point of view. This is contributed by the reduced number of light paths and router port savings. Needless to mention here that, the more is the mismatch between the client traffic rates and the line-side rate(s), more savings could be achieved by exploiting a multilayer optimization approach. Furthermore, exploiting the higher capacity light paths enabled by flexible-rate line interfaces and combining the same with universal OTN switching, the better of both worlds can be achieved. But it is important to mention that flexible-rate line interfaces supporting 16-QAM, for example, will be less important when the network links are very long due to the limited transparent reach with this modulation format. On the other hand, the role of these state-of-the-art node architectures in the context of protection and restoration was also visited and different alternatives were presented. It could be inferred that OCh restoration would be preferable from a CAPEX point of view, with lesser resources being required, while opting for a better QoS (e.g., in terms of recovery time), ODU restoration can be a better choice. At the end, this chapter overviews the present day metro-to-core network scenario which enforces all-opaque traffic engineering and also prospects the economic and technological feasibility of transparent handovers. Recent technological developments, for example, metro core spectral efficiency gap narrowing, alien wavelength/black link standardization aided by state-of-the-art multivendor and multilayer resource provisioning platforms will overcome the hurdles of transparent handovers, and thus exploiting the big saving potentials in terms of interfaces at these boundary nodes.

Author details

Bodhisattwa Gangopadhyay[1*], João Pedro[1,2] and Stefan Spälter[3]

*Address all correspondence to: bodhisattwa.gangopadhyay@coriant.com

1 Coriant Portugal, Amadora, Portugal

2 Instituto de Telecomunicações, Lisboa, Portugal

3 Coriant GmbH, Munich, Germany

References

[1] Coriant. The Role of OTN Switching in 100G & Beyond Transport Networks [Internet]. 2016. Available from: http://www.lightwaveonline.com/content/lw/en/whitepapers/2016/03/the-role-of-otn-switching-in-100g-beyond-transport-networks.whitepaperpdf.render.pdf

[2] ITU. G.709: Interfaces for the Optical Transport Network [Internet]. Available from: http://www.itu.int/rec/T-REC-G.709/en, 2009

[3] Cinkler T, Coudert D, Flammini M, Monaco G, Moscardelli L, Munoz X, Sau I, Shalom M, Zaks S. Traffic grooming: Combinatorial results and practical resolutions. In: Koster A, Muñoz X, editors. Graphs and Algorithms in Communication Networks: Studies in Broadband, Optical, Wireless and Ad Hoc Networks. 28th ed. Springer, Heidelberg; 2010. p. 426, p.97. DOI: 10.1007/978-3-642-02250-0_2

[4] Eira A, Pedro J, Pires J. Optimized design of fixed/flex-rate line-cards and transceivers over multiple planning cycles. In: Optical Society of America, editor. Optical Fiber Communication Conference, 2014. San Francisco: Optical Society of America; 2014. p. W2A-47. DOI: 10.1364/OFC.2014.W2A.47

[5] Gerstel O. Opportunities for optical protection and restoration. In: IEEE, editor. Optical Fiber Communication Conference and Exhibit. IEEE; 22–27 February 1998, San Jose, California, USA. DOI: 10.1109/OFC.1998.657388

[6] Pedro J, Gangopadhyay B. On the trade-offs between ODU and OCh preplanned shared restoration in transport networks. In: IEEE, editor. Telecommunications Network Strategy and Planning Symposium (Networks), 2014 16th International. Madeira: IEEE; 2014. pp. 1–6. DOI: 10.1109/NETWKS.2014.6959252

[7] Schmitt A. Infonetics Research. Integrated OTN Switching Virtualizes Optical Networks [Internet]. 2012. Available from: https://www.infinera.com/wp-content/uploads/2015/07/Infinera-WP-Infonetics-Integrated-OTN-Switching.pdf

[8] Schmitt A. Infonetics Research Excerpt. Global Service Provider Survey Excerpts: OTN, MPLS, and Control Plane Strategies [Internet]. 2013. Available from: https://www.

infinera.com/wp-content/uploads/2015/07/Infonetics-Global_Service_Provider_Survey_Excerpts-OTN_MPLS_Control_Plane_Strategies.pdf

[9] Coriant. Universal Switching and Transport [Internet]. Available from: http://www.coriant.com/solutions/technologies/universal-switching-and-transport.asp, 2015

[10] Pedro J, Santos J, Morais R. Dynamic setup of multi-granular services over next-generation OTN/DWDM networks: Blocking versus add/drop port usage. In: IEEE, editor. Proceedings of IEEE ICTON 2012. IEEE; 2012, Coventry, United Kingdom. DOI: 10.1109/ICTON.2012.6253705

[11] Coriant. Flexing Next-Generation Optical Muscles [Internet]. Available from: http://www.coriant.com/spotlights/flexi-rate-white-paper.asp, 2016

[12] Gangopadhyay B, Pedro J, Spälter S. Cost-effective next-generation information highways leveraging universal OTN switching and flexible-rate. In: Optical Society of America, editor. Optical Fiber Communication Conference; 19–23 March, 2017. Los Angeles: Optical Society of America; 2017

[13] Costa N, Pedro J, Quagliotti M, Serra L. Preplanning framework to evaluate the potential of different modulation formats in DWDM networks. In: 10th Conference on Telecommunications (ConfTele).; TECHDAYS, Aveiro. 2015

[14] Gangopadhyay B, Pedro J, Spälter S. Comparative assessment of network architectures for transporting Packet and TDM traffic. In: IEEE, editor. 21st European Conference on Networks and Optical Communications (NOC), 1–3 June, 2016. Lisbon: IEEE; 2016. DOI: 10.1109/NOC.2016.7507000

[15] Telecom Italia. Sparkle [Internet]. Available from: http://www.tisparkle.com/

[16] Cantono M, Gaudino R, Curri V. A statistical analysis of transparent optical networks comparing merit of fiber types and elastic transceivers. In: IEEE, editor. 18th International Conference on Transparent Optical Networks (ICTON), 10–14 July, 2016, Trento, Italy. IEEE; 2016. DOI: 10.1109/ICTON.2016.7550511

[17] Eira A, Pedro J, Pires J. On the impact of optimized guard-band assignment for superchannels in flexible-grid optical networks. In: Optical Society of America, editor. Optical Fiber Communication Conference/National Fiber Optic Engineers Conference 2013. Optical Society of America; 2013, Anaheim, California, USA. DOI: 10.1364/OFC.2013.OTu2A.5

[18] Rambach F, Konrad B, Dembeck L, Gebhard U, Gunkel M, Quagliotti M, Serra L, López V. A multilayer cost model for metro/core networks. IEEE/OSA Journal of Optical Communications and Networking. 2013;**5**(2013):210–225. DOI: 10.1364/JOCN.5.000210

[19] Pedro J, Pato S. Impact of add/drop port utilization flexibility in DWDM networks [Invited]. IEEE/OSA Journal of Optical Communications and Networking. 2012;**4**(11):B142–B150. DOI: 10.1364/JOCN.4.00B142

[20] Woodward S, Feuer M, Kim I, Palacharla P, Wang X, Bihon D. Service velocity: Rapid provisioning strategies in optical ROADM networks. IEEE/OSA Journal of Optical Communications and Networking. 2012;**4**:92–98. DOI: 10.1364/JOCN.4.000092

[21] Pedro J, Gangopadhyay B. Prospects for transparent handover between the metro and core segments of next-generation transport networks. In: IEEE, editor. 17th International Conference on Transparent Optical Networks (ICTON), 2015; 5–9 July, 2015; Budapest, Hungary: IEEE; 2015. pp. 1–5. DOI: 10.1109/ICTON.2015.7193475

Chapter 12

Novel MT/MPO Single-Mode Multifiber Connector Technologies for Optical Fiber Communications

Mitsuru Kihara

Additional information is available at the end of the chapter

http://dx.doi.org/10.5772/intechopen.68863

Abstract

This chapter reports the latest mechanically transferable (MT) and multifiber push-on (MPO) multifiber connector technologies. Low insertion-loss and high return-loss angled physical contact (APC)-MPO single-mode multifiber connectors are developed. For these multifiber connectors to achieve a high performance, it is necessary to reduce deviations of all components and to control fiber offset due to shear force caused by compressing two obliquely ended MT ferrules. MT single-mode 84-fiber connectors are developed to realize higher density multifiber connectors. We have developed a novel high-density optical cable joint that has a 24-connector array consisting of new 84-fiber MT connectors. The new cable joining technique significantly shortens the time needed to join the optical fiber cables. Novel optical fiber switches are developed on the basis of MT multifiber connector technology. The switches use a manual guide-pin slide method in special guide holes in a novel multifiber array. Because high precision multifiber connectors need to be measured and inspected, a new inspection technique for MT ferrules is developed and equipment for it is fabricated. The fabricated inspection equipment can measure MT ferrules with high speed, cost effectiveness, and high accuracy. These single-mode multifiber connectors can be used as key technologies for advanced optical fiber communication systems.

Keywords: optical connector, insertion loss, return loss, physical contact, fiber-to-the-home, data center

1. Introduction

The number of subscribers to broadband services in Japan now exceeds 38 million, and about 29 million subscribers were using fiber-to-the-home (FTTH) services as of September 2016 [1]. The FTTH networks use many various single-mode optical fiber connection technologies.

One such connection is a field mountable optical fiber connector (FMC) or field installable connector (FIC), and another type is a manufactured optical fiber connector. The mechanically transferable (MT) or multifiber push-on (MPO) connectors, which belong to the manufactured connectors, are particularly expected to be used in upcoming FTTH networks because these multifiber connectors are superior to single-fiber connectors with high-density optical fiber cable joints [2–6].

Optical fiber communication systems have been used in data centers in the USA. Multimode single-fiber connectors such as a little connector (LC) were mainly used previously, and multimode MPO connectors are also used in current data centers. In addition, high precision MPO connectors with single-mode fibers are expected to be used for communicating more information over long distances in huge data centers.

This chapter reports the latest MT/MPO multifiber connector technologies. Current MT/MPO connectors used in Japan and the USA are explained in Section 2, and our developed innovative single-mode multifiber connectors are reported in Section 3. The low insertion-loss and high return-loss angled physical contact (APC)-MPO single-mode multifiber connectors are described in Section 3.1. Next, MT single-mode 84-fiber connectors for realizing higher density multifiber connectors are explained in Section 3.2. Novel optical fiber switches based on MT multifiber connector technology are reported in Section 3.3. These connectors need to be measured and inspected for high precision multifiber connectors. In Section 4, a new inspection technique for MT ferrules and equipment using the technique are introduced. These single-mode multifiber connectors can be used as key technologies for advanced optical fiber communication systems.

2. Overview of MT/MPO multifiber connectors

Figure 1 shows the configuration of a typical FTTH network in Japan. It is mainly composed of an optical line terminal (OLT) in the central office, underground and aerial optical fiber cables, and an optical network unit (ONU) inside a customer's home and building. The network requires various fiber connections at office, outdoor, and home sites. With the fiber connections at aerial and home sites in particular, field installable connectors or field mountable connectors and mechanical splices are used to fit the best wiring depending on aerial conditions and room arrangement. Field assembly small (FAS) connectors and field assembly (FA) termination connectors are field installable connectors [7–9]. In contrast, manufactured connectors, such as miniature-unit (MU) coupling optical fiber and single-fiber coupling (SC) optical fiber connectors, are used in central offices and homes. MT connectors are also used in central offices for multifiber ribbon joints. MPO connectors are used in customers' buildings. These MT and MPO connectors are particularly expected to be used in the upcoming FTTH networks with high-density optical fiber cable joints.

Optical fiber communication systems have been used in data centers in the USA. Previously, multimode single-fiber connectors such as an LC connector were mainly used, and multimode MPO connectors are also used in current data centers. This is because multifiber

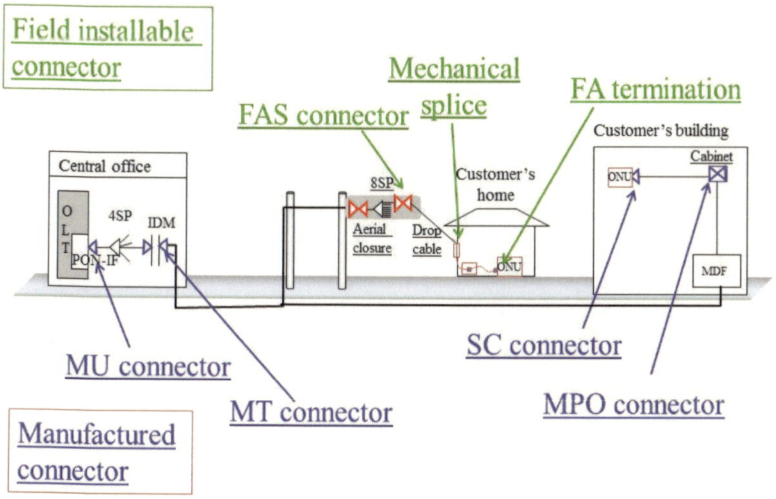

Figure 1. Typical FTTH network and various optical fiber connections.

connectors are superior to other single-fiber connectors with high-density optical fiber cable joints. **Figure 2** shows examples of optical fiber wiring (a) with LC single-fiber connectors and (b) with MPO multifiber connectors. Optical fiber wiring with the single-fiber connectors is slightly complicated, whereas that with the multifiber connectors is very simple and well ordered. The MPO multifiber connectors are consequently expected to be used in data center networks with high-density optical fiber cable joints. In addition, high precision MPO connectors with single-mode fibers are expected to be used for communicating more information over long distances in huge data centers. **Figure 3** shows the structure of the MT multifiber connector [2, 3], which consists of two plastic ferrules with two guide holes, two guide pins, and a clamp spring. The multifibers are positioned in a row between two guide holes. The ferrules are aligned by the two guide pins and two guide holes and then held with the clamp spring to achieve a low connection loss for multifiber ribbon connections. Refractive index matching material is used between the ferrule endfaces to reduce the

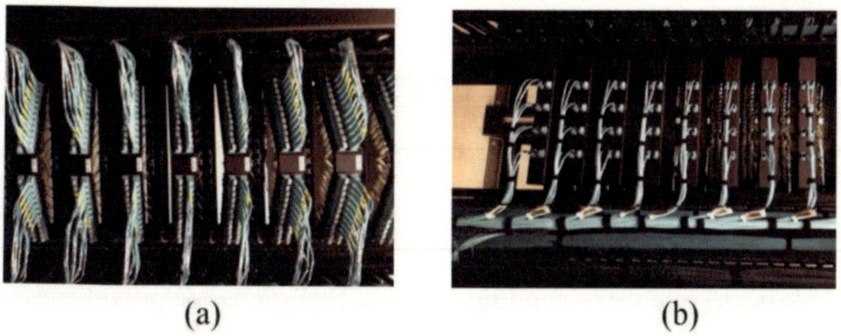

Figure 2. Example of optical fiber wiring (a) with single-fiber connectors and (b) with multifiber connectors.

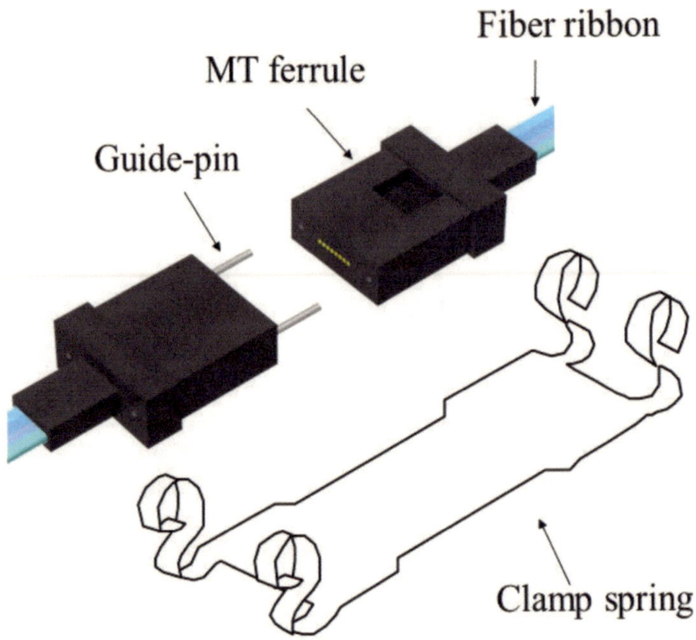

Figure 3. MT multifiber connector.

Fresnel reflection caused by an air-gap [10, 11]. The device is disconnected simply by removing the clamp spring and guide pins. The 4-, 8-, and 16-fiber MT connectors are currently used in the FTTH services in Japan.

Figure 4(a) shows the endface of an MT ferrule. In the MT ferrule endface, several fiber holes are positioned between two guide holes. The origin O is designated as the middle point of two guide hole centers. Each fiber hole is designed to be arranged with the designated fiber pitch on the basis of the origin. **Figure 4(b)** shows the fiber hole eccentricity of the MT ferrule. The actual fiber hole positions on the fabricated MT ferrule endface are different from the designated ideal positions because the molds are not perfectly accurate and/or plastic might deform during the MT ferrule fabrication. This large fiber hole eccentricity might result in large insertion loss of an MT connector, so it must be minimized for low insertion-loss MT connectors. Therefore, a technique has been developed to inspect for fiber eccentricities in MT ferrules, which is described in Section 4.

One application of the MT connector is as a multifiber push-on (MPO) connector [4–6, 12]. **Figure 5** shows the structure of the MPO connector, which is composed of two plugs and an adaptor. The plug contains an MT ferrule that has an endface that is obliquely polished with a slight fiber protrusion to enable physical contact. It is important for physical contact type connectors to eliminate an air gap between fiber ends [13–16]. Two guide pins are fitted into two guide holes in one ferrule to align the ferrules. The plug and adaptor are engaged by fitting a pair of elastic hooks into corresponding grooves. This connector provides easy push-pull reconnections without the need for refractive index matching material.

Figure 6 shows the fabrication process of an MPO plug. First, the mold is fabricated by using high precision mechanical engineering. A high precision mold is used, and an MT ferrule is

http://dx.doi.org/10.5772/intechopen.68863

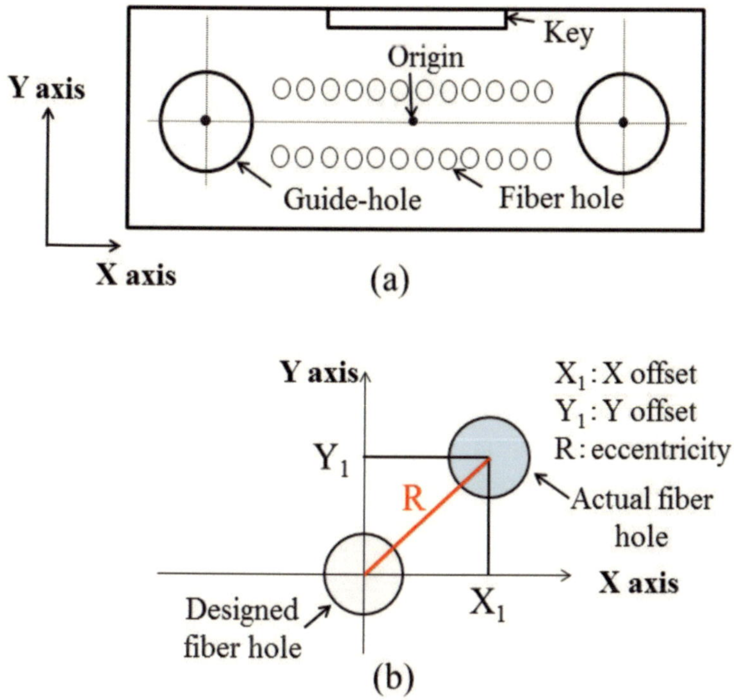

Figure 4. (a) Endface and (b) fiber hole eccentricity of MT ferrule.

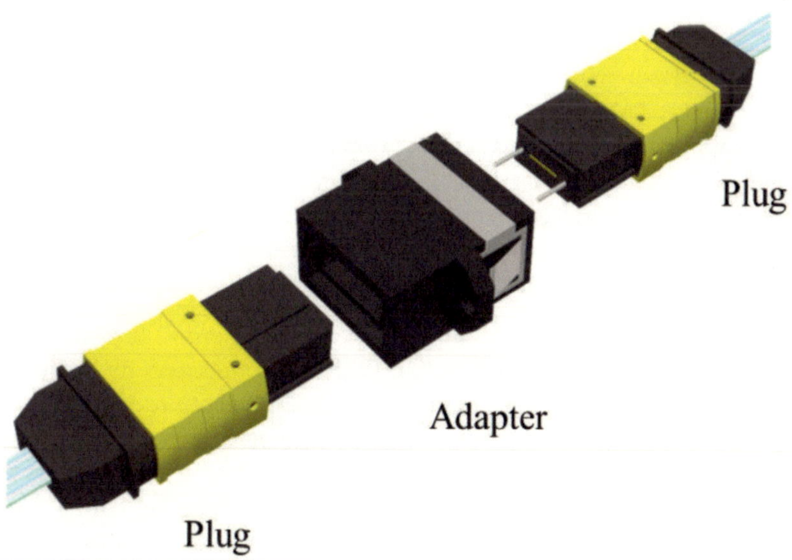

Figure 5. MPO multifiber connector.

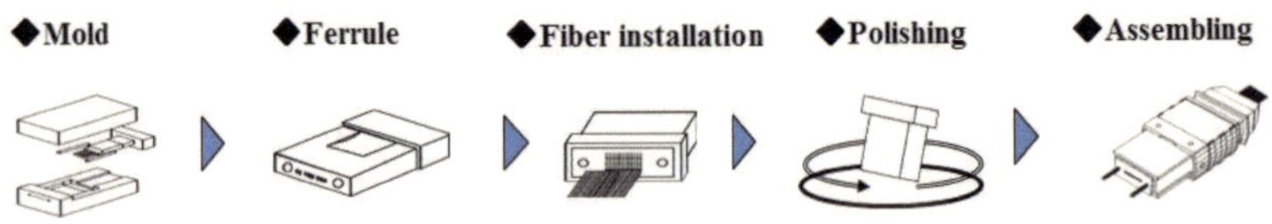

Figure 6. Fabrication process of MPO plug.

fabricated from plastic resin. This MT ferrule is a key device for MT/MPO connectors. Next, the multifiber ribbon is installed and fixed with adhesive in the MT ferrule. The MT ferrule containing the multifiber ribbon is then obliquely polished. Finally, an MPO plug houses the obliquely ended MT ferrule with multifibers. This MPO fabrication process contains two steps: ferrule making and connector assembling. Currently, fewer than 10 companies make ferrules, but more than 100 companies assemble connectors. More companies will make ferrules and assemble connectors in the future.

3. Innovative multifiber connectors based on MT/MPO connectors

This section reports our innovative multifiber connectors based on MT/MPO connector technologies. **Figure 7** shows three target single-mode multifiber connectors. The first is an extremely lower insertion-loss multifiber connector, which is detailed in Section 3.1. The second is a higher density multifiber connector that enables two high-density optical fiber cables to be connected as soon as possible, which is explained in Section 3.2. The third is a functional multifiber connector that enables multifibers to be switched, which is described in Section 3.3.

3.1. Lower insertion-loss multifiber connectors

We have investigated extremely great performances of multifiber connectors and developed angled physical contact (APC) MPO connectors that have an insertion loss lower than 0.2 dB and return loss higher than 60 dB. This section explains these APC-MPO connectors. To create lower insertion-loss connectors, fiber core offsets from the ideal positions must be decreased. APC-MPO connectors have two types of fiber core offset. One is due to deviation of components and the other is due to shear force caused by compressing two obliquely ended MT ferrules. These two fiber core offsets are detailed below.

Optical loss in a fiber connection has four causes: fiber-core offset, fiber-axis tilt, fiber-end separation, and mode field mismatch [17, 18]. With MT connectors, fiber-core offset is the dominant factor generating optical loss, and so we have limited our considerations to this factor. **Figure 8** shows four deviations: d_1, d_2, d_3, and d_4. d_1 is the fiber hole position deviation. d_2 is the deviation due to the clearance between a fiber and a fiber hole. d_3 is the deviation due to the clearance between a guide pin and a guide hole. d_4 is the fiber-core eccentricity [19]. The four deviations dependently affect total fiber core deviation. To realize an insertion loss of lower than 0.2 dB, the total deviation must be less than 0.5 μm.

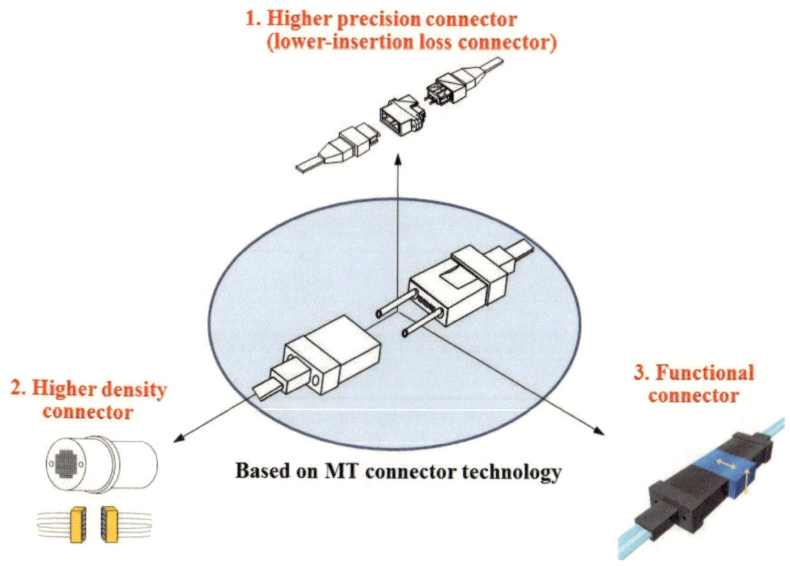

Figure 7. Three targets of single-mode multifiber connector technologies.

Figure 9(a) and **(b)** shows side views of connected MT ferrules and an MT ferrule endface in an MPO plug. The MPO connector uses MT ferrules obliquely polished at 8 degrees and can provide high return-loss characteristics. In contrast, the shear force occurs in the connected obliquely ended MT ferrules when two MT ferrules are compressed. This shear force leads to the Y offset of connected fibers in the MT ferrule. By considering the Y offset, fiber positions are designed to be previously offset, as shown in **Figure 9(b)**. This offset quantity is dependent on clearance between a guide hole and a guide pin and deformed material of an MT ferrule. The fiber offset must be controlled for low insertion-loss MPO connectors.

On the basis of the above design, we fabricated three pairs of high precision single-mode 12-fiber APC-MPO connectors by using three different fabrication molds. The fabricated APC-MPO connectors used high precision ferrules with d_1 of less than 0.5 μm and severely

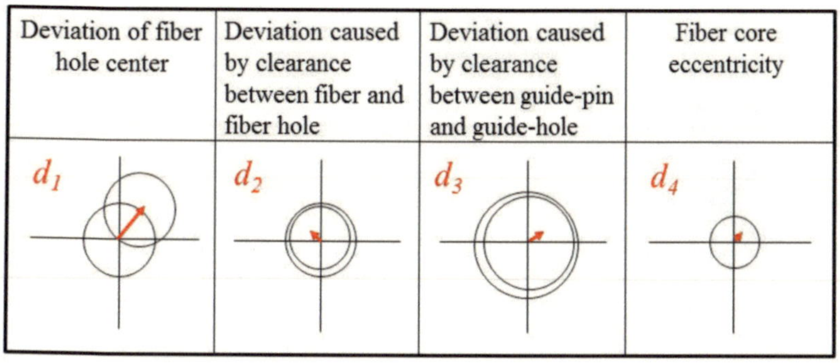

Figure 8. Four deviations resulting in fiber core offset.

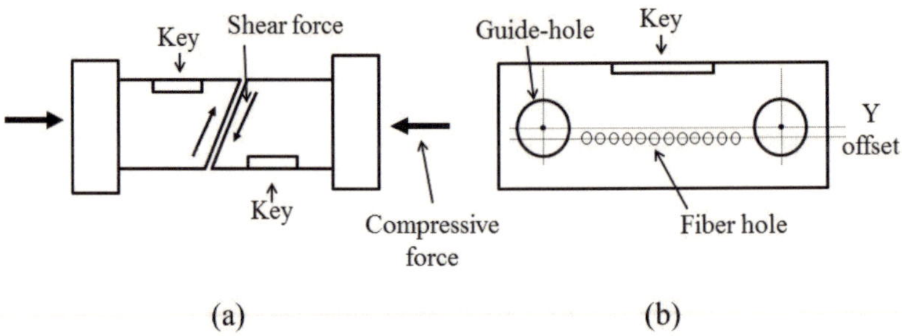

Figure 9. (a) Side view of connected MT ferrules and (b) MT ferrule endface in MPO plug.

controlled guide pins and optical fibers with lower d_2, d_3, and d_4. The fabricated APC-MPO connectors also used the controlled Y offset quantity. We measured the insertion losses and return losses using 1.31 μm LD, a power meter, and a backreflection meter. **Figure 10** shows histograms of the insertion losses and return losses of the connectors. The average insertion loss was 0.08 dB with a maximum of 0.17 dB. The return losses were greater than 64 dB. Consequently, we have revealed that the fabricated APC-MPO has an excellent performance.

3.2. Higher density multifiber connectors

We have developed a higher density multifiber connector to enable two high-density optical fiber cables to be connected as soon as possible [20–23]. **Figure 11(a)** and **(b)** shows the novel high-density preconnectorized cable joint and endface of high-count fiber MT connectors, respectively. The cable joint consists of an arrayed high-count fiber MT connector and a cable joint housing and can connect high-density optical fiber cables such as 2000-fiber cables at one time. When it is connected, each MT connector is aligned with two guide pins and pushed toward the opposite ferrule by a spring. The MT connectors are aligned and connected individually. This structure is the same as those of a conventional MT connector and MPO connector. The connector endface is polished at a right angle. The connectors need an index matching material to eliminate the Fresnel reflection caused by an air gap. We used a conventionally sized 2.5-mm-thick MT ferrule that has guide holes of 4.6 mm in pitch and 0.7 mm in diameter.

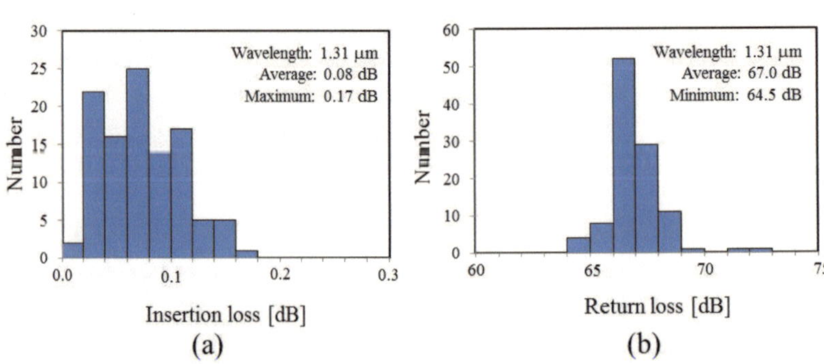

Figure 10. (a) Insertion loss and (b) return loss of fabricated high precision APC-MPO connectors.

This is because this type of ferrule is utilized widely in optical cable components such as MPO connectors. The optical fiber ribbons are generally installed into fiber holes of a ferrule. We used a conventional 12-fiber ribbon with a 0.3 thickness and 0.25 fiber pitch to assemble our fabricated connector. By considering the thickness of the fiber ribbon and that of the MT connector ferrule, we figured that seven is the maximum number of fiber ribbons installable in an MT ferrule. Therefore, we decided that 84 (7 rows of 12) is the target number of fibers installed in our high-count fiber MT connector as shown in **Figure 11(b)**.

On the basis of the above design, we fabricated an 84-fiber MT connector and high-density preconnectorized cable joint for 2000 fiber cables. **Figure 12(a)** and **(b)** shows endface photographs of the fabricated high-count fiber MT connector and high-density preconnectorized cable joint, respectively. The high-count fiber MT connector ferrule consists seven rows of 12 fiber ends between the two guide holes and the high-density preconnectorized cable joint has a 4 × 6 array of a high-count 84-fiber MT connector. Consequently, the fabricated high-density preconnectorized cable joint can potentially connect 2016 fiber cables at one time.

The fabricated high-count 84-fiber MT connector was measured at a wavelength of 1.31 μm. We connected six pairs of high-count fiber MT connectors using the same cramp spring as the conventional MT connectors. The average and maximum insertion losses were 0.40 and 1.8 dB, which are slightly larger than the average (0.3 dB) and maximum (1.0 dB) of the conventional 8-fiber MT connectors. This is because the fabricated fiber hole positions of 84-fiber MT ferrules are thought to have a large offset from the designed positions. However, we think that these offsets can be decreased with the same precision as with the conventional MT connector. The assembly of the high-density preconnectorized cable joint must not generate any excess loss. We demonstrate the insertion-loss changes of the 84-fiber MT connectors in the center and the corner of the installed cable joint. **Figure 13(a)** and **(b)** shows the measured insertion-loss change of the high-count 84-fiber MT connectors located at the center and corner of the cable joint, respectively. In **Figure 13**, there is no loss change of the high-count fiber MT connector before and after assembly. Consequently, there turned out to be no excess loss in the cable joint. Although it takes about 20 hours to connect two sets of

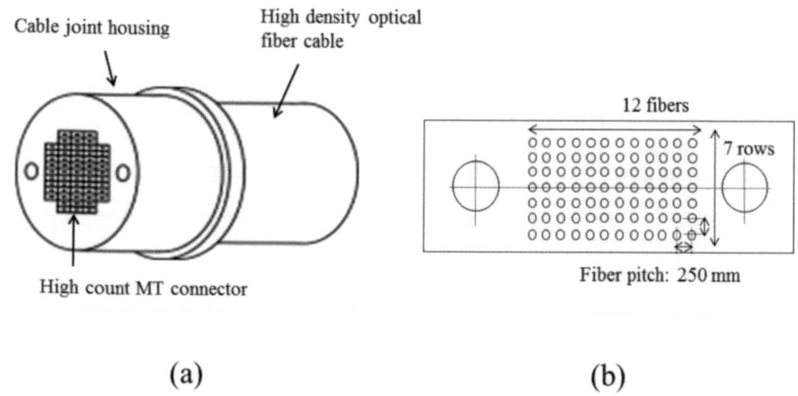

Figure 11. (a) High-density preconnectorized cable joint and (b) high-count fiber MT connector.

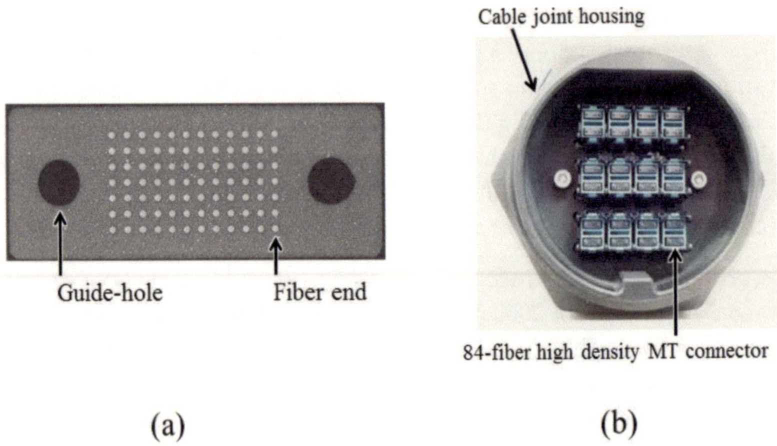

Figure 12. Endface pictures of (a) fabricated high-count fiber MT connector and (b) high-density preconnectorized cable joint.

2000 single-mode optical fiber cables using a fusion splice, it took about 5 minutes to connect two sets of 2000 single-mode optical fiber cables using the fabricated cable joint. The cross-sectional area of the cable joint was less than one-fifth that of a conventional optical closure. We have found that the fabricated cable joint can significantly reduce the time and space needed to join high-density optical fiber cables.

3.3. Functional multifiber connector for switching fibers

We have designed new kinds of functional multifiber connectors on the basis of MT connector technologies [24]. One is a device for switching fibers [25], which is described in this section. Figure 14 shows the basic structure and the switching mechanism of the proposed

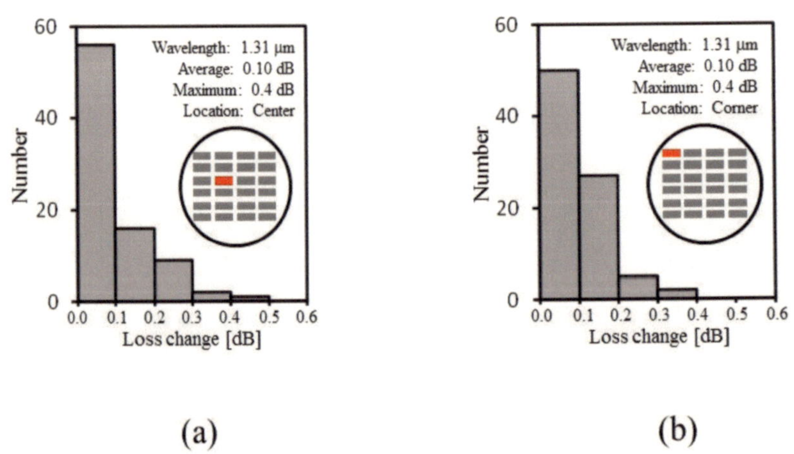

Figure 13. Measured insertion-loss change of the high-count 84-fiber MT connectors located at (a) center and (b) corner of cable joint.

switch. **Figure 14(a)–(c)**, respectively, shows the fiber connection states before, during, and after switching from one fiber to another. First, two guide pins are fixed at the bottom of the guide holes with the elastic materials in the multifiber array, as shown in **Figure 14(a)**. Next, when the conventional MT ferrule is pushed up with a certain force in a perpendicular direction, the two guide pins begin to slide in the same direction because the elastic materials change to the horizontal direction, as shown in **Figure 14(b)**. Finally, two guide pins are fixed at the top of the guide holes in the multifiber array by the restored elastic materials again after two guide pins have finished sliding, as shown in **Figure 14(c)**. Consequently, a certain optical fiber in the conventional MT ferrule can be switched from one optical fiber to another in the multifiber array.

We discuss the various applications of the proposed optical switch. **Figure 15** shows four types of switches. Types a and b are based on the conventional MT ferrule and multifiber array. In type a, two guide pins slide parallel to the multifiber arrangement between two guide holes. With this switch, a fiber from a row in the MT ferrule can be switched from one optical fiber to another in the same row in the multifiber array. In type b, two guide pins slide vertically in relation to the multifiber arrangement with two rows. In this switch, the optical fibers can be switched by the unit of the row. Type c can be inserted and removed between the two endfaces of conventional MT ferrules. With this switch, two guide pins can slide either parallel or vertical to the multifiber arrangement. This switch can be applied to conventional MT connectors and is useful both for installing MT connectors and for existing MT connectors in current optical fiber systems. Type d is an insertion and removal switch in which two guide pins slide vertically in relation to the multifiber arrangement. Therefore, the optical fibers are switched by the unit of the MT ferrule. Consequently, by using the proposed switching method where two guide pins slide in special guide holes, various types of switches with multifiber arrays can be realized for various uses.

We fabricated types a and c for an optical switch. The structure and performances of the type-a optical fiber switch are described elsewhere [25]. This section explains the type c optical fiber

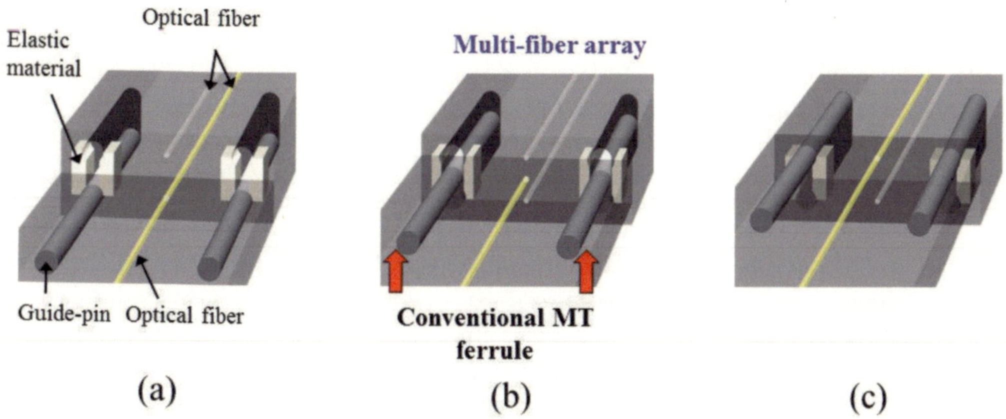

Figure 14. Basic switch structure and the switching mechanism (a) before, (b) during, and (c) after switching.

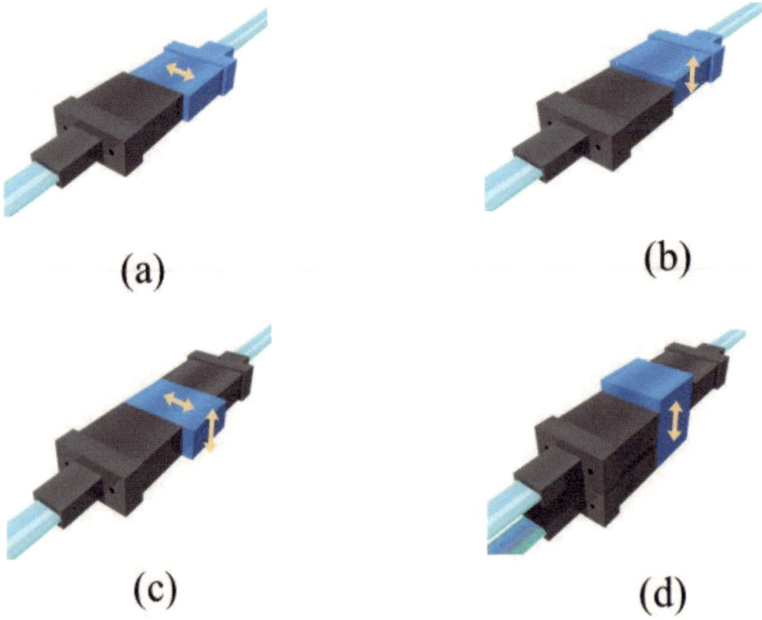

Figure 15. Various types of switches. (a) Fiber switch in a row, (b) fiber row switch, (c) insertable and removable switch, and (d) MT ferrule switch.

switch. **Figure 16(a)** and **(b)** shows the structure and a photograph of the fabricated multifiber-ribbon switch, which is composed of two MT ferrules and a multifiber array. The multifiber array has a novel guide hole structure in which two guide pins slide vertically to the four-fiber arrangement. The MT ferrules have two rows of four fibers. **Figure 17** shows the switching mechanism. In condition A before switching (a), the upper fiber ribbon in MT ferrule 1 is connected to optical fibers in the multifiber array and is also connected to the upper fiber ribbon in MT ferrule 2. When MT ferrule 1 is pushed up manually with a certain force in a perpendicular direction, the two guide pins slide in the opposite direction. In condition B after switching (b), where two guide pins in the multifiber array slide to the 0.25 mm length vertically to the arrangement direction, the lower fiber ribbon in MT ferrule 1 is connected to optical fibers in the multifiber array and is also connected to the upper fiber ribbon in MT ferrule 2. Therefore, the upper fiber ribbon in MT ferrule 1 is switched to the lower fiber ribbon in MT ferrule 1.

We measured the connection losses of the fabricated multifiber-ribbon switch at a wavelength of 1.3 μm using an LD and an optical power meter. The conventional two MT ferrules and multifiber array are connected with all four guide pins and a special clamp spring. Refractive index matching material is used between the ferrule and the multifiber array endfaces (both connections 1 and 2). **Figure 18** shows the measured connection losses. The average connection loss is 1.1 dB. These connection losses are higher than those of the conventional MT connectors because the fabricated fiber positions are thought to have large misalignments of the offset from the designed positions. However, we think that these offset misalignments can be minimized with the same precision as with the conventional MT connector. We also measured

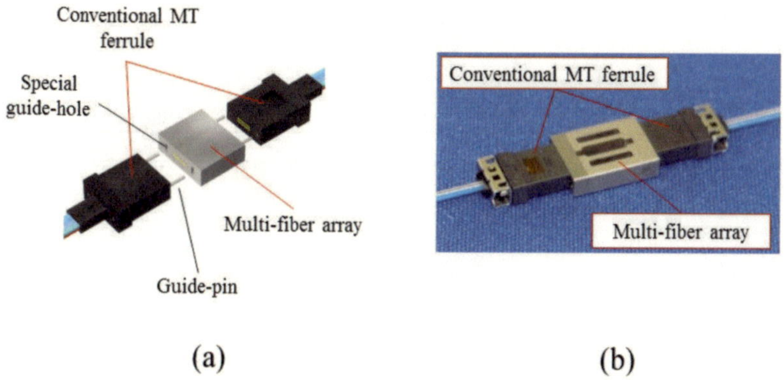

Figure 16. Fabricated fiber-ribbon switch. (a) Structure and (b) photograph of insertable and removable switch.

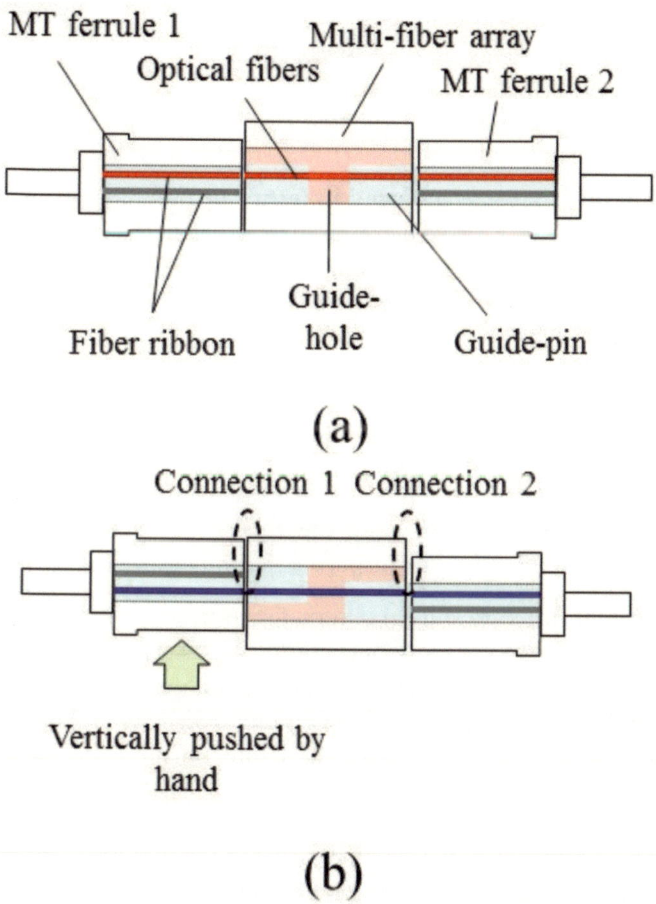

Figure 17. Switching mechanism of fiber-ribbon switch. (a) Before switching (condition A) and (b) after switching (condition B).

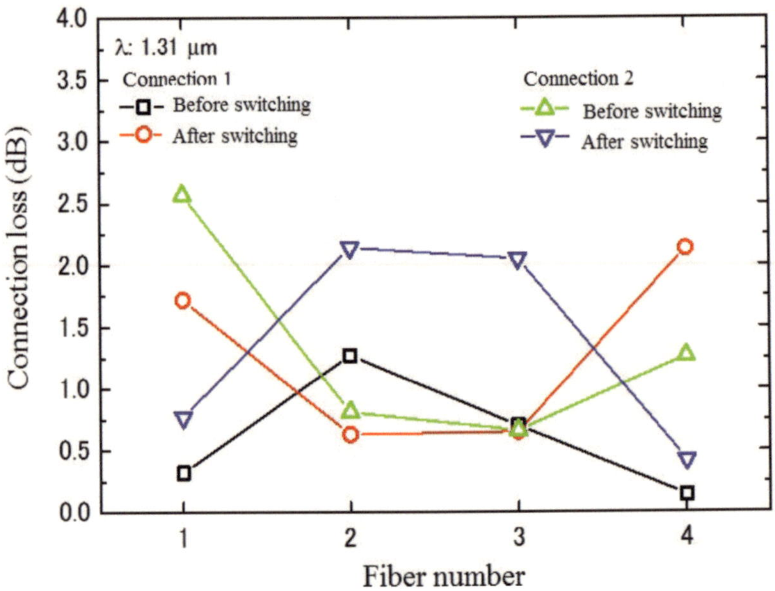

Figure 18. Connection losses of fabricated fiber-ribbon switch.

the force needed to push the conventional MT ferrule or multifiber array by hand in order to switch fibers. The measured forces were about 9.5 N. Consequently, we revealed that the fabricated switch can provide ease of handling.

4. Novel inspection technique for MT ferrule

To fabricate a high precision multifiber connector, MT ferrules need to be inspected. To shorten the time it takes to measure MT ferrules using conventional inspection equipment, we have created a novel inspection technique that has both fast measurement and low cost features. This section describes the new inspection technique.

Figure 19 shows the **(a)** structure and **(b)** observation area of conventional inspection equipment for MT ferrules [26]. The conventional equipment consists of a light source, high precision sample stage, an objective lens with 20 times, a charge-coupled device (CCD) camera with 0.4 M pixels, an image processor, and a computer. As shown in **Figure 19 (b)**, the observation area of conventional equipment is 0.2 × 0.17 mm, which corresponds to covering area of a fiber hole. To inspect whole fiber holes and two guide holes on an MT ferrule, the MT ferrule sample must be scanned on the high precision sample stage. Therefore, the conventional inspection equipment takes a long time to measure MT ferrules. We have thus designed and fabricated new inspection equipment to shorten the measurement time.

Figure 20 shows the **(a)** structure and **(b)** observation area of new inspection equipment for MT ferrules [27]. The novel equipment consists of a light source, an objective lens with 1 time, a CCD camera with 10 M pixels, an image processor, and computer. To reduce measurement

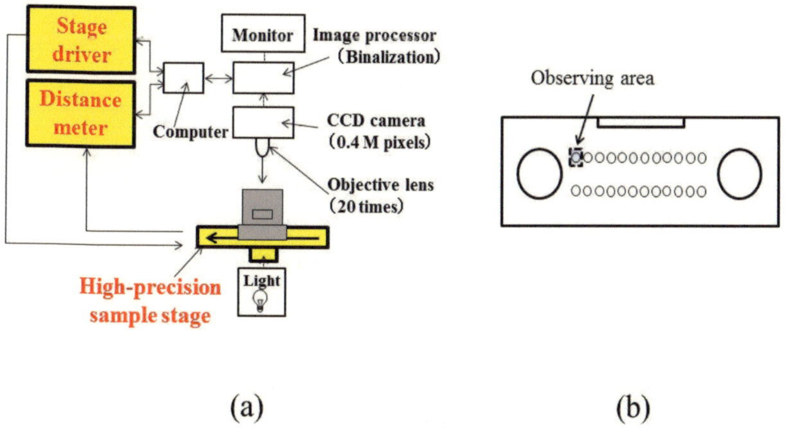

Figure 19. (a) Structure and (b) observation area of conventional inspection equipment for MT ferrules.

time, the scanning measurement must be eliminated. Therefore, the objective lens is changed from the conventional 20-time lens to a 1-time lens. Consequently, the observation area is expanded to be covered by the whole endface of an MT ferrule: 6.4 × 4.6 mm. In addition, the new inspection equipment is given a simpler structure and made more cost effective by eliminating the conventional high precision sample stage.

Our goal is to give the new inspection equipment the same measurement accuracy as the conventional equipment: less than 0.1 μm. The measurement accuracy of the new inspection equipment generally worsens because the objective lens is changed. To improve this worsened measurement accuracy, two methods were used. One is to change the pixels of the CCD camera and the other is to use multivalue processing with an image processor.

Figure 21 shows a photograph of the fabricated inspection equipment for MT ferrules. The fabricated inspection equipment is simpler and more cost-effective because it does not need a high precision sample stage, a stage driver, or a laser distance meter. We have experimentally demonstrated the multivalue processing using an image processor. **Figure 22** shows

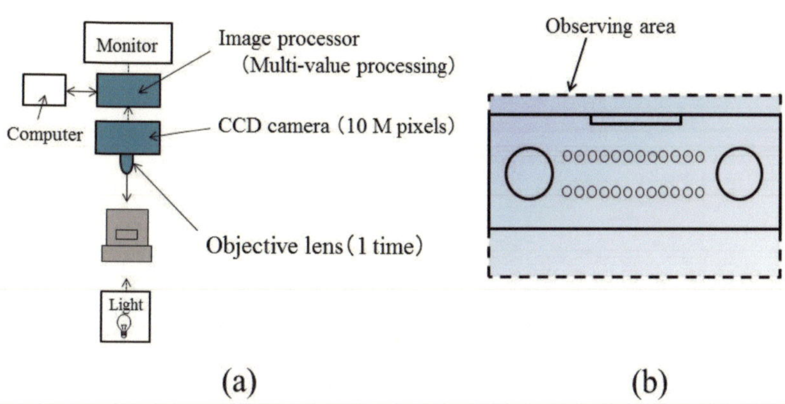

Figure 20. (a) Structure and (b) observation area of novel inspection equipment for MT ferrules.

experimental results for measurement time and accuracy for a 24-fiber MT ferrule. The measurement accuracy is the standard deviation of nine measurements. As the multivalue with the image processor becomes larger, the normalized standard deviation decreases whereas the normalized measurement time increases. These results indicate that there are optimal conditions for both high speed and high accuracy measurement. **Figure 23** shows measured fiber hole offsets for the same 24-fiber MT ferrule using the conventional and the fabricated inspection equipment. The offset results for the conventional and fabricated equipment are almost the same. The measured accuracy with the fabricated inspection equipment is less

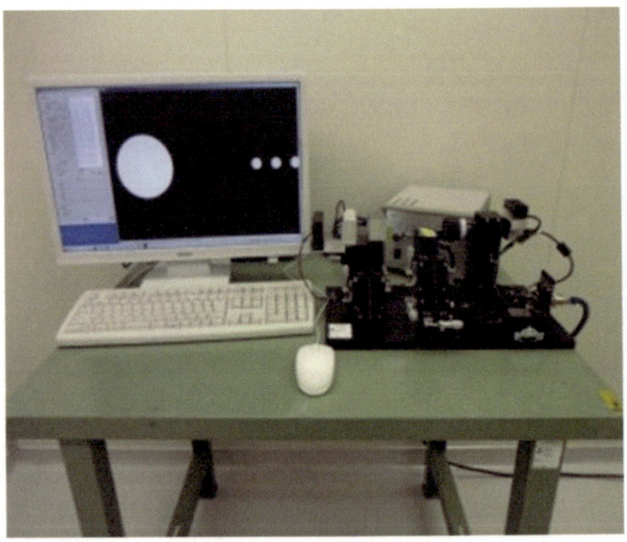

Figure 21. Photograph of the fabricated inspection equipment for MT ferrules.

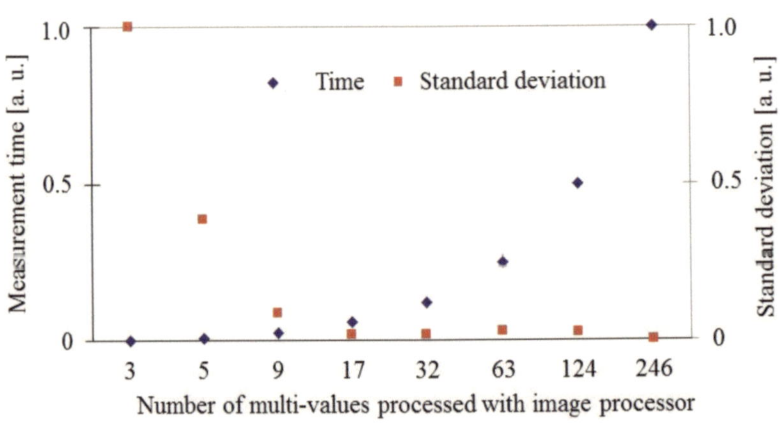

Figure 22. Experimental results of measurement time and accuracy to 24-fiber MT ferrule.

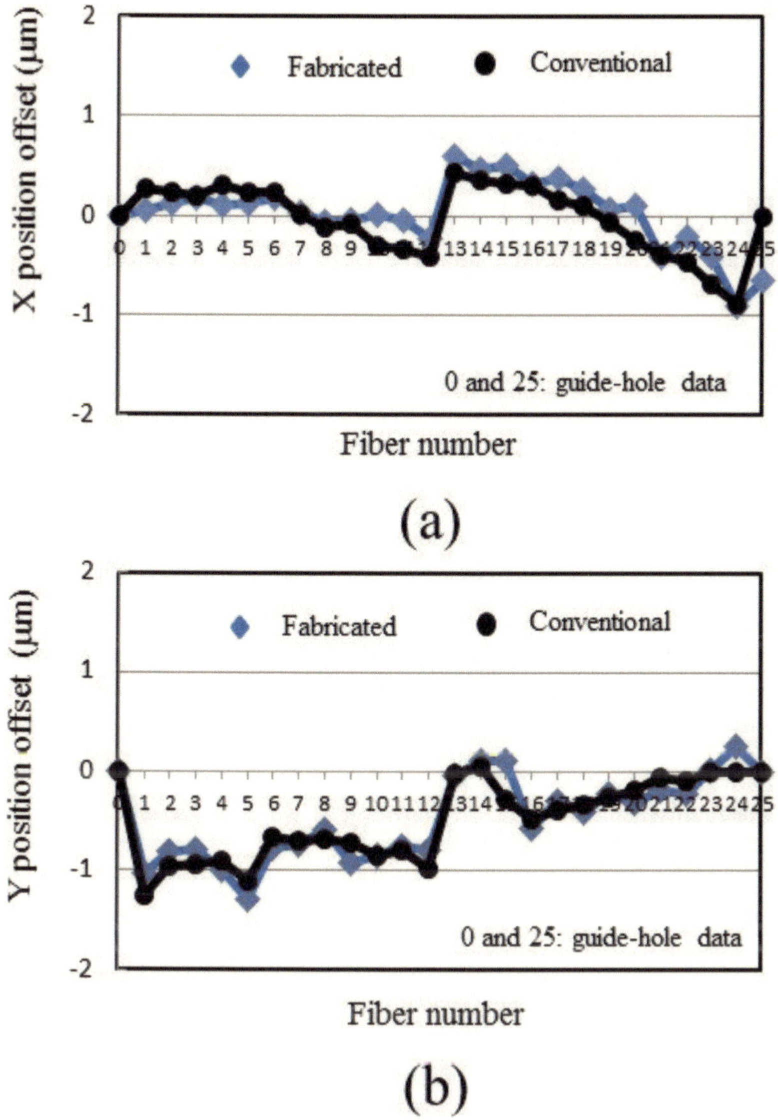

Figure 23. Measured fiber hole offsets of same 24-fiber MT ferrule using the conventional and fabricated inspection equipment. (a) X position offset and (b) Y position offset.

than 0.1 μm, which is as good as that of the conventional inspection equipment. It takes 3 minutes 40 seconds and 11 minutes 30 seconds to measure 24-fiber and 84-fiber MT ferrules using the conventional inspection equipment, respectively. In contrast, it takes 10 seconds each to measure 24-fiber and 84-fiber MT ferrules using the fabricated inspection equipment. Consequently, we reveal that the fabricated inspection technique can measure MT ferrules with high speed, cost effectiveness, and high accuracy.

5. Conclusion

This chapter reported the latest mechanically transferable (MT) and multifiber push-on (MPO) multifiber connector technologies.

Low insertion-loss and high return-loss angled physical contact (APC)-MPO single-mode multifiber connectors were developed. For these multifiber connectors to achieve a high performance, it is necessary to reduce deviations of all components and to control fiber offset due to shear force caused by compressing two obliquely ended MT ferrules.

Next, MT single-mode 84-fiber connectors were developed to realize higher density multifiber connectors. We have developed a novel high-density optical cable joint that has a 24-connector array consisting of new 84-fiber MT connectors. The new cable joining technique significantly shortens the time needed to join the optical fiber cables.

Novel optical fiber switches were developed on the basis of MT multifiber connector technology. The switches use a manual guide-pin slide method in special guide holes in a novel multifiber array. The fabricated fiber switches can be easily pushed up at a force of less than 10 N by hand.

Because high precision multifiber connectors need to be measured and inspected, a new inspection technique for MT ferrules was developed and equipment for it was fabricated. The fabricated inspection equipment can measure MT ferrules with high speed, cost effectiveness, and high accuracy.

These single-mode multifiber connectors can be used as key technologies for advanced optical fiber communication systems.

Acknowledgements

The author thanks Drs. K. Shiraki and T. Kurashima at NTT Access Service Systems Laboratories (ANSL) for their encouragement and helpful suggestions. The author is deeply grateful to R. Koyama, K. Saito, Y. Yamaguchi, and Y. Machida at ANSL for discussing the low insertion APC-MPO connectors and high-density MT connectors; T. Shimizu and C. Fukai at ANSL for discussing the novel optical switch based on MT connector technology; K. Noda and K. Nakashoji for helpful discussions regarding the inspection technique for MT ferrules; and Dr. T Satake at USConec Ltd. for providing the pictures of **Figure 2**.

Author details

Mitsuru Kihara

Address all correspondence to: kihara@osakac.ac.jp

NTT Access Network Service Systems Laboratories, Nippon Telegraph and Telephone Corporation, Hanabatake, Tsukuba, Ibaraki, Japan (He is currently with Osaka Electro-Communication University)

References

[1] Ministry of Internal Affairs and Communications. Subscriber Number and Share of Telecommunications [Internet]. 2016. Available from: http://www.soumu.go.jp/main_content/000454424.pdf [Accessed: February 12, 2017]

[2] Satake T, Nagasawa S, Arioka R. A new type of demountable plastic-molded single-mode multifiber connector. Journal of Lightwave Technology. 1986;**LT-4**(8):1232–1236. DOI: 10.1109/JLT.1986.1074860

[3] Satake T, Nagasawa S, Hughes M, Lutz S. MPO-type single-mode multi-fiber connector: Low-loss and high-return-loss intermateability of APC-MPO connectors. Optical Fiber Technology. 2011;**17**:17–30. DOI: 10.1016/j.yofte.2010.09.004

[4] Nagasawa S, Yokoyama Y, Ashiya F, Satake T. A high-performance single-mode multifiber connector using oblique and direct endface contact between multiple fibers in a plastic ferrule. IEEE Photonics Technology Letters. 1991;**3**:937–939. DOI: 10.1109/68.93269

[5] Nagasawa S, Ashiya F, Satake T. Multifiber optical connector plug with low reflection and low insertion loss. 1993. US Patent 5214730

[6] Kihara M, Nagasawa S, Tanifuji T. Design and performance of an angled physical contact type multifiber connector. Journal of Lightwave Technology. 1996;**14**(4):542–547. DOI: 10.1109/50.491391

[7] Nakajima T, Terakawa K, Toyonaga M, Kama M. Development of optical connector to achieve large-scale optical network construction. In: Proceedings of the International Wire and Cable Symposium (IWCS). 2006. pp. 439–443

[8] Kihara M, Koyama R, Abe Y, Kobayashi, Tomita S. Novel field installable optical fiber splice and connector for optical access networks. Optical Fiber Technology. 2013;**19**:269–278. DOI: 10.1016/j.yofte.2013.02.007

[9] Kihara M, Okada M, Hosoda M, Iwata T, Yajima Y, Toyonaga M. Tool for inspecting faults from incorrectly cleaved fiber ends and contaminated optical fiber connector end surfaces. Optical Fiber Technology. 2012;**18**:470–447. DOI: 10.1016/j.yofte.2012.07.004

[10] Kihara M, Nagasawa S, Tanifuji T. Temperature dependence of return loss for optical fiber connector. IEEE Photonics Technology Letters. 1995;**7**:795–797. DOI: 10.1109/68.393209

[11] Kihara M, Nagasawa S, Tanifuji T. Return loss characteristics of optical fiber connectors. Journal of Lightwave Technology. 1996;**14**:1986–1991. DOI: 10.1109/50.536966

[12] Takaya M, Kihara M, Nagasawa S. Design and performance of a multifiber backpanel type connector. IEEE Photonics Technology Letters. 1996;**8**:655–657. DOI: 10.1109/68.491570

[13] Kihara M, Tomita S, Haibara T. Influence of wavelength and temperature changes on optical performance of fiber connections with small gap. IEEE Photonics Technology Letters. 2006;**18**:2120–2122. DOI: 10.1109/LPT.2006.883256

[14] Kihara M, Uchino M, Omachi X, Watanabe H. Fault analysis: Optical performance investigation of fiber connections with imperfect physical contact. OSA Optics Letters. 2011;**36**:4791–4793. DOI: 10.1364/OL.36.004791

[15] Kihara M, Uchino M, Omachi M, Watanabe H. Investigation into optical performance of fiber connections with imperfect physical contact. Journal of Lightwave Technology. 2013;**31**:967–974. DOI: 10.1109/JLT.2013.2240256

[16] Kihara M, Watanabe H. Corrections to "Investigation into optical performance of fiber connections with imperfect physical contact". Journal of Lightwave Technology. 2016;**34**:3592–3595. DOI: 10.1109/JLT.2016.2579650

[17] Marcuse D. Loss analysis of single-mode fiber splices. Bell System Technical Journal. 1997;**56**(5):703–718. DOI: 10.1002/j.1538-7305.1977.tb00534.x

[18] Kihara M. Optical performance analysis of single-mode fiber connections. In: Yasin M, Harun SW, Arof H, editors. Optical Fiber Communications and Devices. InTech; 2011. pp. 239–256. ISBN: 978-953-307-954-7

[19] Shibata K, Takaya M, Nagasawa S. Design and performance of high-precision MT-type connector for 1.55-μm zero-dispersion-shifted fiber-ribbon cables. IEEE Photonics Technology Letters. 2001;**13**:136–138. DOI: 10.1109/68.910513

[20] Haibara T, Tomita S, Matsumoto M, Yabuta T. High-speed, low-loss connection techniques for high-count pre-connectorized cables. In: Proceedings of the International Wire and Cable Symposium (IWCS). 1991. pp. 296–302

[21] Tomita S, Matsumoto M, Haibara T, Nakashima T, Kihara M. Optical fiber cable and connector technology for FTTH network. The Institute of Electronics, Information and Communication Engineers. 1992;**E75-B**:862–870. ISSN: 0916–8516

[22] Takaya M, Shibata K. Design and performance of very high-density 60-fiber ferrules. Journal of Lightwave Technology. 2003;**21**:1549–1556. DOI: 10.1109/JLT.2003.812460

[23] Koyama R, Saito K, Fukai C, Kihara M, Kurashima T. A novel high-density pre-connectorized cable joining technique for single-mode optical fiber cables using new 84-fiber MT connectors. In: Proceedings of the International Wire & Cable, Connectivity Symposium (IWCS). 2016. pp. 572–575. Available from: http://www.iwcs.org/polopoly_fs/1.3285734.1474652248!/fileserver/file/687359/filename/099_0779-000072.pdf [Accessed: March 26, 2017].

[24] Kihara M, Nagasawa S, Takaya M, Tanifuji T. A new type of demountable optical device for coupling single-mode planar waveguides to a multifiber array using passive alignment. IEEE Photonics Technology Letters. 1995;**7**:1453–1455. DOI: 10.1109/68.477280

[25] Shimizu T, Kihara M, Izumita H, Kurashima T. Novel optical switch using manual guide-pin slide method based on MT multifiber connector technique. In: Proceedings of the OFC/NFOEC. 2013. pp. JTh2A.24. DOI: 10.1364/NFOEC.2013.JTh2A.24

[26] Measurement method for the hole location of a multiway connector plug. IEC 61300-3-27. JIS C5961 6.10

[27] Kihara M, Koyama R, Fukai C, Katayama K, Kurashima T, Novel equipment for inspecting fiber eccentricities in MT/MPO connector to enable high speed measurement. In: Technical Reports of the OFT (Japanese); 2017. OFT2016-176. Print edition: ISSN 0913-5685. Online edition: ISSN 2432–6380

Chapter 13

Power-Over-Fiber Applications for Telecommunications and for Electric Utilities

Joao Batista Rosolem

Additional information is available at the end of the chapter

http://dx.doi.org/10.5772/68088

Abstract

Beyond telecommunications, optical fibers can also transport optical energy to powering electric or electronic devices remotely. This technique is called power over fiber (PoF). Besides the advantages of optical fiber (immunity to electromagnetic interferences and electrical insulation), the employment of a PoF scheme can eliminate the energy supplied by metallic cable and batteries located at remote sites, improving the reliability and the security of the system. Smart grid is a green field where PoF can be applied. Experts see smart grid as the output to a new technological level seeks to incorporate extensively technologies for sensing, monitoring, information technology, and telecommunications for the best performance electrical network. On the other hand, in telecommunications, PoF can be used in applications, such as remote antennas and extenders for passive optical networks (PONs). PoF can make them virtually passives. We reviewed the PoF concept, its main elements, technologies, and applications focusing in access networks and in smart grid developments made by the author's research group.

Keywords: power over fiber, PoF, fiber powering, sensing, monitoring, photovoltaic converters, smart grid, fiber to the antennas

1. Introduction

Optical fiber (OF) technology has many decades of research and development mainly focused in telecommunication applications. Other classical applications for optical fibers include sensors, endoscopic imaging, and illumination.

Decades ago, in the 1970s, scientists of American Telephone and Telegraph (AT&T) had the idea to use the fiber to supply energy using optical fiber to make a sound alert in telephones, instead of the electric option. The concept of power over fiber (PoF) [1] was born this way. In other words, beyond their classical applications, optical fibers can also be used to transport optical energy to powering electric or electronic devices remotely. This is the concept of PoF.

Since that time in the 1970s, many applications have emerged for PoF mainly in two different areas: telecommunications and utilities.

Why PoF is interesting? The reasons depend on the applications but they are associated to the optical fiber characteristics.

For telecom applications, the key factors are (i) PoF eliminates the necessity of batteries, solar panels, and long cupper feeder wires in the remote sites, improving the reliability and the security of the system; (ii) PoF permits the reduction of the space and the installation cost in remote sites, which is very important to telecom companies.

For electrical utilities, there are at least four key factors: (i) PoF uses optical fiber, which is made of nonconductive material. This characteristic is important because in most applications in electrical world, the sensors are placed in high voltage. Any conductance in high-voltage elements can create current leaks; (ii) optical fiber is immune to electromagnetic interferences. The electrical world environment is polluted of electromagnetic interferences; therefore, optical fiber can transmit signals without quality degradation; (iii) the optical fiber eliminates the need to run conductive copper wire into a high-ground potential rise (GPR) zone. GPR arises when lightning strikes occur in substations and can cause severe interference problems in electronic equipment and systems. Considering that the sensor using PoF has a complete galvanic isolation to the ground potential, it is practically immune to the GPR effects; and (iv) there are many low-cost/low-power/high-efficiency electronic sensors available for transmission lines and substations monitoring that can be supplied by PoF without incurring in the described problems.

This chapter describes a revision of PoF, its technical principles, main elements, technologies, and the applications in telecom and in utilities, developed by the author's group.

2. Power over fiber

This section presents the technical principles of PoF including a short history, the main elements (high-power lasers, fibers, and photovoltaic converters (PVs)), and their main limitations (laser power, converter efficiency, and fiber-fuse effect), circuits, topologies and networks, application on multimode (MM), single mode (SMF) and special fibers, and finally the future perspectives.

2.1. Historical overview

2.1.1. Telecom

The concept of PoF was born together with the first optical fiber developments. Optical fiber with low attenuation (less than 20 dB/km) was developed in 1970, and the first fiber optic

http://dx.doi.org/10.5772/68088

link was installed in Chicago in 1976, USA. To the best of author's knowledge, DeLoach et al. [1] published the first work about PoF in 1978. They proposed to use PoF to operate a sound alert of a telephone remotely. Continuing the previous experiment, in 1979, Miller and Lawry [2] implemented a two-way speech communication between an electrically powered station and an optically powered station, and in 1982, Miller et al. [3] demonstrated a bidirectional speech-television communication over a single optical fiber, with emergency optical powering of the remote station telephone. A great result of those researches was the development of a highly efficient photovoltaic converter based on GaAlAs.

In 1993, Banwell et al. investigated the issues of PoF application in fiber-in-the-loop (FITL) [4]. In this work, many technical and cost aspects have been considered in the analysis. Almost a decade after that, Miyakawa et al. [5–7] presented fiber optic power and signal transmission systems considering the application of DC (direct current) powering to information equipment such as personal computers.

Werthen and Cohen [8] in 2006 and 2008 [9] proposed the use of PoF for driving low-power switches or actuators to assuring path diversity in upstream data delivery in passive optical network (PON) architectures and for providing battery backup power in case of power outages.

In 2007, Wake et al. [10] studied the combination of radio over fiber and PoF, and in [11], they described optically powered radio-over-fiber remote units designed and constructed for distributed antenna system applications.

Sato and Matsuura [12] and Matsuura and Sato [13] demonstrated in 2013 the transmission of RF signal and high-power light using a double-clad fiber (DCF). A DCF has a single-mode core and a multimode inner clad. The RF signal and the optical powering signal travel, respectively, through core and in the inner clad of DCF.

In 2014, Penze et al. [14] proposed powering PON extenders using PoF. In this research, SOAs (semiconductor optical amplifier) were powered using PoF. The SOAs were used in a bidirectional amplification of signals of Gigabit PON (GPON) and 10 Gigabit PON (XGPON). In the same year, Ikeda [15] proposed a PoF instead of metal waveguides in order to protect microwave radio equipment from lightning.

Yan et al. [16], in 2015, developed a wireless sensor system based on PoF to realize a flexible distributed sensing over a middle distance, in environments of high voltage, strong magnetic field, flammable, and explosives. In the same year, Matsuura and Sato [17] demonstrated experimentally a bidirectional radio over fiber using a DCF for optically powered remote antenna units. The feasibility of the technique was demonstrated by bidirectional RoF transmission over a 100-m DCF optically feeding with 4.0 W, and in [18], Sato et al. demonstrated a bidirectional radio-over-fiber transmission using a double-clad fiber with 40-W optical power feeding. In the same year, Lee et al. [19] described the concept of cloud radio access network (CRAN) based on PON exploiting PoF, andSuto et al. [20] detailed the challenges of QoE-guaranteed and power-efficient network operation for CRAN based on PON exploiting PoF.

In 2016, Minamoto and Matsuura [21] and Matsuura and Minamoto [22] presented an optically controlled beam steering system using 60-W PoF in the fiber to feed remote antenna units, and in [23], Yoneyama et al. demonstrated a 1.3-μm dual-channel radio-over-fiber system using a 300-m double-clad fiber feed with 30-W optical power of a PoF system. In the same year, Umezawa et al. [24] reported a high-conversion gain in a high-speed optical receiver based on PoF supply and designed to work with radio over fiber at 100-GHz region. In [25, 26], the same authors proposed the use of a multicore-based radio and PoF in a transmission using a 100-GHz photo receiver.

2.1.2. *Utilities and other industries*

In 1980, Caspers and Neumann [27] first described a PoF method used to supply active electronic circuits at high potential. In the experiment, they used four infrared light-emitting diodes (LEDs) emitting at a wavelength of 940 nm which were connected in parallel and coupled to an uncladded glass rod of 40-mm diameter and a length of 1 m with an attenuation smaller than 200 dB/km at a wavelength of 850 nm. Four square silica solar cells of 20 mm length were connected in series in this experiment.

Ohte et al., in 1984 [28], demonstrated a transducer with optical-fiber data link to provide electrical isolation. A pulse-position–modulated optical signal was used in transmitted signal by optical fiber from a remote converter to mainframe. Wavelength division multiplexing (WDM) was used to transmit the modulated signal and the optical feed power in the same optical fiber.

In 1988, Trisno and Wobschall [29] described a method for improving the efficiency of the conversion of pulsed optical power to electrical power by photovoltaic. The system required a pulsed power source with a storage capacitor to hold energy for a time after the optical power is turned off. In the same year, Bjork et al. [30] demonstrated the ability to provide data communication a variety of off-the-shelf electronic transducers in a single fiber with proper attention to the synchronization issues and in [31] Lenz and Bjork presented a comparative study of four concepts for standardizing and multiplexing of fiber optic sensors for aircraft applications and compared them with PoF.

Kirkham and Johnston [32] presented in 1989 an optical frequency–modulated (FM) data link whose remote electronic sensors were optically powered. An application to current measurement is described in a high-voltage line by means of a linear coupler.

In 1991, Yamagata et al. [33] described a PoF system to measuring gas density of extra high-voltage substation gas-insulated switchgear. This system employed a method of matching the impedances of load and photodiode by optical electrical conversion of pulsed light with a photodiode and boosting the voltage with a transformer. In [34], Sai presented an optimization of this method. In the same year, Nieuwkoop et al. [35] described an alternative system to increase the voltage to the remote sensor using a low-cost coil.

Spillman and Crowne [36] presented in 1995 a throttle level angle (TLA)-sensing system applied to aircrafts utilizing a capacitance-based rotary position transducer which was powered and interrogated via light from a single MM optical fiber. The system used a unique GaAs device that served as both a power converter and optical data transmitter. In the same year, Tardy et al. [37] described a PoF system designed to measure current in high voltage;

Pember et al. [38] demonstrated a PoF network concept; and Dubaniewicz and Chilton [39] described a PoF system to gas monitoring in mines.

In 1996, Werthen et al. [40] presented an optically powered current transducer based on shunt technology using PoF and detailed a practical use of this technique. Wang et al. [41] and Zhijing [42] detailed the use of PoF to measure the pressure of liquids in 1998. In 2005, Turán et al. [43] review the basic properties of PoF, their key elements, and its industrial applications.

Bottger et al. [44], in 2008, presented an optically powered video camera link that allows acquiring and communicating a 100-Mb/s video stream over a distance of hundreds of meters. In [45], M. Roger demonstrated an optically powered sensor network with subscribers consuming less than 1-μW average power, and an optically powered high-speed video link transmitting data at a bit rate of 100 Mbit/s.

In 2010, Rosolem et al. [46] described the results of a PoF sensing system for monitoring partial discharges on hydro generators and de Nazaré and Werneck [47] proposed a monitoring system based on PoF to measure the temperature and current of transmission lines.

Audo et al. [48], in 2011, proposed PoF-based architecture to extend multidisciplinary-cabled networks or to create a dedicated submarine hydrophone or seismometer network, and in 2012, Lau et al. [49] reported a prototype optically remote-powered subsea video monitoring system that provides an alternative approach to powering subsea video cameras. In the same year, Tanaka and Kurokawa [50] presented a review of fiber optic sensor network that employs PoF. A hybrid sensor network with wireless sensors is also achievable by introducing wireless/optical interface nodes.

In 2014, Rosolem et al. [51] presented a PoF system to long-distance applications based on charge/discharge of super-capacitors. The system was installed in a transmission line tower, and it was connected to the substation using an optical ground wire cable (OPWG), and in [52], Silva et al. proposed a new MM fiber optic cable to transmit optical energy for long reach in PoF systems.

Rosolem et al. [53], in 2015, presented the results of a PoF system to monitor high-voltage switchgear using video cameras and free space optics (FSO), and in [54], Rosolem et al. proposed the use of PoF and FSO to measure current in high-voltage transmission lines.

In 2016, Zhang et al. [55] proposed a 15-kV silicon carbide (SiC) MOSFET gate drive using PoF and replaced the traditional design based on isolation transformer.

2.2. Technical principles of PoF

A generic PoF system is shown in **Figure 1**. In the left side, the control unit composed by the high-power optical source (HPOS) unit together with the optical reception unit (ORU) is shown, which receives signals from the remote sensor unit, shown on the right side of **Figure 1**. Two optical fibers connect the local control unit to the remote unit. The optical fibers may be standard SMF optical fibers or MM fibers. In the remote unit, a photovoltaic converter

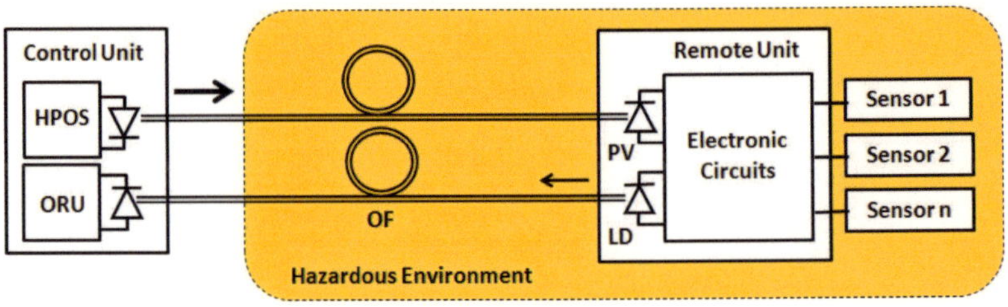

Figure 1. Generic PoF system diagram showing its main elements HPOS, OF, PV, LD, electronic circuit, and sensors.

detects the power transmitted by the HPOS. The electrical energy produced by the photovoltaic converter is used to power up a low-threshold laser (LD), electronic circuits, and sensors of the remote unit. The optical fibers and the remote unit can be installed in a hazardous environment, such as high-voltage substations, oil refineries, mines, oil tanks, water reservoirs, and sea depth.

The biggest challenge for a generic PoF system is to provide to a load the highest possible power, at the greatest possible distance, with the highest reliability. There are limits today to PoF applications and the limits are attributed to technological, physical, and cost aspects of PoF elements.

The amount of power that the PoF system can deliver is determined to a great extent by its components: laser, fiber, and photovoltaic cell. The following parameters can be used to evaluate the delivered electric power (P_{Load}) to an electronic load: the maximum transmitted optical power without causing damages in the optical fiber ($P_{MaxFiber}$); the optical power of the HPOS (P_{HPOS}); the optical power in the PV input (P_{In}); the total loss on the fiber (α_{Fiber}); the total loss of the optical connectors (α_{Conn}); the link distance (L); and the efficiency of the PV (η_{PV}). The power delivered to the extender can be expressed by

$$P_{Load} = P_{In} \cdot \eta_{PV} \tag{1}$$

$$P_{In} = P_{HPOS} \cdot \alpha_{Fiber} \cdot \alpha_{Conn} \tag{2}$$

$$\alpha_{Fiber} = 10^{(-L.\alpha_{FiberdB}/10)} \tag{3}$$

$$\alpha_{Conn} = 10^{(-\alpha_{ConndB}/10)} \tag{4}$$

where $\alpha_{FiberdB}$ is the fiber attenuation in dB/km and α_{ConndB} is the total connector loss in dB. Notice that

$$P_{HPOS} \leq P_{MaxFiber} \tag{5}$$

In the following subsections, the main elements of PoF will be described, and **Figure 2** will be used as a reference for further discussion. **Figure 2** shows a qualitative graph of some parameters of the PoF elements, such as the loss of the optical fibers and the responsivity of PV cells. The main spectral bands for HPOS are also shown in **Figure 2**.

http://dx.doi.org/10.5772/68088

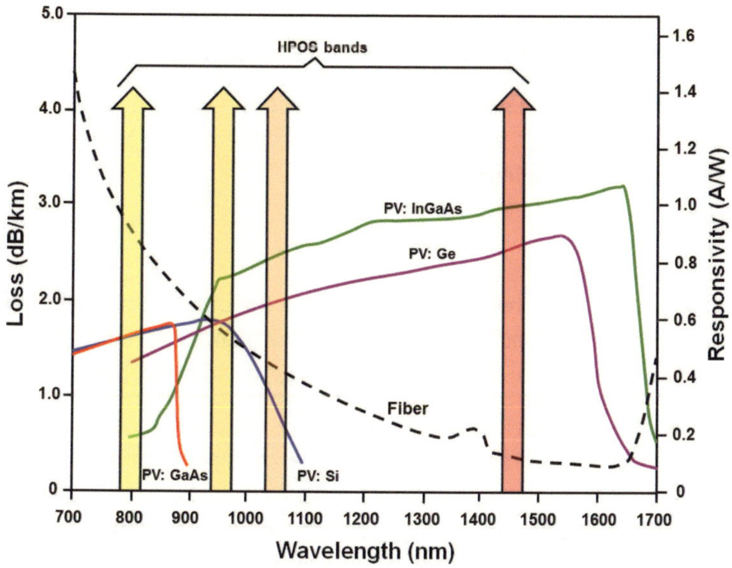

Figure 2. Qualitative graph of some PoF parameters.

2.3. Main elements of PoF

2.3.1. High-power optical source

Figure 2 shows the main spectral bands of HPOS. There are four main bands: 800, 950, 1050, and 1480 nm. The HPOSs for these bands have applications in medicine, pumping, thermal printing, and industrial applications. The power of these HPOSs can reach from 2 to 10 W for TO220 packaging depending on the wavelength band and from 6 to 650 W for laser modules. The output fiber core diameter for these HPOSs ranges from 50 to 600 μm depending on the HPOS output power. These fibers are MM types. For SMF fibers, the output power changes from 0.2 to 1.0 W.

Normally, an SMA (Sub-Miniature A) high-power type with metal ferrule is used in the HPOS output optical fiber. This metal ferrule is used for thermal dissipation in the HPOS connection to the fiber link. The high-power SMA connector utilizes air-gap-ferrule technology that eliminates the materials near the fiber end face that absorbs energy (e.g., epoxy). This absorption can damage the connector end face [56].

HPOS can also transmit information to the remote unit. The information signal is modulated in the continuous optical power generated by the HPOS. In general, this signal transmits commands to be executed in the remote unit, such as requesting the sensor parameters transmission, switching the sensor, and so on.

The most common modulation formats used to modulate the HPOS are on-off keying (OOK), pulse-width modulation (PWM), or frequency modulation. The extinction ratio of this signal can be high (100%) for few milliseconds like burst signals or low (less than 5%) during all the time of optical power sending. The extinction ratio is the ratio of the power used to transmit a logic level "1," to the power used to transmit a logic level "0." **Figure 3** shows an oscilloscope trace (voltage vs. time) of a telecommand signal transmitted over the optical supply power.

The main HPOS parameters required for a PoF project are maximum output power (P_{HPOS}), the operation wavelength (λ_{op}), the output fiber core diameter (D_{core}), the numerical aperture of the fiber (NA), and connector loss α_{ConndB}.

2.3.2. Optical fiber

The next element in the PoF system after the HPOS is the optical fiber, which is responsible to send the optical power from the HPOS to the PV. The fiber has two main parameters in the PoF system design: its attenuation and its limit of transmitted optical power.

The attenuation in modern optical fiber is basically caused by the intrinsic loss due to Rayleigh backscattering. Silica fibers are glasses that have materials with microscopic variations of density and refractive index. This effect gives rise to energy losses due to the scattered light. The loss due to the Rayleigh backscattering has high dependence with the operation wavelength ($1/\lambda_{op}^4$). The typical loss in the HPOS bands is 2.8 (800 nm), 1.8 (950 nm), 1.4 (1050 nm), and 0.25 dB (1480 nm).

The loss of optical fiber is not a big problem for the major PoF applications that has dozens of meters of fibers. It can be a problem for distances longer than 1 km. The major problem is regarding the maximum transmitted optical power. Next, the physical mechanism that limits the power in optical fiber will be detailed. After this limit, a degradation of the fiber core and the fiber coating occurs.

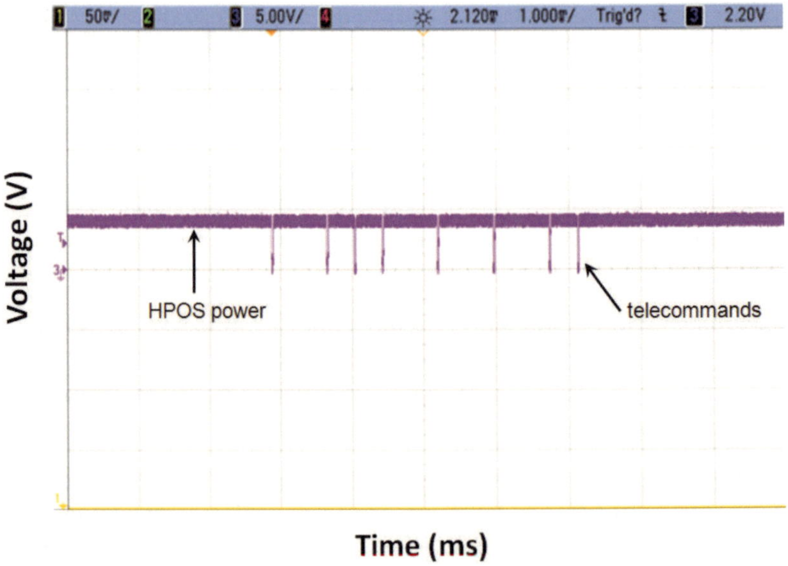

Figure 3. Oscilloscope trace showing a telecommand signal transmitted over the optical supply power.

Optical fibers when subjected to reduced diameter curvatures exhibit further attenuation of the optical signal. This attenuation may be significant when the bending diameter reaches a critical value. The attenuation of the optical signal in the bending zone of the fiber is related to the leakage of the fiber core region signal to the cladding region. The material of the outer-coating layer of the optical fiber absorbs the signal lost to the cladding region. When a high-power optical signal is propagated in the fiber, the energy absorbed by the coating in the zone of curvature is high, generating a local increase in temperature in the coating [56].

The heating of the coating of the optical fiber leads to its degradation, implying a consequent reduction of the useful life of the fiber. This degradation of the optical fiber is related to the reduction of the volume of the coating in the zone of curvature.

The second problem resulting from the propagation of high-power signals is the occurrence of the "fiber-fuse" effect on the optical fibers. This phenomenon was first observed in 1987 in a standard SMF fiber with optical intensity exceeding 5 MW/cm^2 at 1064 nm [57]. The fiber-fuse effect was initiated at a point of the fiber with a high-temperature value, then propagating toward the optical source with a velocity of about 1 m/s and emitting an optical signal in the spectral region of the visible. The propagation of this high-temperature zone is similar to the burning of a wick, which gave rise to the name of the "fiber-fuse" effect. After propagation of the fiber-fuse effect, the optical-fiber core presents a periodic bubble chain (see **Figure 4**) and is permanently destroyed, being unable to guide an optical signal.

Currently, the most accepted general explanation for the fiber-fuse effect relates to the ignition of this effect with the increase of fiber optic absorption at a point with a high-temperature value [56]. In turn, the increase in optical signal absorption is responsible for the catastrophic temperature increase in the fiber core, reaching values higher than the vaporization temperature of the silica. Through the thermal diffusion mechanism, this high-temperature zone is transmitted to neighboring regions and the process evolves toward the optical source. Thus, the "fiber-fuse" effect is related to the increase in fiber optic absorption and the thermal diffusion process.

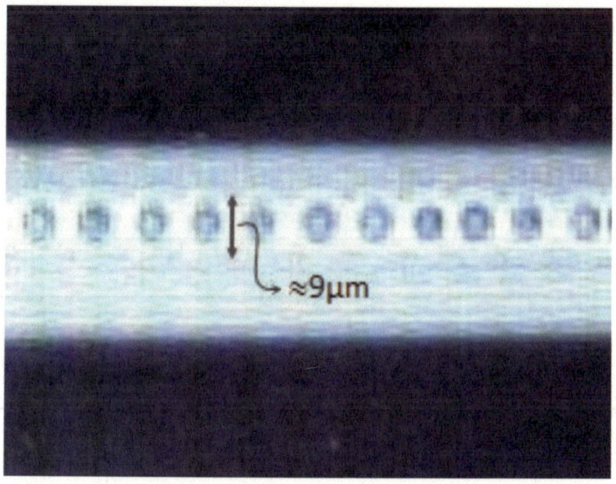

Figure 4. Schematic of a fiber-fuse effect and a photo of an optical fiber core presenting a periodic bubble chain due to fiber-fuse effect.

It should be noted that ignition and propagation of the fiber-fuse effect occurs only if the power of the optical signal in the fiber is maintained above a threshold value. According to [56], this threshold is proportional to the diameter of the fiber modal field (MDF) and dependent on the wavelength of the optical signal. A typical SMF fiber (MDF = 7.8 μm) presents a threshold of 1.5 W at 1467 nm. In other study [58], it was observed that an SMF fiber type 28F (MDF = 7.8 μm) presents a threshold of 1.0 W, and an MM fiber of 62.5-μm core (MDF = 12.0 μm) presents a threshold of 4.0 W at 1060 nm.

In order to increase the optical power in SMF fibers, many efforts have been made to transmit the high power by the cladding of the SMF fiber. In [21, 22], an optically controlled beam steering system using 60 W transmitted in the cladding of the fiber to feed remote antenna units is presented.

The propagation of signals with optical power exceeding the threshold required for ignition and propagation of the fiber-fuse effect is not sufficient to trigger it, and an ignition point characterized by having a high-temperature value is also required. This point of ignition occurs in contaminated and/or degraded connectors or in optical fibers subject to tight bending. As described previously, in optical fibers subjected to curvatures of reduced diameters, an additional attenuation of the optical signal occurs, which in combination with high-power signals generates a considerable localized heating.

Optical connectors can be easily contaminated with dust or organic particles that are common in outdoor environments. This additional attenuation in the connectors, contaminated and/or degraded, may be considerable and when associated with high-power signals also generate localized heating.

PoF is generally used in hazardous environment requiring rugged optical cables for this kind of environment. It is desirable that such cables containing fiber from 100- to 1000-μm core diameter be available on the market, since PoF requires fibers with large core diameter. The development of a rugged PoF optical cable using 100-μm fiber core for long-distance applications is described in [52].

The main OF parameters required for a PoF project are maximum transmitted optical power without causing damages (P_{MaxFiber}), fiber attenuation (α_{FiberdB}) in dB/km at the operation wavelength (λ_{op}), the fiber core diameter (D_{core}), and the numerical aperture of the fiber (NA).

2.3.3. Photovoltaic converter

A photovoltaic power converter is the element where the light will be finally transformed in electricity to supply the control unit and the sensors. PV is one (or more) photodiode that operates in photovoltaic mode. When the supply optical power reaches the PV, electron-hole pairs (carriers) are generated creating a photocurrent. The total current is the photocurrent produced by the supply optical power minus the dark current, which is related to the reverse saturation current. This photocurrent (I_{ph}) is proportional to the incident supply optical power according to Eq. (6)

$$I_{\text{ph}} = P_{\text{In}} \cdot R(\lambda_{\text{op}}) \quad (6)$$

http://dx.doi.org/10.5772/68088

$R(\lambda_{op})$ is the spectral responsivity of the cell, which is dependent on the HPOS operation wavelength (**Figure 2**).

The most important photovoltaic parameter is the conversion efficiency (η). It is defined as the maximum electric power produced by the PV ($P_{elecmax}$) divided by the incident light power under standard light conditions or as defined by Eq. (7):

$$\eta = P_{elecmax}/P_{Pin} = (V_{OC} \cdot I_{SC} \cdot FF)/(P_{Pin}) \tag{7}$$

where V_{oc} is the open-circuit voltage, I_{SC} is the short circuit current, and FF is the fill factor of PV.

Many PV types are actually made of micro-PV arrays [59, 60]. In this way, connecting the micro-PVs in parallel to the output voltage can reach an adequate level. **Figure 5** shows a schematic arrangement of a commercial PV array with four elements compared with 100-μm-core optical-fiber diameter. This type of fiber is normally used in fiber-packaged PVs.

The fill factor is the parameter that evaluates the junction quality and series resistance of the PV array. The fill factor is the ratio of the maximum electric output power to the product of VOC and ISC.

A comparative study of PV efficiency for the materials Si, GaAs, and Ge is founded in [61]. PV devices made of Si or GaAs proper for PoF uses can be founded in market. Unfortunately, these devices operate in the spectrum region where the fiber does not have the minimum attenuation (**Figure 2**). For operation in 1310 or 1550 nm, the PV arrays can be obtained easily connecting many InGaAs or Ge photodiodes in series or in parallel using an optical splitter. This is a topic for the next section.

The main PV parameters required for a PoF project are the maximum output electric power ($P_{elecmax}$), the maximum optical input power without damaging the PV (P_{Inmax}), the operation spectral band (λ_{op}), the efficiency (η), open-circuit voltage (V_{oc}), and the short-circuit current (I_{SC}).

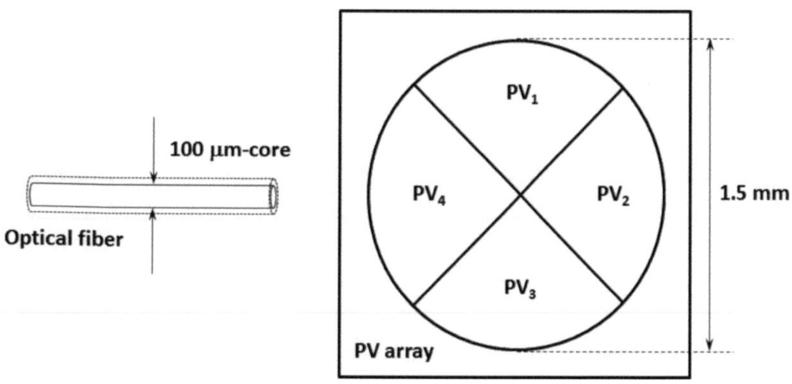

Figure 5. Schematic arrangement of a commercial PV array with four elements compared with 100-μm-core optical fiber diameter.

2.3.4. Basic remote unit circuits

The electronic circuits of unit control including LD driver and sensors need to work in a specific voltage (V_{cc}), and they consume an electrical current (I_{load}). To provide the required voltage and current, some basic circuits can be designed. The most basic circuits use just one PV and a DC-DC converter to stabilize the voltage, as shown in **Figure 6(a)**. In this case, the PV should be an array type to provide voltage higher than 3 V, although there are many integrated circuits that work below 3 V. If the PV chosen has low-output voltage (lower than 2V), the circuit shown in **Figure 6(b)** can be used. In this circuit, the DC-DC converter was substituted by a step-up converter, which elevates the PV output voltage to the voltage required by the circuit. Other possibility to elevate the voltage is using an optical splitter (with *n* output ports) connected to the optical fiber and the PVs (or photodetectors) connected in series (**Figure 6(c)**). Typically, splitters with $n = 4–8$ have been used, to provide V_{cc} ranging from 2.4 to 4.8 V. Each splitter output port is connected to one PV. This is an easy solution to use in spectral bands where there is no commercial PV array available, such as in 1310 and 1550 nm. **Figure 6(d)** shows a different version of **Figure 6(c)** circuit. In this case, the association of the PVs is made in parallel in order to increase the current available to the circuit. Certainly, the association of PVs can be made simultaneously in parallel and in series.

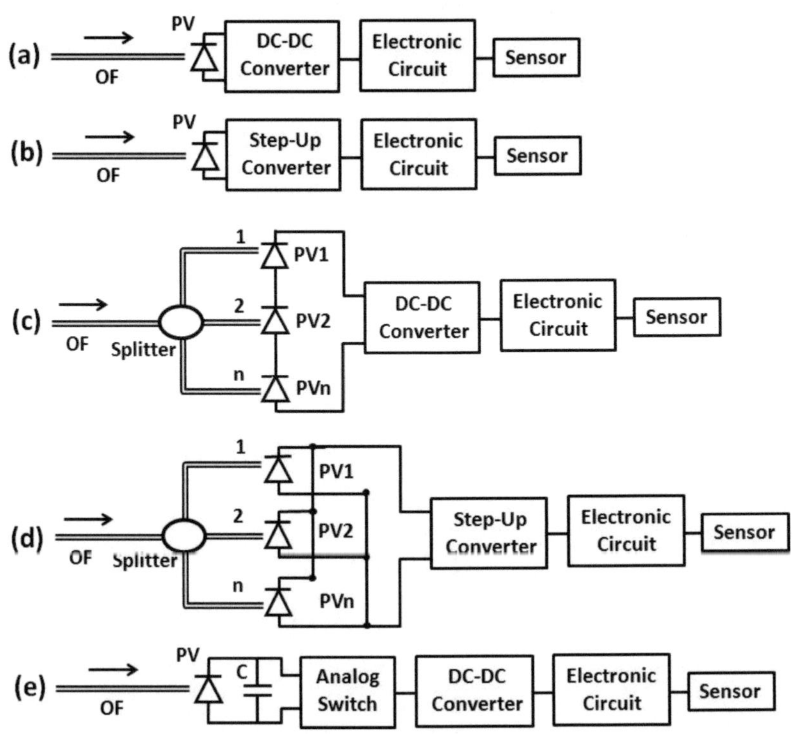

Figure 6. Some types of circuits used in the design of PoF remote units, (a) direct conversion, (b) conversion using a step-up converter, (c) PVs connected in series, (d) PVs connected in parallel, and (e) charging circuit using a capacitor.

Figure 6(e) shows a charging circuit type. This type of circuit is used to supply loads in long-distance PoF links where the available input optical power level in the PV is not enough to provide a necessary electric power. In this circuit, when the capacitor C connected in the PV output is charged in the maximum PV voltage output, an analog switch connects the capacitor to a DC-DC converter that stabilizes its output voltage for the load for a specific time period. This type of circuit should be used only if the sensors do not require measuring their parameters all the time.

2.4. PoF topologies

The connection between the control units and remote units in PoF systems is usually made in the same way of telecommunications links. **Figure 7(a)** shows a typical PoF link using two fibers. However, in some cases the quantity of fibers is a limiting factor in specific application. WDM can be used to multiply the supply optical power transmitted by the HPOS and the sensor signal transmitted by the LD in the remote unit. The WDM device must support the high power of the HPOS. WDM devices for high-power applications can support around 2 W of handling power. The wavelength pairs used in a WDM PoF system as the one shown in **Figure 7(b)** are as follows:

- λ_1 = 808–830 (only for MM fibers), 980, and 1480 nm;
- λ_2 = 1310 and 1550 nm.

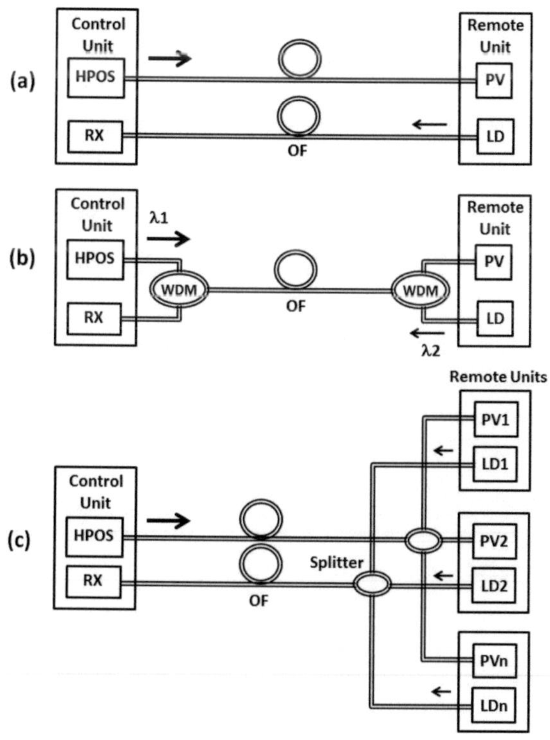

Figure 7. Illustrations of some topologies for PoF systems, (a) two fibers topology, (b) bidirectional WDM topology, and (c) tree topology.

PoF networks have also been reported. A typical tree topology (**Figure 7(c)**) permits the connection of a single control unit to many remote units. This type of network works in a similar way as a passive optical network used in telecommunications.

2.5. Example of PoF link calculations

Using Eqs. (1)–(5), it is possible to calculate the maximum PoF link for two cases, using SMF and MM fibers. **Table 1** shows the parameters used in this calculation.

Figure 8 shows a plot of the delivered power (P_{Load}) to an electronic load as a function of link distance. It can be observed that MM fiber option is better until 2.2 km. After this distance, the SMF option is better.

Parameter/fiber	SMF	MM
HPOS optical power	1.0 W	2.0 W
Wavelength	1480 nm	830 nm
Total connectors attenuation	1.0 dB	1.0 dB
Fiber attenuation	0.25 dB/km	3.0 dB/km
PV efficiency	0.2	0.4
PV material	InGaAs	GaAs

Table 1. Parameters used to calculate the maximum PoF link distance.

In utilities, the major applications of PoF occur in the substations, which the maximum internal distances are less than 0.2 km. For telecommunications, the distances in the major application are longer than 2 km.

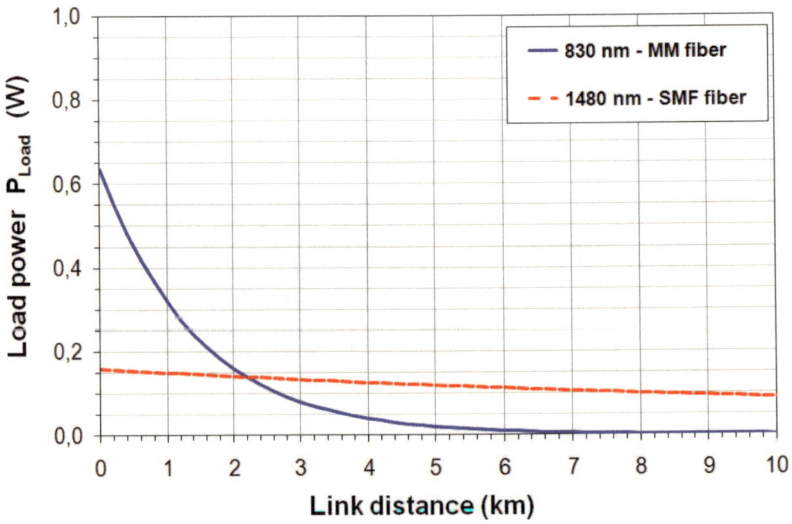

Figure 8. Plot of the delivered power to an electronic load as a function of link distance.

http://dx.doi.org/10.5772/68088

2.6. PoF applications developed at CPqD

This section presents some PoF systems and sensors developed by the author's group for telecommunications and for utilities applications. Most PoF systems for utilities have been tested in field trials.

2.6.1. PoF applications in telecommunications

In telecommunication, we developed a new approach to powering PON extenders using PoF [14]. PON extenders are elements that recover the downstream and upstream signals. These signals are highly attenuated by the splitters and by link length in any PON. However, the use of extenders in PON modifies its passive characteristic since extenders require electrical supply to work. Using PoF to supply, the extenders turns them virtually passives.

Using PoF to eliminate the batteries located at telecom remote sites, the reliability and the security of the system are improved. Furthermore, PoF can reduce the copper cables theft, used to power the extenders in many cases. We demonstrated the performance of a fiber-powered extender using semiconductor optical amplifiers in a 10-gigabit passive optical network system and a gigabit passive optical network system (G-PON) setup using a 1:32 splitter and 50-km reach. The extender was powered from a remote site placed in the access area using a 62.5-μm MM fiber at 830 nm. **Figure 9** shows a schematic of the PON using the fiber-powered extender and the photo of the extender.

2.6.2. PoF applications in utilities

In utilities, we developed applications for monitoring hydrogenerators, power transformers, switchgears, and for transmission lines.

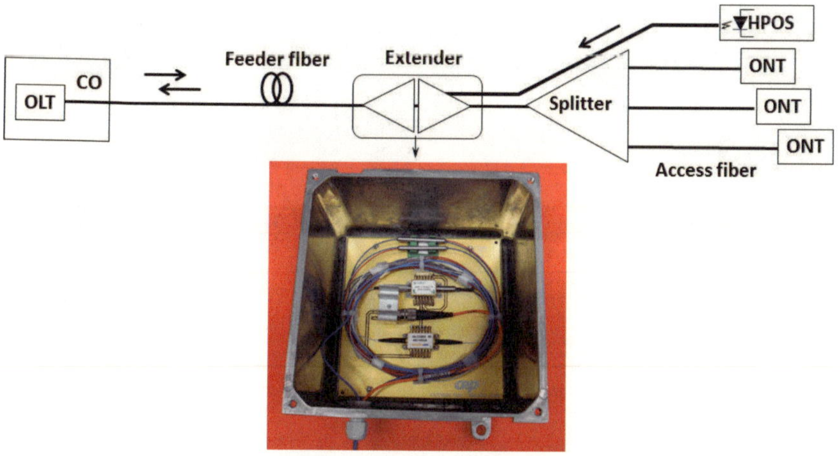

Figure 9. Schematic of a PON using extender powered by a PoF system and the extender photo.

Figure 10 shows the applications for partial discharges monitoring in hydrogenerators [46] and for 500-kV power transformers [62]. Partial discharges on high-voltage equipment insulation are a symptom of fragility of the dielectric capacity. The growing of partial discharges can cause serious consequences for the equipment and the electrical system. Partial discharges generate some physical effects, such as conducted and radiated electromagnetic pulses, light, acoustic noise, localized temperature variation, and chemical reactions. In the developed PoF-monitoring systems, we used a powered antenna to detect the radiated electromagnetic pulses of discharges.

For hydrogenerator monitoring (**Figure 10(a)**), the sensor is composed by one dipole meander antenna, one photovoltaic converter, and one semiconductor laser. Some passive components such as resistors, capacitors, and inductors are omitted in the schematic of sensor circuit to simplification. This system was installed in the Eletrobrás' Coaracy Nunes power plant, which is located in Amapá state, in the north of Brazil.

The sensor for monitoring the high-voltage power transformers (**Figure 10(b)**) is composed by one monopole antenna, one photovoltaic converter, one field-effect transistor (FET) amplifier, and one semiconductor laser. The FET was used in this sensor to increase the sensitivity of the sensor in this particular application. This system was installed in the Cemig's Neves substation, which is located in Minas Gerais state, in the southeast of Brazil.

Figure 10 shows the schematic circuit and sensor photo of antenna powered by fiber for partial discharges monitoring in (a) hydrogenerators and (b) high-voltage power transformers.

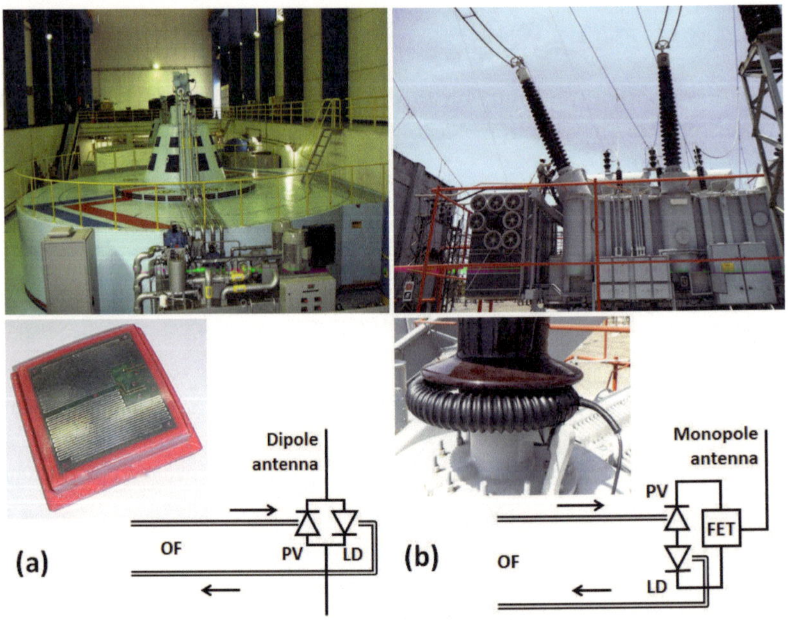

Figure 10. Schematic circuit and sensor photo of antenna powered by fiber for partial discharges monitoring in (a) hydrogenerators and (b) high-voltage power transformers.

Figure 11 shows some applications using more complex electronic circuits with PoF. **Figure 11(a)** shows a 138-kV switchgear-monitoring application using fiber-powered video cameras to inspect the quality contact of the switchgear [53]. In this application, the three phases of 138-kV switchgear were monitored using three sensors. The connection of the sensors to the control unit used the tree topology shown in **Figure 7(c)**. This system was installed in the Cemig's Bonsucesso substation, which is located in Minas Gerais state, in the southeast of Brazil.

Figure 11(b) shows a fiber-powered camera installed in a 138-kV transmission line tower [51]. The camera was used to monitor possible invasions in the security area of transmission line. The circuit of the camera works in a noncontinuous regime (circuit of **Figure 6(e)**) since the power transmission was done using an SMF fiber embedded in an optical ground wire cable. This system was installed in the Cemig's Bonsucesso/Gutierrez transmission line, which is located in Minas Gerais state, in the southeast of Brazil.

Remembering that for electrical utilities, there are key factors to use PoF. Optical fiber is made of nonconductive material. In high-voltage environment, any conductance can

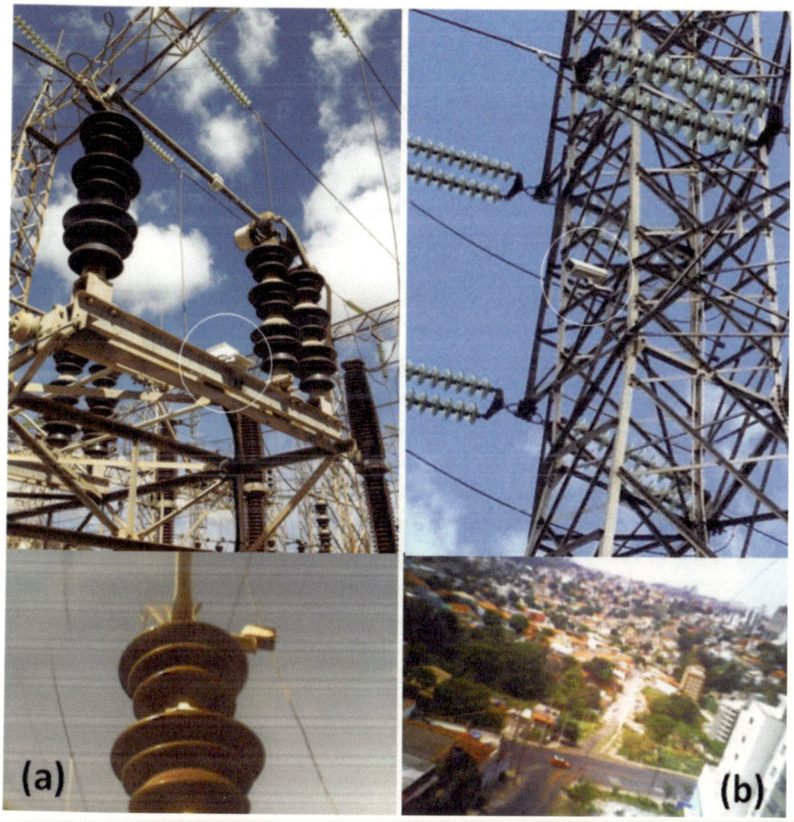

Figure 11. PoF applications with fiber-powered video cameras, (a) 138-kV switchgear sensor shown in the circle (above) and a sensor image (below) and (b) 138-kV transmission line sensor shown in the circle (above) and a sensor image (below).

create current leaks. Optical fiber is immune to electromagnetic interferences. The electrical world environment is polluted of electromagnetic interferences; therefore, optical fiber can transmit signals without quality degradation. The optical fiber eliminates the need to run conductive copper wire into a GPR zone. GPR arises when lightning strikes occur in substations and can cause severe interference problems in electronic equipment and systems. PoF sensors have a complete galvanic isolation to the ground potential, then they are practically immune to the GPR effects and finally, as we show, there are many low-cost/low-power/high-efficiency electronic sensors available for transmission lines and substations monitoring that can be supplied by PoF increasing the monitoring capacity for utilities companies.

3. Conclusion and future perspectives

In conclusion, this chapter described a revision of PoF, its technical principles, main elements, technologies, and the applications focusing in telecom and in utilities, developed by the author's group.

The applications for PoF have evolved over the last years mainly in terms of power availability for the load in the remote units. Many publications describe incredible dozen watts power transmitted in the SMF fiber cladding. This evolution allows applications to become more complex using PoF, such as video cameras powering, antennas powering, or PoF networks.

The increase of supplier's options for PoF devices and fibers has also been occurring and it can reduce the cost of this interesting optical fiber technique.

Acknowledgements

The author wishes to thank his CPqD colleagues Claudio Floridia, Danilo C. Dini, Rivael S. Penze, Fabio R. Bassan, and João P. V. Fracarolli. Many PoF projects cited in this work were funded by ANEEL (Brazilian Electricity Regulatory Agency). CNPq (National Counsel of Technological and Scientific Development) sponsors the author under scholarship DT.

Author details

Joao Batista Rosolem

Address all correspondence to: rosolem@cpqd.com.br

CPqD—Research and Development Center in Telecommunications, Campinas, SP, Brazil

References

[1] DeLoach BC, Miller RC, Kaufman S. Sound alerter powered over an optical fiber. Bell System Technical Journal. 1978;**57**:3309–3316. DOI: 10.1002/j.1538-7305.1978.tb02205.

[2] Miller RC, Lawry RB. Optically powered speech communication over a fiber lightguide. Bell System Technical Journal 1979;**58**:1735–1741. DOI: 10.1002/j.1538-7305.1979.tb02280.x

[3] Miller RC, DeLoach BC, Stakelon TS, Lawry RB. Wideband, bidirectional lightguide communication with an optically powered audio channel. Bell System Technical Journal 1982;**61**:1538–7305. DOI: 10.1002/j.1538-7305.1982.tb04349.x

[4] Banwell TC, Estes RC, Reith LA, Shumate Jr PW, Vogel EM. Powering the fiber loop optically-a cost analysis. Journal of Lightwave Technology. 1993;**11**:481–494. DOI: 10.1109/50.219583

[5] Miyakawa H, Hyodo R, Tanaka Y, Kurokawa T. Photovoltaic cell characteristics for high-intensity laser light in fiber optic power transmission systems. In: Proceedings of Conference Record of the Twenty-Ninth IEEE Photovoltaic Specialists Conference; 19–24 May 2002; New Orleans, USA; **29**:1653–1655. DOI: 10.1109/PVSC.2002.1190934

[6] Miyakawa H, Herawaty E, Yoshimoto M, Tanaka Y, Kurokawa T. Power-over-optical local area network systems. In: Proceedings of 3rd World Conference on Photovoltaic Energy Conversion; 11–18 May 2003; Osaka, Japan; **3**:2466–2469.

[7] Miyakawa H, Tanaka Y, Kurokawa T. Design approaches to power-over-optical local-area-network systems. Applied Optics 2004;**43**:1379–1389. DOI: 10.1364/AO.43.001379

[8] Werthen JG, Cohen M. Photonic power: delivering power over fiber for optical networks. In: Proceedings of 2006 International Conference on Photonics in Switching; 16–18 October 2006; Heraklion, Crete; pp. 1–3. DOI: 10.1109/PS.2006.4350205

[9] Werthen JG. Powering next generation networks by laser light over fiber. In: Proceedings of 2008 Conference on Optical Fiber Communication/National Fiber Optic Engineers (OFC/NFOEC 2008); 24–28 February 2008; San Diego; USA, pp. 1–3. DOI: 10.1109/OFC.2008.4528749

[10] Wake D, Nkansah A, Gomes N. Optical powering of remote units for radio over fiber links. In: Proceedings of 2007 International Topical Meeting on Microwave Photonics; 3–5 Oct. 2007; VIC, Canada; pp. 29–32. DOI: 10.1109/MWP.2007.4378127

[11] Wake D, Nkansah A, Gomes NJ, Lethien C, Sion C, Vilcot J P. Optically powered remote units for radio-over-fiber systems. Journal of Lightwave Technology. 2008;**26**:2484–2491. DOI: 10.1109/JLT.2008.927171

[12] Sato J, Matsuura M. Radio-over-fiber transmission with optical power supply using a double-clad fiber. In: Proceedings of 18th OptoElectronics and Communications

Conference held Jointly with 2013 International Conference on Photonics in Switching (OECC/PS); 30 June–4 July 2013; Kyoto, Japan; pp. 1–2.

[13] Matsuura M, Sato J. Power-over-fiber using double-clad fibers for radio-over-fiber systems. In: Proceedings of 19th European Conference on Networks and Optical Communications (NOC); 4–6 June 2014; Milano, Italy; pp. 126–131. DOI: 10.1109/NOC.2014.6996840

[14] Penze RS, Rosolem JB, Duarte UR, Paiva GER, Filho RB. Fiber powered extender for XG-PON/G-PON applications. IEEE/OSA Journal of Optical Communications and Networking. 2014;**6**:250-258. DOI: 10.1364/JOCN.6.0002

[15] Ikeda K. Lightning protection of microwave radio equipment using radio on fiber and power over fiber experimental demonstration of communication quality. In: Proceedings of Microwave Photonics (MWP) and the 9th Asia-Pacific Microwave Photonics Conference (APMP) International Topical Meeting on; 20–23 October 2014; Sendai, Japan; pp. 185–188. DOI: 10.1109/MWP.2014.6994526

[16] Yan J, Wang J, Lu Y, Jiang J, Wan H. Novel wireless sensor system based on power-over-fiber technique. In: Proceedings of 14th International Conference on Optical Communications and Networks (ICOCN); 3–5 July 2015; Nanjing, China; pp. 1–3. DOI: 10.1109/ICOCN.2015.7203710

[17] Matsuura M, Sato J. Bidirectional radio-over-fiber systems using double-clad fibers for optically powered remote antenna units. IEEE Photonics Journal. 2015; **7**:1–9. DOI: 10.1109/JPHOT.2014.2381669

[18] Sato J, Furugori H, Matsuura M. 40-watt power-over-fiber using a double-clad fiber for optically powered radio-over-fiber systems. In: Proceedings of Optical Fiber Communications Conference and Exhibition (OFC); 22–26 March 2015; Los Angeles, USA; pp. 1–3. DOI: 10.1364/OFC.2015.W3F.6

[19] Lee Y, Suto K, Nishiyama H, Kato N, Ujikawa H, Suzuki K. I. A novel network design and operation for reducing transmission power in cloud radio access network with power over fiber. In: Proceedings of IEEE/CIC International Conference on Communications in China (ICCC); 2–4 November; Shenzhen, China; pp. 1–5. DOI: 10.1109/ICCChina.2015.7448616

[20] Suto K, Miyanabe K, Nishiyama H, Kato N, Ujikawa H, Suzuki KI. QoE-guaranteed and power-efficient network operation for cloud radio access network with power over fiber. IEEE Transactions on Computational Social Systems;**2**:127–136. DOI: 10.1109/TCSS.2016.2518208

[21] Minamoto Y, Matsuura M. Optically controlled beam steering system with 60-W power-over-fiber feed for remote antenna units. In: Proceedings of Optical Fiber Communications Conference and Exhibition (OFC), Anaheim, USA; 20–24 March 2016; pp. 1–3.

[22] Matsuura M, Minamoto Y, Optically powered and controlled beam steering system for radio-over-fiber networks. Journal of Lightwave Technology. 2017; 35:979-988. DOI: 10.1109/JLT.2016.2631251

[23] Yoneyama A, Minamoto Y, Matsuura M. Power-over-fiber transmission using 1.3-μm dual-channel radio-over-fiber signals in a double-clad fiber. In: Proceedings of 21st OptoElectronics and Communications Conference (OECC) held jointly with 2016 International Conference on Photonics in Switching (PS); 3–7 July 2016; Niigata, Japan; pp. 1–3.

[24] Umezawa T, et al. 11-Gbps 16-QAM OFDM radio over fiber demonstration using 100 GHz high-efficiency photoreceiver based on photonic power supply. In: Proceedings of 21st OptoElectronics and Communications Conference (OECC) held jointly with 2016 International Conference on Photonics in Switching (PS); 3–7 July 2016; Niigata, Japan; pp. 1–33.

[25] Umezawa T, et al. Multi-core based 94-GHz radio and power over fiber transmission using 100-GHz analog photoreceiver. In: Proceedings of 42nd European Conference on Optical Communication (ECOC 2016), 18–22 Sept. 2016; Dusseldorf, Germany; pp. 1–3. DOI: 10.1109/JSTQE.2016.2611638

[26] Umezawa T, et al. 100-GHz fiber-fed optical-to-radio converter for radio- and power-over-fiber transmission. IEEE Journal of Selected Topics in Quantum Electronics. 2017; **23**:1–8. DOI: 10.1109/JSTQE.2016.2611638

[27] Caspers F, Neumann EG. Optical power supply for measuring or communication devices at high-voltage levels. IEEE Transactions on Instrumentation and Measurement. 1980; 29:73–74. DOI: 10.1109/TIM.1980.4314866

[28] Ohte A, Akiyama K, Ohno I. Optically-powered transducer with optical-fiber data link. In: Proceedings of SPIE 0478, Fiber Optic and Laser Sensors II; 17 September 1984; Arlington, USA. DOI: 10.1117/12.942655

[29] Trisno YS, Wobschall D. Optimization of an optically pulsed photocell array as a sensor power source. IEEE Transactions on Instrumentation and Measurement. 1988; **37**:142–144. DOI: 10.1109/19.2683

[30] Bjork P, Lenz J, Fujiwara K. Optically powered sensors. In: Proceedings of Optical Fiber Sensors, OSA Technical Digest Series; Mar 1988; New Orleans, USA. DOI: 10.1109/19.2683

[31] Lenz J, Bjork P, Optically powered sensors: a systems approach to fiber optic sensors. In: Proceedings of SPIE 0961, Industrial Optical Sensing, 29 November 1988; Dearborn, USA. DOI: 10.1117/12.947855

[32] Kirkham H, Johnston AR. Optically powered data link for power system applications. IEEE Transactions on Power Delivery. 1989; **4**:1997–2004. DOI: 10.1109/61.35623

[33] Yamagata Y, Kumagai T, Sai Y, Uchida Y, Imai K. A sensor powered by pulsed light (for gas density of GIS). In: Proceedings of Solid-State Sensors and Actuators, International Conference on Transducers '91, 24–27 June 1991; San Francisco, USA; pp. 824–827. DOI: 10.1109/SENSOR.1991.149011

[34] Sai Y. Optimization of optically powered sensors. In: Proceedings of International Conference on Industrial Electronics, Control and Instrumentation, IECON '91; 28 October–1 November 1991, Kobe, Japan, pp. 2439–2443. DOI: 10.1109/IECON.1991.238962

[35] Nieuwkoop E, Kapsenberg T, Steenvoorden GK, Bruinsma AJ. Optically powered sensor system using conventional electrical sensors. In: Proceedings of SPIE 1511, Fiber Optic Sensors: Engineering and Applications; 1 August 1991; The Hague, Netherlands. DOI:10.1117/12.45999

[36] Spillman Jr WB, Crowne DH. Optically powered and interrogated rotary position sensor for aircraft engine control applications. Optics Lasers Engineering. 1992; **16**:105–118. DOI:10.1016/0143-8166(92)90003-P

[37] Tardy A, Derossis A, Dupraz JP. A Current sensor remotely powered and monitored through an optical fiber link. Optical Fiber Technology; 1995; **1**:181–185. DOI: 10.1006/ofte.1995.1010

[38] Pember SJ, France CM, Jones BE. A multiplexed network of optically powered, addressed and interrogated hybrid resonant sensors. Sensors and Actuators A: Physical; 1995; **47**:474–477. DOI:10.1006/ofte.1995.1010

[39] Dubaniewicz Jr TH, Chilton JE. Fiber optic powered remote gas monitor. In: Proceedings of SPIE 2622, Optical Engineering Midwest '95; 18 August 1995; Chicago, USA. DOI:10.1117/12.216819

[40] Werthen JG, Andersson G, Weiss ST, Bjorklund HO. Current measurements using optical power. In: Proceedings of Transmission and Distribution Conference and Exposition; 15–20 Sept. 1998; Los Angeles, USA, pp. 213–218. DOI: 10.1109/TDC.1996.545937

[41] Wang L, Wang Y, Shi J, Zheng L. Optically powered hydrostatic tank gauging system with optical fiber link. In: Proceedings of SPIE 3555, Optical and Fiber Optic Sensor Systems; 13 August 1998; Beijing, China; DOI:10.1117/12.318165

[42] Zhijing Y. Optically powered oil tank multichannel detection system with optical fiber link. In: Proceedings of SPIE 3555, Optical and Fiber Optic Sensor Systems; 13 August 1998; Beijing, China. DOI:10.1117/12.318152

[43] Turán J, Ovseník L, Turán Jr J. Optically powered fiber optic sensors. Acta Electrotechnica et Informatica. 2005; **5**:1–7.

[44] Bottger G, et al. An optically powered video camera link. IEEE Photonics Technology Letters; 2008; **20**:39–41. DOI: 10.1109/LPT.2007.912695

[45] Röger M, Böttger G, Dreschmann M, Klamouris C, Huebner M, Bett AW, Becker J, Freude W, Leuthold J. Optically powered fiber networks. Optics Express. 2008; **16**:21821–21834. DOI:10.1364/OE.16.021821

[46] Rosolem JB, Floridia C, Sanz JPM. Field and laboratory demonstration of a fiber-optic/RF partial discharges monitoring system for hydrogenerators applications. IEEE Transactions on Energy Conversion. 2010; **25**:884–890. DOI: 10.1109/TEC.2010.2044507

[47] de Nazaré FVB, Werneck MM. Temperature and current monitoring system for transmission lines using power-over-fiber technology. In: Proceedings of IEEE Instrumentation & Measurement Technology Conference; 3–6 May 2010; Austin, USA; pp. 779–784. DOI: 10.1109/IMTC.2010.5488198

[48] Audo F, Guegan M, Quintard V, Perennou A, Bihan JL, et al. Quasi-all-optical network extension for submarine cabled observatories. Optical Engineering. 2011; 50:1-8. DOI: 10.1117/1.3560542

[49] Lau FK, Stewart B, McStay D. An optically remote powered subsea video monitoring system. In: Proceedings of SPIE 8372, Ocean Sensing and Monitoring IV, 837209; 12 June 2012; Baltimore, USA. DOI:10.1117/12.920834

[50] Tanaka Y, Kurokawa T. A fiber sensor network using fiber optic power supply. In: Proceedings of SPIE 8421, OFS2012 22nd International Conference on Optical Fiber Sensors; 84211M; 14 October 2012; Beijing, China. DOI:10.1117/12.981524

[51] Rosolem JB, Bassan FR, Pereira FR, Penze RS, Leonardi AA, et al. Fiber powered sensing system for a long reach single mode fiber link and non-continuous applications. In: Proceedings of SPIE 9157, 23rd International Conference on Optical Fibre Sensors, 9157AE; 02 June 2014; Santander, Spain. DOI:10.1117/12.2058672

[52] Silva JCV, Souza ELAS, Garcia V, Rosolem JB, Floridia C., Sanches MAB. Design of a multimode fiber optic cable to transmit optical energy for long reach in PoF systems. In: Proceedings of IWCS International Cable, Connectivity Symposium, 9–12 November 2014, Rhode Island, USA.

[53] Rosolem JB. Quantitative and qualitative monitoring system for switchgear with full electrical isolation using fiber-optic technology. IEEE Transactions on Power Delivery. 2015; **30**:449–1457 DOI: 10.1109/TPWRD.2014.2377122

[54] Rosolem JB, Bassan FR, Penze RS, Leonardi AA, Fracarolli JPV, Floridia C. Optical sensing in high voltage transmission lines using power over fiber and free space optics. Optical Fiber Technology. 2015; **26**:180–183. DOI: 10.1016/j.yofte.2015.09.003

[55] Zhang X, et al. A gate drive with power over fiber-based isolated power supply and comprehensive protection functions for 15-kV SiC MOSFET. IEEE Journal of Emerging and Selected Topics in Power Electronics. 2016; **4**:946–955. DOI: 10.1109/JESTPE.2016.2586107

[56] Seo K, Nishimura N, Shiino M, Yuguchi R, Sasaki H. Evaluation of high-power endurance in optical fiber links. Furukawa Review. 2003; **24**:17–22.

[57] Kashyap R, Blow KJ. Observation of catastrophic self-propelled self-focusing in optical fibers. Electronics Letters. 1988; **24**:47–49. DOI: 10.1049/el:19880032

[58] Davis DD, Mettler SC, DiGiovanni DJ. Experimental data on the fiber fuse", In: Proceedings of SPIE 2714, 27th Annual Boulder Damage Symposium: Laser-Induced Damage in Optical Materials; 1995; Boulder, USA; pp. 202–210. DOI:10.1117/12.240382

[59] Andreev V., et al. High current density GaAs and GaSb photovoltaic cells for laser power beaming. In: Proceedings of 3rd World Conference on Photovoltaic Energy Conversion; 11–18 May 2003; Osaka, Japan; pp. 761–764.

[60] Sohr S, Rieske R, Nieweglowski K, Wolter KJ. Laser power converters for optical power supply. In: Proceedings of 34th International Spring Seminar on Electronics Technology (ISSE); 11–15 May 2011; Tratanska Lomnica, Slovakia. DOI: 10.1109/ISSE.2011.6053563

[61] Allwood G, Wild G, Hinckley S. Power over fibre: material properties of homojunction photovoltaic micro-cells. In: Proceedings of Sixth IEEE International Symposium on Electronic Design, Test and Application; 17–19 Jan. 2011; Queenstown, New Zealand, pp. 78–82. DOI: 10.1109/DELTA.2011.66

[62] Rosolem JB, et al. A fiber optic powered sensor designed for partial discharges monitoring on high voltage bushings. In: Proceedings of SBMO/IEEE MTT-S International Microwave and Optoelectronics Conference (IMOC); 3–6 Nov. 2015; Porto de Galinhas, Brazil, pp. 1–5. DOI: 10.1109/IMOC.2015.7369072

Chapter 14

Ultra-Fast All-Optical Memory based on Quantum Dot Semiconductor Optical Amplifiers (QD-SOA)

Yossef Ben Ezra and Boris I. Lembrikov

Additional information is available at the end of the chapter

http://dx.doi.org/10.5772/intechopen.68527

Abstract

All-optical signal processing is characterized by high bit-rate, power efficiency, high bandwidth, and transparency. All-optical logic gates are basic logic units for the all-optical signal processing implementation. Typically, all-optical gates are based on strong optical nonlinearities related in particular to semiconductor optical amplifiers (SOA). We briefly review the state of art in the field of all-optical logic gates and all-optical memory. In the original part, we discuss the ultrafast all-optical memory loop based on the Mach-Zehnder interferometer (MZI) with quantum dot (QD) semiconductor optical amplifier (SOA) in each arm.

Keywords: all-optical signal processing, quantum dots (QD), semiconductor optical amplifier (SOA)

1. Introduction

Optical signal processing is based on the using of linear and nonlinear optical techniques in order to manipulate and process digital, analogue, and quantum information [1]. Optical signal processing increases the processing speed of devices and reduces the energy consumption and latency of communication systems [1]. In particular, ultrafast optical nonlinearities provide a substantial speed advantage as compared to electronic techniques for simple logic: switching, regeneration, wavelength conversion (WC), performance monitoring, and analog-to-digital conversion (ADC) [1]. Silicon photonics and highly nonlinear nanophotonic devices are providing strong optical nonlinearities for ultrafast processing on millimeter length scales [1].

The recent progress in optical signal processing is based on the combination of the advanced modulation techniques, coherent detection, and digital signal processing (DSP) [2]. The interface of all-optical techniques and advanced DSP will enhance electronic processing capabilities [1]. Optical signal processing is essentially based on the following advanced technologies: coherent detection, high-speed electronics for DSP, advances in strongly nonlinear materials and devices, photonic integrated circuits (PIC), and access to four optical domains of amplitude, phase, polarization, and wavelength [2]. A simple digital modulation scheme is the on-off keying (OOK) referred to as intensity modulation with direct detection (IM/DD) [3]. In such a case, an electrical binary bit stream modulates the intensity of an optical carrier inside the optical transmitter, and the resulting optical signal is converted to the original signal in the electrical domain in an optical receiver [4]. The phase modulation combined with the coherent detection increases the spectral efficiency (SE) of optical communication systems and improves the sensitivity of optical receivers [4]. In general case, amplitude-shift keying (ASK), phase-shift keying (PSK) or M-ary quadrature amplitude modulation (QAM) can be realized [3, 4]. Polarization-division multiplexing (PDM), advanced multilevel modulation formats such as M-ary QAM, digital spectral shaping at the transmitter, coherent detection and advanced forward error correction (FEC) can increase SE of the communication system [1]. Typically, DSP must overcome deterministic signal distortions, while FEC overcomes stochastic impairments caused by noise and interference [1]. At the transmitter, DSP together with digital-to-analog converters (DAC) and FEC converts the incoming data bits into a set of analogue signals [1]. An optical coherent receiver recovers the amplitude and phase of the signal by mixing it with the local oscillator (LO) which is typically a continuous-wave (CW) laser [3, 4]. DSP, ADC, and FEC recover the data from the set of analogue electrical signals [1]. The main functions of the receiver-based DSP are equalization and synchronization [1]. Equalization must realize the polarization rotation tracking and dispersion compensation including both the chromatic dispersion and polarization-mode dispersion (PMD) [1]. Synchronization carries out the transmitter and receiver electrical and optical signal frequency and phase matching [1].

The all-optical signal processing is implemented by using the nonlinear optical phenomena such as self-phase modulation (SPM), cross-phase modulation (XPM), four-wave mixing (FWM) related to the third-order susceptibility and sum frequency, difference frequency, second harmonic generation (SHG) related the second-order susceptibility [1–4]. The typical nonlinear elements used in optical communication systems are highly nonlinear optical fibers (HNLF), silicon waveguides, chalcogenide waveguides, photonic crystals, nonlinear optical loop mirrors (NOLM), parametric amplifiers, and semiconductor optical amplifiers (SOA) [2, 3]. SOA are characterized by the extremely strong third-order optical nonlinearity and fast response and can be integrated monolithically with other devices on the same chip [3].

Optically assisted signal processing combines optics and electronics for what each one of them does best [2]. Optical components can perform some functions very fast, while electronic components carry out complex computations with buffers and memory [2]. For instance, optically assisted network routing technique uses optical correlation on headers of Internet data packets [2]. Optically assisted signal processing can be also used for a target pattern search in large amounts of data [2]. In such cases, the data information is encoded on an optical carrier at Tb/s speed and sent to an optical correlator for pattern recognition [2]. The output at Gb/s speed is searched and processed electronically with high accuracy before being sent to the user [2].

http://dx.doi.org/10.5772/intechopen.68527

In optical networks, the bandwidth mismatch between optical transmission and electronic routers results in the development of different optical signal processing and the investigation of optical packet switching (OPS) [5]. Some applications require selective switching of one or more bits to a different port [3]. The packet switching takes place when a packet of tens or hundreds of bits is selected from a bit stream [3]. The flip-flop memory is an essential component of the packet switch [3]. Typically, such a memory is implemented using two coupled lasers switching the output signal between two wavelengths λ_1 and λ_2 [3]. Recently, we proposed a novel architecture of an all-optical memory loop combining the ultrafast all-optical signal processor based on the Mach-Zehnder interferometer (MZI) with quantum dot (QD) SOA and the DSP block for the mitigation of dispersion and nonlinearity impairments [6].

The chapter is organized as follows. OPS in optical communication systems is discussed in Section 2. The different types of all-optical logic gates used in OPS are briefly reviewed in Section 3. The operation principle of the novel all-optical memory is described in Section 4. The QD SOA theoretical model is briefly discussed in Section 5. The numerical simulation results and conclusions are presented in Sections 6 and 7, respectively.

2. Optical packet switching (OPS)

OPS process requires many components for buffering, header processing, and switching [3]. Each packet begins with a header containing the destination information [3]. When a packet arrives at a node, a router reads the header and sends it toward its destination [3]. The basic element of an optical router is a packet switch directing incoming packets to the corresponding output ports depending on the information in the header [3].

Consider the architecture and operation principle of the all-optical packet switch. The scheme of the 1 × 2 all-optical packet switch is shown in **Figure 1** [3, 5, 7]. The all-optical packet switch consists of three functional blocks: the all-optical header-processing block, the all-optical flip-flop memory block, and the WC block [7]. All-optical header processing can be realized by using the different methods such as tunable Bragg gratings, FWM in a SOA, terahertz optical asymmetric demultiplexers (TOAD), two-pulse correlation in a semiconductor laser amplifier in a loop optical mirror (SLALOM) [5, 7].

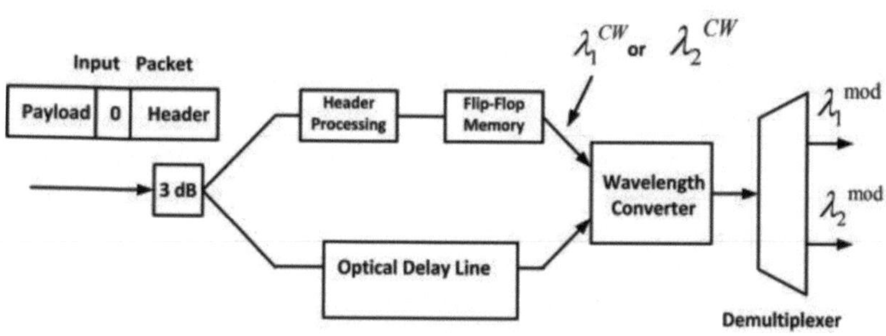

Figure 1. System concept for 1 × 2 all-optical packet switches.

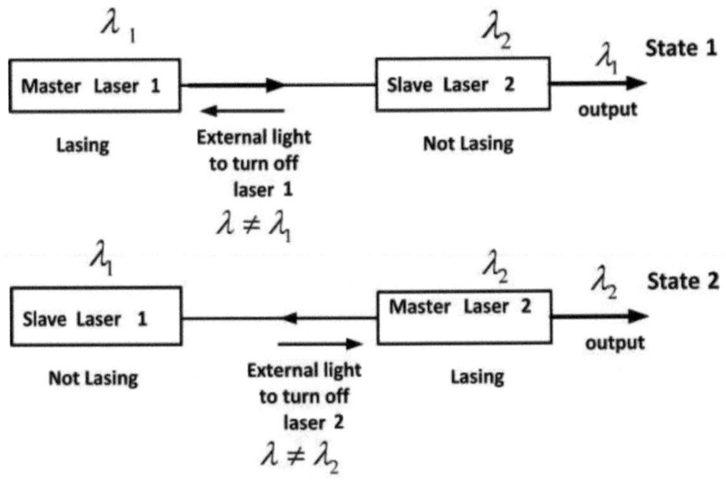

Figure 2. The all-optical flip-flop memory based on two coupled lasers.

The high-speed memory is necessary for OPS networks in order to avoid the packet collisions during packet routing [8]. The all-optical flip-flop memory is based on two coupled lasers with separate laser cavities and can have two states [7]. It is shown in **Figure 2** [7].

In state 1, light from laser 1 suppresses lasing in laser 2 emitting CW light at wavelength λ_1, while in state 2, light from laser 2 suppresses lasing in laser 1 emitting CW light at wavelength λ_2 [7]. The output pulse of the optical header processor is used to set the optical flip-flop memory into the desired wavelength [7]. The amount of light P_{sw} which is necessary for the change of states, the threshold carrier number N_{th}, and photon lifetime τ_p are given by, respectively [7]:

$$P_{sw} = E\frac{v_g R \tau_p}{L(1-R)}\ln\left(\frac{1}{R}\right)\left(1-\frac{2R}{\delta(1-R)}\right)\left(\frac{I}{q}-\frac{N_{th}}{\tau_e}\right) \tag{1}$$

$$N_{th} = \frac{V}{\tau_p \Gamma v_g a} + N_0; \ \frac{1}{\tau_p} = v_g\left(\alpha_{\text{int}} + \frac{1}{L}\ln\left(\frac{1}{R}\right)\right) \tag{2}$$

Here, E is the photon energy, R is the reflectivity at the end facets of lasers, δ is the coupling constant between the two laser cavities, v_g is the group light velocity, L is the length of the active region in the laser, I is the injection current, q is the electron charge, τ_e is the carrier lifetime, V is the volume of the laser cavity active region, Γ is the confinement factor, a is the gain factor, N_0 is the carrier number at transparency, and α_{int} is the internal laser cavity losses factor. Note that the outputs of the lasers on the left side are defined by the reflectivity at the end facets of lasers R and do not influence the memory states.

WC component converts the incoming data packet wavelength to the output wavelength of the flip-flop memory [3]. The demultiplexer directs output at different wavelengths to different ports depending on the header information [3].

3. All-optical logic gates with SOA

In this section, we briefly discuss the scheme and operation principle of all-optical gates which are core logic units for the all-optical signal processing system implementation [9]. All-optical gates may be divided into two groups: without SOA and with SOA [9]. The all-optical gates without SOA are based on the change in nonlinear refraction index in silica fiber [9]. The intensity-dependent refractive index of silica results in the following nonlinear optical effects: SPM, cross gain modulation (XGM), and FWM [2, 9]. The all-optical gates without SOA based on these nonlinear optical phenomena can be realized in the following configurations: dispersion shifted fiber/high nonlinear fiber (DSF/HNLF) configuration; waveguide configuration; circular configuration; optical channel-dropping (C/D) filter configuration; multilayer waveguide configuration; double heterostructure optical thyristor (DHOT) configuration; and acousto-optical tunable filter (AOTF) configuration [9]. The detailed description and comparison between non-SOA gates are presented in Ref. [9]. For instance, DSF/HNLF, waveguide, circular, and AOTF configurations are polarization sensitive; DSF/HNLF, waveguide, circular configurations are characterized by bad or moderate integration capacity [9].

On the other hand, the SOA-based devices are mainly polarization non-sensitive and possess compact integration capacity [9]. They are highly competitive due to the high nonlinearity, low switching power, wide gain bandwidth, and compact size [8]. Recently, novel two inputs optical logic gates (NOT, AND, OR and NOR) based on a traveling wave SOA (TW-SOA) operating at 40 Gb/s had been demonstrated [10].

The implementation of all-optical gates with SOA is based on the different interferometer techniques such as ultra-high nonlinear interferometer (UNI), Sagnac interferometer (SI), MZI, and delay interferometer (DI) [9]. In these techniques, the XPM-induced phase shift is used for optical switching [4]. Typically, a weak signal pulse is divided equally between two arms of the interferometer and is undergoing identical phase shift in each arm [4]. In such a case, it is transmitted through constructive interference [4]. Consider now the situation when a pump pulse at a different wavelength as compared to the signal pulse is injected into one arm of the interferometer. As a result, the signal phase in that arm would be changed due to XPM. If the XPM-induced phase shift is close to π, the signal pulse will not be transmitted due to the destructive interference at the interferometer output [4]. An intense pulse pump can switch the signal pulse through the XPM-induced phase shift [4].

In particular, all-optical gate based on SOA-MZI can be realized with the copropagating, counterpropagating, and copropagating push-pull configurations [9]. Copropagation MZI operates on the principle of phase change caused by the light propagating through the 3 dB coupler [9]. MZI copropagating gates consist of a symmetrical MZI with two SOA placed in the upper and lower arm of the interferometer [9]. Data and clock pulses of different wavelengths are inserted into SOA operated under the gain saturation condition, where the optical gain is distributed between wavelengths according to their relative photon densities [9]. The data are transferred in the clock pulse in the inverted form in both arms of MZI [9]. After passing through the first 3 dB coupler, the phase difference $\pi/2$ is created between the upper and lower arms of clock pulse, after passing through the second 3 dB coupler the total phase shift becomes π [9]. Then, if both data have the same value, they will cancel, and at the T-port 0 will appear, if data have different value, then it will not cancel and 1 will appear at the T-port [9]. In MZI counter-propagating gates, the clock and data pulse propagate in opposite directions through MZI [9]. If any of the data is 1, then XPM between the clock and data pulse inside SOA creates the differential phase shift between the two clock components, MZI becomes unbalanced, and the clock pulse exits at T-port [9]. If both the data are the same, the total phase shift will become π, and the clock pulse is cancelled at T-port [9].

Consider now a typical all-optical logic element based on transforming of XPM into an intensity modulation and implemented as the MZI copropagating push-pull gate with SOA in the two arms shown in **Figure 3** [9]. The optical fibers are used as interconnects. The SOA-based MZI with couplers at the input and output is shown in **Figure 4**. At the SOA-based MZI block output, there is a coupler shown in **Figure 4**. The outputs at the right side of this coupler are connected to the T-port and R-port shown in **Figure 3**.

Note that the co-propagating data streams configuration permits to avoid the SOA length restriction, and the MZI with push-pull configuration allows increasing the memory bit-rate beyond the limitation of the SOA carrier recovery time [8]. The copropagating data streams A and B of the same wavelengths are inserted into upper and lower arm of MZI shown in **Figure 4**.

The data A in the upper arm is ahead of one bit period to data B traveling in the lower arm of MZI, and the lower arm data B is one bit period ahead to upper arm data A [9]. As a result, a

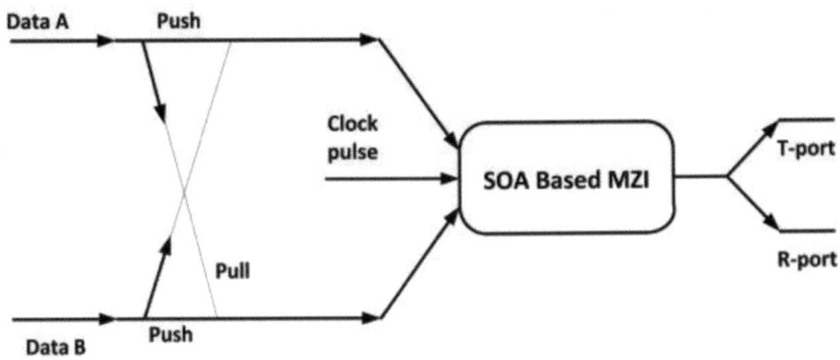

Figure 3. MZI with push-pull configuration.

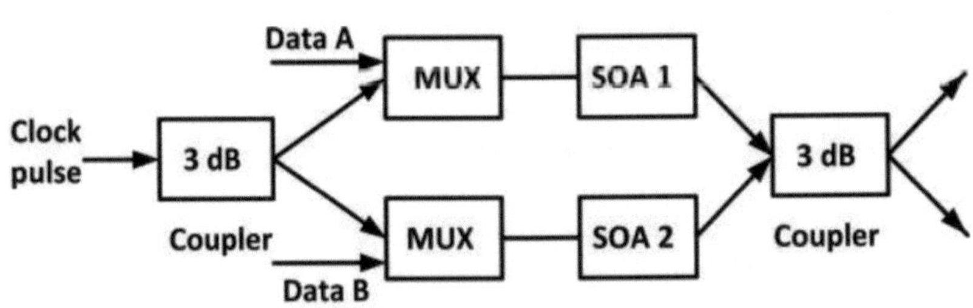

Figure 4. SOA–based MZI.

switching window for data streams occurs [9]. The clock pulse copropagating with the data streams A and B is inserted into the 3 dB coupler. Assume that the data A is 1 and data B is 0. Then, the pulse from data A splits into two parts in such a way that one pulse is pushed to the upper SOA 1 and other is delayed by the switching window. Consequently, the upper SOA 1 is switched before the lower SOA 2 [9]. The MZI is unbalanced, and clock wave is switched to the T-port. In the opposite case, when data A is 0 and data B is 1, then lower SOA 2 is switched and wave also appears at T-port [9]. Assume now that data A and data B are the same. As a result, SOA 1 and SOA 2 are equally influenced by the injected pulse. The respective push and pull pulse temporarily coincide with each other, the phase difference between the two arms of MZI equals to zero, and no switching occurs at T-port [9].

The disadvantages of the all-optical logic gates discussed above are twofold: (i) the operating speed of SOA is limited by the carrier recovery time of the order of magnitude of 100 ps; (ii) the scheme can be used only for OOK modulation format [8]. However, the SOA operation rate can be significantly increased up to 100 Gb/s by using the quantum dot semiconductor optical amplifiers (QD-SOA) [8, 11]. A theoretical model of an ultrafast all-optical signal processor based on QD-SOA MZI has been developed with limiting bit rates of 100 and 200 Gb/s at the injection currents of I = 30 mA and I = 50 mA, respectively [12].

4. All-optical memory loop based on SOA MZI

Consider now an all-optical memory loop consisting of an AND gate and a regenerator based on the two push-pull co-propagating MZI with a SOA in each arm and couplers at the input and the output shown in **Figure 4** and discussed in the previous section [8]. The scheme of the all-optical memory loop is shown in **Figure 5** [8]. The input data burst at the wavelength λ_2 is inserted into the memory through the AND gate is converted into the output data burst at the wavelength λ_1 [8]. A regenerator is used into the loop in order to improve the quality of the data burst at the wavelength λ_1 eliminating the signal degradation caused by dispersion, nonlinearity, and other physical impacts [8]. The output of the regenerator is fed back to the AND gate as a pump at the wavelength λ_3 [8]. The data can be stored in the loop for a long term [8]. The length of the data burst nT must be less than or equal to the length of the memory loop NT where T is the bit period [8].

The data format of the all-optical memory shown in **Figure 5** is OOK, while for the PSK modulation format the memory should be modified [8].

Recently, we proposed a novel architecture of an all-optical memory loop combining the ultrafast all-optical signal processor based on QD-SOA MZI [12] and the DSP block for the mitigation of the dispersion and nonlinearity impairments [6]. The proposed all-optical memory loop is shown in **Figure 6**. The advantages of the novel all-optical memory loop are following [6].

1. It includes only one MZI with two QD-SOA reducing the complexity of the electronic synchronization scheme.
2. It can operate at the bit rates up to 100 Gb/s due to the fast gain recovery time of QD-SOA.

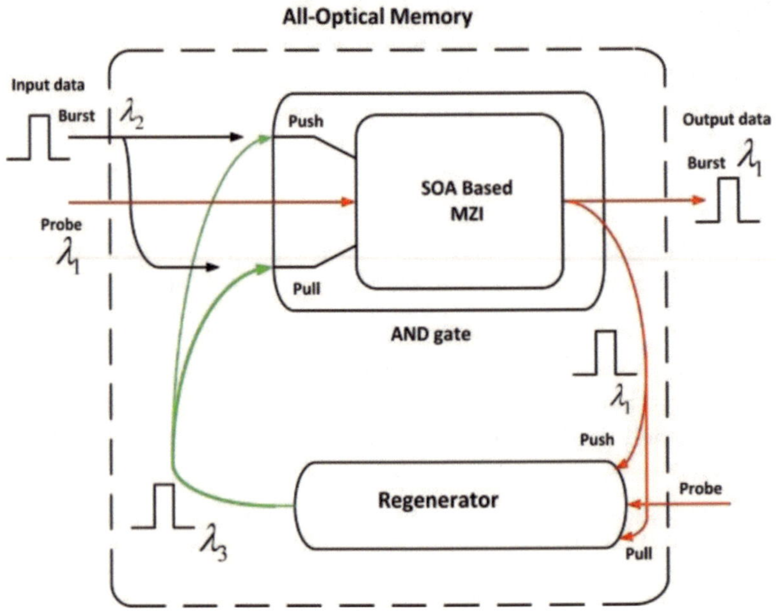

Figure 5. Schematic setup of the all-optical memory based on SOA MZI.

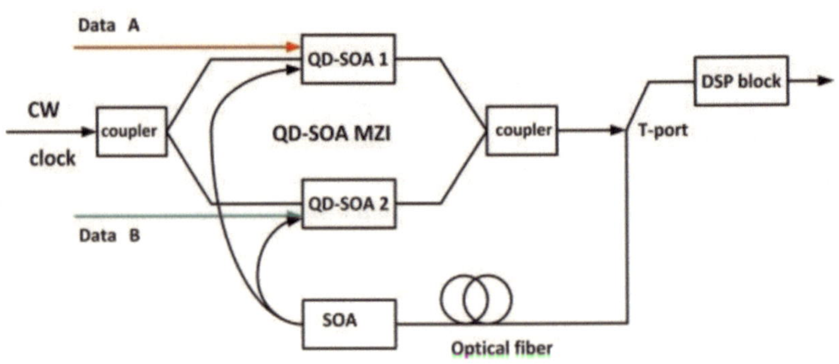

Figure 6. The architecture of the all-optical memory loop based on QD-SOA–based MZI.

3. DSP block can improve the signal quality as compared to an MZI-based regenerator.
4. An additional SOA is inserted into the loop in order to compensate the optical fiber losses and to increase the loop length.

In our case, the phase difference at the output of the QD-SOA–based MZI is caused by the signal power difference in the upper and lower MZI arms, unlike the MZI copropagating push-pull gate mentioned above [9]. Typically, 80 and 20% of the input signal power were fed through the coupler into the upper and lower arm of the MZI, respectively [6]. As a result, dynamic processes in QD-SOA placed into the upper and lower arms of MZI are characterized by different carrier relaxation time and gain recovery time, and the flatness of the switching window is significantly improving [6].

http://dx.doi.org/10.5772/intechopen.68527

The QD-SOA MZI output light intensity P_{out} and phase difference $\varnothing_1(t) - \varnothing_2(t)$ are given by Refs. [12–14]:

$$P_{out} = \frac{P_{in}}{4}\{G_1(t) + G_2(t) - 2\sqrt{G_1(t)G_2(t)}\cos[\phi_1(t) - \phi_2(t)]\} \tag{3}$$

$$\phi_1(t) - \phi_2(t) = -\frac{\alpha}{2}\ln\left(\frac{G_1(t)}{G_2(t)}\right) \tag{4}$$

where the P_{in} is continuous wave (CW) clock stream optical signal divided and introduced into the two QD-SOA, $G_{1,2} = exp(g_{1,2}L)$ and $\varnothing_{1,2}(t)$ are the gain and phase shift, respectively, in the two arms of QD-SOA MZI, α is the line width enhancement factor (LEF), $g_{1,2}$ is the SOA gain, and L is the active medium length. In order to evaluate the QD-SOA gain, we must use the theoretical model of the QD-SOA, which will be briefly discussed in Section 5.

5. The theoretical model of the QD-SOA

QD-SOA had been thoroughly investigated both theoretically and experimentally (see, for example, [15, 16] and references therein). The promising features of QD-SOA such as high saturation power, broad gain bandwidth, pattern-free amplification of single- and multi-channel signals, efficient WC can provide high-performance amplifiers and all-optical switches for optical networks [16].

QD is a nanostructure where the electron and hole movement is confined in the three dimensions, and these dimensions are of a few nanometers [15]. In QD, the charge carriers occupy only a restricted number of discrete energy levels like the electrons in an atom [15]. The density of states in QD is quantized, and the number of carriers necessary to fill these states decreases as compared to the structures with higher dimensionality. Consequently, the threshold current density in QD lasers substantially reduces, while the transparency and the population inversion necessary for the optical gain can be achieved more easily [15]. The QD grown by using the Stranski-Krastanov technology typically has a pyramidal shape with a base of about 15–20 nm and a height of about 5 nm [15]. The QD has a significant size dispersion, which results in the inhomogeneous broadening of the QD laser and SOA spectrum [15]. QD structures exhibit ultrafast gain recovery time which results in the ultrafast carrier dynamics [15]. The active layer of a QD-SOA contains one or several quantum wells (QW) referred to as a wetting layer (WL) with a continuous carrier energy band [15]. The electrons in the Stranski-Krastanov grown QD are typically characterized by two energy levels: the ground state (GS) situated about 100 meV below the band gap of WL and the excited state (ES), or the first excited level [15]. The energy level structure of a QD laser or QD-SOA is presented in **Figure 7**.

The carrier dynamics in QD is described by the system of the rate equations taking into account the following electron transitions: the fast electron transitions from WL to ES with the relaxation time $\tau_{wE} \sim 3$ ps; the fast electron transitions between ES and GS with the corresponding relaxation times$\tau_{EG} \approx 0.16$ ps; $\tau_{GE} \sim 1.2$ ps; the slow electron transitions from ES to WL with the escape time $\tau_{Ew} \sim 1$ ns. The QD-SOA rate equations have the form [12, 17]:

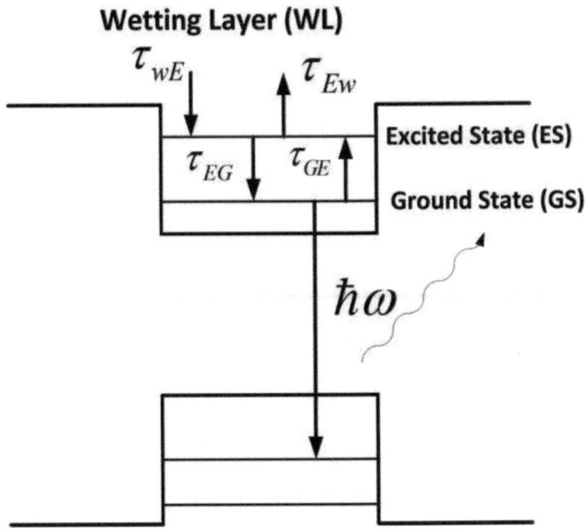

Figure 7. The energy level structure and the electron transitions in a QD.

$$\frac{\partial N_w}{\partial t} = \frac{J}{eL_w} - \frac{N_w(1-h)}{\tau_{wE}} + \frac{N_w}{\tau_{Ew}} - \frac{N_w}{\tau_{wR}} \tag{5}$$

$$\frac{\partial h}{\partial t} = \frac{N_w L_w (1-h)}{N_Q \tau_{wE}} - \frac{N_w L_w h}{N_Q \tau_{Ew}} - \frac{(1-f)h}{\tau_{EG}} + \frac{f(1-h)}{\tau_{GE}} \tag{6}$$

$$\begin{aligned} \frac{\partial f}{\partial t} = & \frac{(1-f)h}{\tau_{EG}} - \frac{f(1-h)}{\tau_{GE}} - \frac{f^2}{\tau_R} - \frac{g_p L}{N_Q}(2f-1) S_p \frac{c}{\sqrt{\varepsilon_r}} \\ & - \frac{g_s L}{N_Q}(2f-1) S_s \frac{c}{\sqrt{\varepsilon_r}} \end{aligned} \tag{7}$$

where J is injection current density, N_w is the WL carrier density per unit volume, f is the electron occupation probability of GS, h is the electron occupation probability of ES, $S_{p,s}$ are the pump (data A or data B) and signal (clock stream) wave photon densities averaged over the length of SOA L, $g_{p,s}$ are the pump and signal wave modal gains, respectively, e is the electron charge, $N_Q \sim (10^{10} - 10^{12})\ \mathrm{cm}^{-2}$ is the QD density per unit area, L_w is the effective thickness of the active layer, ε_r is the SOA material permittivity, and c is the free space light velocity. The average photon densities $S_{p,s}$ are given by Ben Ezra [12]:

$$S_{p,s}(\tau) = \frac{[S_{p,s}(\tau)]_{in}}{L} \int_0^L dz \exp\left(\int_0^z (g_{p,s} - \alpha_{\mathrm{int}}) dz' \right) \tag{8}$$

where $\tau = t - \left(\frac{z}{v_g}\right)$, t, z, v_g are time, coordinate, and the optical wave group velocity, respectively. The pump and signal wave phase $\varnothing_{p,s}$ and modal gain $g_{p,s}$ are given by, respectively [12]:

$$\theta_{p,s}(\tau) = -\frac{\alpha}{2}\int_0^L g_{p,s} dz \tag{9}$$

$$g_{p,s}(\omega) = \frac{2\Gamma N_Q}{a}\int d\omega F(\omega)\sigma(\omega_0)(2f-1) \tag{10}$$

where Γ is the confinement factor, l is the number of QD layers, a is the mean size of QD, $\sigma(\omega_0)$ is the cross section of interaction of photons of frequency ω_0 with carriers in QD at the transition frequency ω including the homogeneous broadening factor, $F(\omega)$ is the Gaussian distribution of the transition frequency in the QD ensemble related to the inhomogeneous broadening. The inhomogeneous broadening is caused by the QD shape and size variations as it was mentioned above. It is given by Ben Ezra [12]:

$$F(\omega) = \frac{1}{\Delta\omega\sqrt{\pi}}\exp\left[-\frac{(\omega-\overline{\omega})^2}{(\Delta\omega)^2}\right] \tag{11}$$

where $\Delta\omega$ is related to the inhomogeneous bandwidth, $\gamma_{inhom} = 2\Delta\omega\sqrt{ln2}$, and $\overline{\omega}$ is the average transition frequency.

6. The simulation results

We solved numerically Eqs. (3)–(11) for the typical values of parameters [12] for the QD-SOA–based all-optical loop shown in **Figure 6** [6]. The numerical simulations have been carried out for the OOK and pulse amplitude modulation 4 (4-PAM) formats [6]. We used the MATLAB environment. Simulation results for the eye diagrams in the case of the OOK modulation format are shown in **Figures 8** and **9**.

We used the QD-SOA–based all-optical memory model with the loop length of L = 2 km and the input OOK modulated signal with the bit rate of 50 Gb/s and the quality factor at the input Q = 15.8932. The comparison of the eye diagrams presented in **Figures 8** and **9** shows that after 4 rounds in the loop, the quality factor of the signal decreases by approximately 18%.

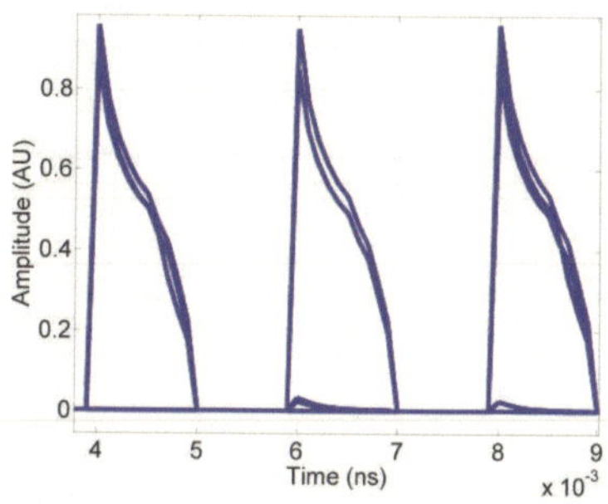

Figure 8. The eye diagram for the OOK modulation format, a bit rate of 50 Gb/s after one round in the QD-SOA–based all optical memory loop. The memory loop length L = 2 km, T = 10 μs, the quality factor Q = 13.3156.

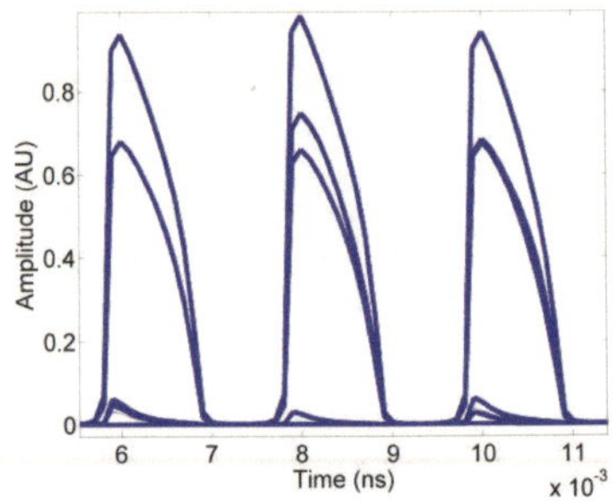

Figure 9. The eye diagram for the OOK modulation format, a bit rate of 50 Gb/s after four rounds in the QD-SOA–based all-optical memory loop. The memory loop length L = 2 km, T = 40 μs, the quality factor Q = 13.0327.

The simulations results for the 4-PAM modulation format, QD SOA–based all-optical memory loop with the length of L = 1 km and bit rates of 50 and 100 Gb/s are shown in **Figures 10–13**, respectively.

Eye diagrams in **Figures 10** and **11** clearly show that for the bit rate of 50 Gb/s, the patterning effect is negligible after two rounds in the memory loop. In the case of the bit rate of 100 Gb/s, the patterning effect is slightly pronounced after the two rounds in the memory loop as it is

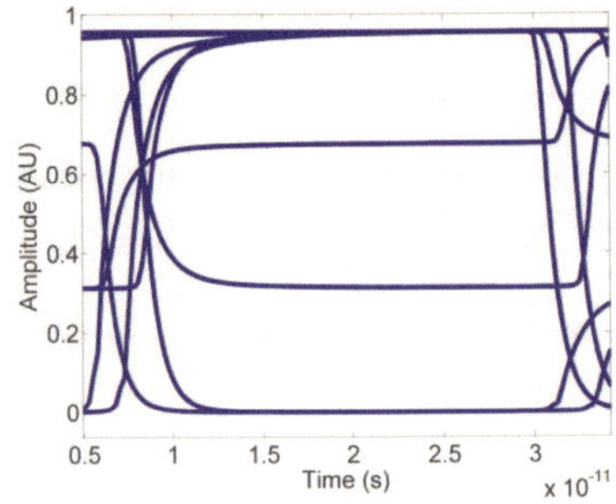

Figure 10. The eye diagram for 4-PAM modulation format and a bit rate of 50 Gb/s, the memory loop length L = 1 km. The signal is at the input of the all-optical memory loop.

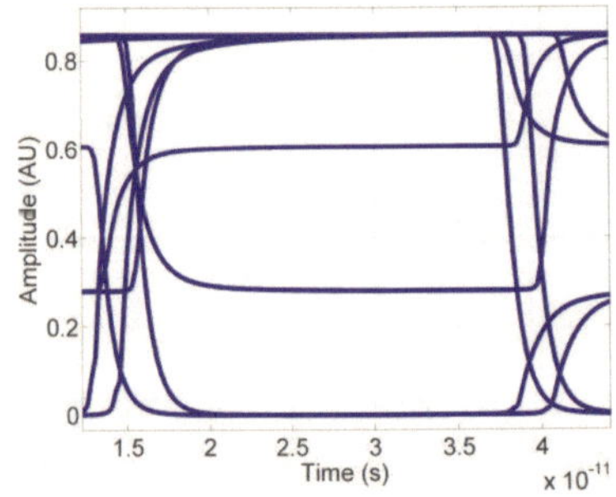

Figure 11. The eye diagram for 4-PAM modulation format and a bit rate of 50 Gb/s, the memory loop length L = 1 km. The signal is after two rounds in the all-optical memory loop.

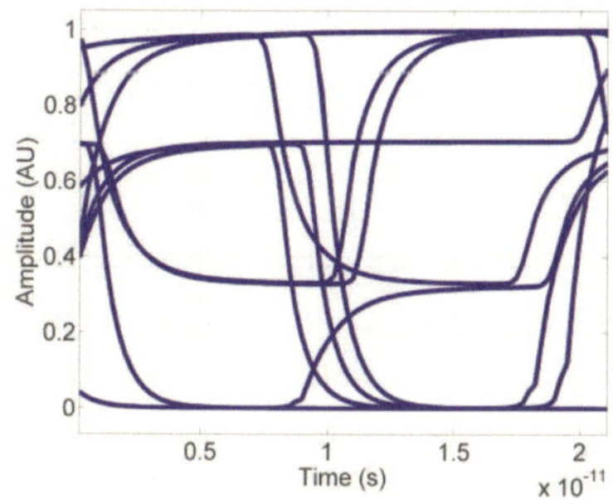

Figure 12. The eye diagram for 4-PAM modulation format and a bit rate of 100 Gb/s, the memory loop length L = 1 km. The signal is at the input of the all-optical memory loop.

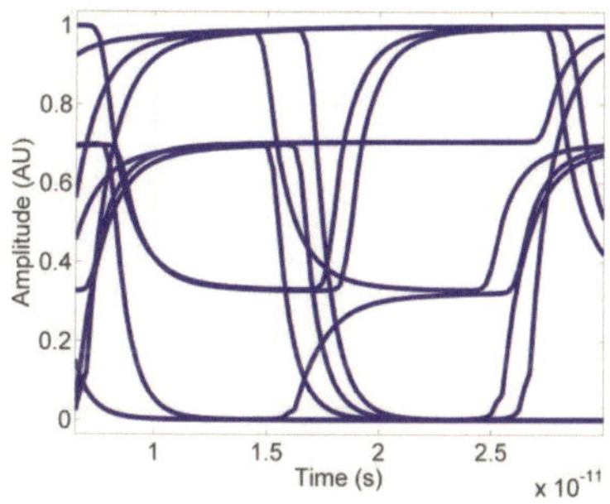

Figure 13. The eye diagram for 4-PAM modulation format and a bit rate of 100 Gb/s, the memory loop length L = 1 km. The signal is after two rounds in the all-optical memory loop.

seen from **Figures 12** and **13**. The QD-SOA–based all-optical memory performance does not deteriorate up to the bit rate of 100 Gb/s.

The numerical estimations show that for loop length $L = 1$ km, the light velocity in the optical fiber $v \approx 2 \times 10^8$ m/s and the bit rates of 50 and 100 Gb/s the all-optical memory storage values are of 0.25 and 0.5 Mb, the storage times are 5 and 10 μs, respectively [6].

7. Conclusions

Optical signal processing substantially increases the device processing speed, provides an alternative to electronic techniques, and, at the same time, can enhance the processing capabilities of electronics when the optical and electronic signal processing techniques are combined. Optical packet-switched networks are promising candidates for the advanced optical telecommunication systems. All-optical signal processing is essential for ultrafast OPS in such networks. Flip-flop memory is a basic component of an all-optical packet switch. We discussed the all-optical memory based on SOA-MZI. It is characterized by the high performance due to the SOA high nonlinearity, low switching power, wide gain bandwidth, compact size, and integration capability with other photonic devices. Unfortunately, the SOA operation rate is limited by its comparatively slow gain recovery time.

We proposed a novel architecture of the ultrafast all-optical memory based on MZI with two QD-SOA. QD-SOA is characterized by high operation rate and low threshold current caused by the 3D carrier confinement and fast electron transitions in QD. For this reason, the operation rate of the proposed all-optical memory loop increases up to 100 Gb/s. The time delay in the typical push-pull scheme of the memory loop is replaced with the output signal phase difference caused by the signal power difference in the MZI arms due to the highly efficient XPM in QD-SOA. The DSP block is inserted for the dispersion and nonlinearity impairment mitigation. The proposed all-optical memory loop includes only one MZI with two QD-SOA and reduces the complexity of electronic synchronization scheme. We carried out numerical simulations of the proposed all-optical memory loop based on the QD-SOA rate equations and the expression for the MZI output power for the OOK and 4-PAMmodulation formats. The simulation results show that the proposed all-optical memory exhibits a high performance up to the bit rate of 100 Gb/s and the corresponding memory storage and storage time values of 0.5 Mb and 10 μs, respectively.

Author details

Yossef Ben Ezra[1,2] and Boris I. Lembrikov[1]*

*Address all correspondence to: borisle@hit.ac.il

1 Department of Electrical Engineering and Electronics, Holon Institute of Technology (HIT), Holon, Israel

2 MER Cellos, Holon, Israel

References

[1] Agrell E, Karlsson M, Chraplyvy AR, et al. Roadmap of optical communications. Journal of Optics. 2016;**18**:1–40. DOI: 10.1088/2040-8978/18/6/063002

[2] Willner AE, Khaleghi S,Chitgarha MR, Yilmaz OF. All-optical signal processing. Journal of Lightwave Technology. 2014;**32**:660–680. DOI: 10.1109/JLT.2013.2287219

[3] Agrawal GP. Fiber-Optic Communication Systems. 4th ed. New York: Wiley; 2010. ISBN: 978-0-470-50511-3

[4] Agrawal GP. Nonlinear Fiber Optics. 5th ed. New York: Academic Press; 2013. ISBN: 978-0-12397-023-7

[5] Liu Y, Hill MT, Calabretta N, Tangdiongga E, Geldenhuys R, Zhang S, Li Z, De Waardt H, Khoe GD, Dorren HJS. All-optical signal processing for optical packet switching networks. In: Proceedings of SPIE; September 15, 2005; San Diego, California, USA; 2005. 59070J-1-12. DOI: 10.1117/12.621346

[6] Ben Ezra Y, Lembrikov BI. All-optical memory based on quantum dot semiconductor optical amplifiers (QD-SOAs) for advanced modulation formats. In: Proceedings of 18th International Conference on Transparent Optical Networks (ICTON 2016); July 10-14, 2016; Trento, Italy; Tu. A5.3 1-3. ISBN: 978-1-5090-1466-8

[7] Dorren HJS, Hill MT, Liu Y, Calabretta N, Srivatsa A, Huijskens FM, De Waardt H, Khoe GD. Optical packet switching and buffering by using all-optical signal processing methods. Journal of Lightwave Technology. 2003;**21**:2–12. DOI: 10.1109/JLT.2002.803062

[8] Xuelin Y, Qiwei W, Weisheng H. High-speed all-optical long-term memory using SOA MZIs: Simulation and experiment. Optics Communications. 2012;**285**:4043-4047. DOI: 10.1016/j.optcom.2012.06.027

[9] Singh P, Tripathi D Kr, Jaiswal S, Dixit HK. All-optical logic gates: Designs, classification and comparison. Advances in Optical Technologies. 2014;**2014**:1–13. DOI: 10.1155/2014/275083

[10] Shehata MI, Mohammed NA. Design and optimization of novel two inputs logic gates (NOT, AND, OR and NOR) based on single commercial TW-SOA operating at 40 Gbit/s. Optical and Quantum Electronics. 2016;**48**:336(1–16). DOI: 10.1007/s1 1082-016-0602-2

[11] Stubkjaer KE. Semiconductor optical amplifier-based all-optical gates for high-speed optical processing. IEEE Journal of Selected Topics in Quantum Electronics. 2000;**6**:1428–1435. DOI: 10.1109/2944.902198

[12] Ben Ezra Y, Lembrikov BI, Haridim M. Ultra-fast all-optical processor based on quantum dot semiconductor optical amplifiers (QD-SOA). IEEE Journal of Quantum Electronics. 2009;**45**:34-41. DOI: 10.1109/JQE.2008.2003497

[13] Sun H, Wang Q, Dong H, Dutta NK. XOR performance of a quantum dot semiconductor optical amplifier based Mach-Zender interferometer. Optics Express. 2005;**13**:1892-1899. ISSN: 1094-4087

[14] Wang Q, Zhu G, Chen H, Jaques J, Leuthold J, Picirilli AB, Dutta NK. Study of all-optical XOR using Mach-Zender interferometer and differential scheme. IEEE Journal of Quantum Electronics. 2004;**40**:703-710. DOI: 10.1109/JQE.2004.828261

[15] Rafailov EU, Cataluna MA, Avrutin EA. Ultrafast Lasers Based on Quantum Dot Structures. Weinheim: Wiley-VCH; 2011. p. 250. ISBN: 978-3-527-40928-0

[16] Sugawara M, Ebe H, Hatori N, et al. Theory of optical signal amplification and processing by quantum-dot semiconductor optical amplifiers. Physical Review B. 2004;**69**:235332-1-2 35332-39. DOI: 10.1103/PhysRevB.69.235332

[17] Qasaimeh O. Optical gain and saturation characteristics of quantum-dot semiconductor optical amplifiers. IEEE Journal of Quantum Electronics. 2003;**39**:793-798. DOI: 10.1109/JQE.2003.810770

Section 4

Optical Signal Processing and Security

Chapter 15

Digital Signal Processing for Optical Communications and Networks

Tianhua Xu

Additional information is available at the end of the chapter

http://dx.doi.org/10.5772/intechopen.68323

Abstract

The achievable information rates of optical communication networks have been widely increased over the past four decades with the introduction and development of optical amplifiers, coherent detection, advanced modulation formats, and digital signal processing techniques. These developments promoted the revolution of optical communication systems and the growth of Internet, towards the direction of high-capacity and long-distance transmissions. The performance of long-haul high-capacity optical fiber communication systems is significantly degraded by transmission impairments, such as chromatic dispersion, polarization mode dispersion, laser phase noise and Kerr fiber nonlinearities. With the entire capture of the amplitude and phase of the signals using coherent optical detection, the powerful compensation and effective mitigation of the transmission impairments can be implemented using the digital signal processing in electrical domain. This becomes one of the most promising techniques for next-generation optical communication networks to achieve a performance close to the Shannon capacity limit. This chapter will focus on the introduction and investigation of digital signal processing employed for channel impairments compensation based on the coherent detection of optical signals, to provide a roadmap for the design and implementation of real-time optical fiber communication systems.

Keywords: optical communications, optical networks, digital signal processing, coherent detection, chromatic dispersion, polarization mode dispersion, laser phase noise, fiber nonlinearities

1. Introduction

The performance of high-capacity optical communication systems can be significantly degraded by fiber attenuation, chromatic dispersion (CD), polarization mode dispersion (PMD), laser phase noise (PN), and Kerr nonlinearities [1–10]. Using coherent detection, the powerful compensation of transmission impairments can be implemented in electrical domain. With the full information of the received signals, the chromatic dispersion, the polarization mode dispersion, the carrier phase noise, and the fiber Kerr nonlinearities can be equalized and mitigated using digital signal processing (DSP) [11–22].

Due to the high sensitivity of the receiver, coherent optical transmission was investigated extensively in the eighties of last century [23, 24]. However, the development of coherent communication has been delayed for nearly 20 years after that period [25, 26]. Coherent optical detection re-attracted the research interests until 2005, since the advanced modulation formats, i.e., *m*-level phase shift keying (*m*-PSK) and *m*-level quadrature amplitude modulation (*m*-QAM), can be applied [27–30]. In addition, coherent optical detection allows the electrical mitigation of system impairments. With the two main merits, the reborn coherent detections brought us the enormous potential for higher transmission speed and spectral efficiency in current optical fiber communication systems [31, 32].

With an additional local oscillator (LO) source, the sensitivity of coherent receiver reached the limitation of the shot-noise. Furthermore, compared to the traditional intensity modulation direct detection system, the multilevel modulation formats can be applied using the phase modulations, which can include more information bits in one transmitted symbol than before.

Meanwhile, since the coherent demodulation is linear and all information of the received signals can be detected, signal processing approaches, i.e., tight spectral filtering, CD equalization, PMD compensation, laser PN estimation, and fiber nonlinearity compensation, can be implemented in electrical domain [33–40].

The typical block diagram of the coherent optical transmission system is shown in **Figure 1**. The transmitted optical signal is combined coherently with the continuous wave from the narrow-linewidth LO laser so that the detected optical intensity in the photodiode (PD) ends can be increased and the phase information of the optical signal can be obtained. The use of LO laser is to increase the receiver sensitivity of the detection of optical signals, and the performance of coherent transmission can even behave close to the Shannon limit [3, 12].

The development of the coherent transmission systems has stopped for more than 10 years due to the invention of Erbium-doped fiber amplifiers (EDFAs) [1, 2]. The coherent transmission techniques attracted the interests of investigation again around 2005, when a new stage of the coherent lightwave systems comes out by combining the digital signal processing techniques [41–46]. This type of coherent lightwave system is called as digital coherent communication system. In the digital coherent transmission systems, the electrical

http://dx.doi.org/10.5772/intechopen.68323

signals output from the photodiodes are sampled and transformed into the discrete signals using high-speed analogue-to-digital convertors (ADCs), which can be further processed by the DSP algorithms.

The phase locking and the polarization adjustment were the main obstacles in the traditional coherent lightwave systems, while they can be solved by the carrier phase estimation and the polarization equalization, respectively, in the digital coherent optical transmission systems [47–55]. Besides, the chromatic dispersion and the nonlinear effects can also be mitigated by using the digital signal processing techniques [56–62]. The typical structure of the DSP compensating modules in the digital coherent receiver is shown in **Figure 2**.

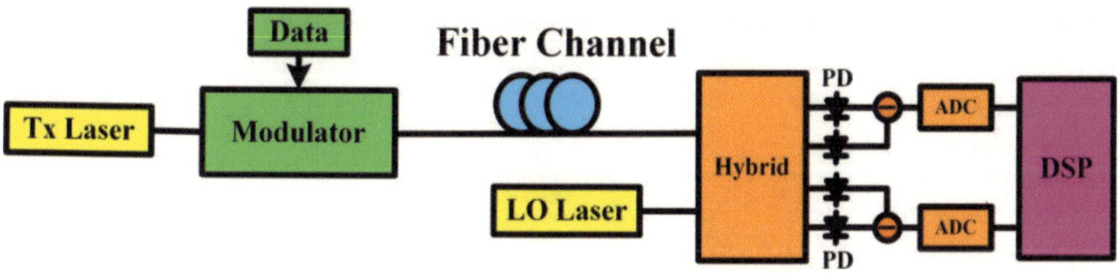

Figure 1. Schematic of coherent optical communication system with digital signal processing.

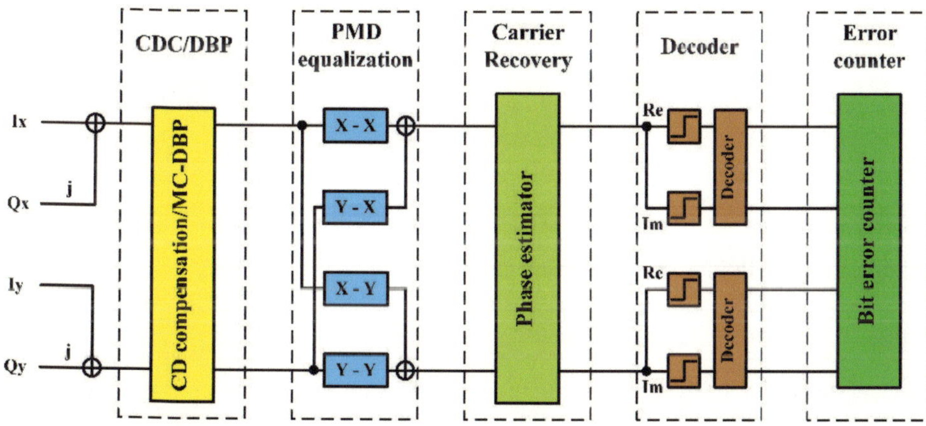

Figure 2. Block diagram of DSP in digital coherent receiver.

2. Digital signal processing for compensating transmission impairments

In this section, the chromatic dispersion compensation, polarization mode dispersion equalization, and carrier phase noise compensation are analyzed and discussed using corresponding DSP algorithms.

2.1. Chromatic dispersion compensation

Digital filters involving the time-domain least-mean-square (TD-LMS) adaptive filter, the static time-domain finite impulse response (STD-FIR) filter, and the frequency-domain equalizers (FDEs) are investigated for CD compensation. The characters of these filters are analyzed based on a 28-Gbaud dual-polarization quadrature phase shift keying (DP-QPSK) coherent transmission system using postcompensation of dispersion. It is noted that the STD-FIR filter and the FDEs can also be used for the dispersion predistorted coherent communication systems.

2.1.1. Time domain least-mean-square equalizer

The TD-LMS filter employs an iterative algorithm that incorporates successive corrections to weights vector in the negative direction of the gradient vector, which eventually leads to a minimum mean square error [34, 38, 63–65]. The transfer function of the TD-LMS digital filter can be described as follows:

$$y_{out}(n) = \vec{W}_{LMS}^{H}(n)\vec{x}_{in}(n) \tag{1}$$

$$\vec{W}_{LMS}(n+1) = \vec{W}_{LMS}(n) + \mu_{LMS}\vec{x}_{in}(n)e_{LMS}^{*}(n) \tag{2}$$

$$e_{LMS}(n) = d_{LMS}(n) - y_{out}(n) \tag{3}$$

where $\vec{x}_{in}(n)$ is the vector of received signals, $y_{out}(n)$ is the equalized output signal, n is the index of signal, $\vec{W}_{LMS}(n)$ is the vector of tap weights, H is the Hermitian transform operator, $d_{LMS}(n)$ is the desired symbol, $e_{LMS}(n)$ is the error between the desired symbol and the output signal, * is the conjugation operator, and μ_{LMS} is the step size. To ensure the convergence of tap weights $\vec{W}_{LMS}(n)$, the step size μ_{LMS} has to meet the condition of $\mu_{LMS} < 1/U_{\max}$, where $U_{\max}$ is the largest eigenvalue of the correlation matrix $R = \vec{x}_{in}(n)\vec{x}_{in}^{H}(n)$ [63]. The TD-LMS dispersion compensation filter can be applied in the "decision-directed" or the "sequence-training" mode [63].

The tap weights in TD-LMS adaptive equalizer for 20 km fiber CD compensation are shown in **Figure 3**. The convergence for 9 tap weights in the TD-LMS filter with step size equal to 0.1 is shown in **Figure 3(a)**, and it is found that the tap weights reach their convergence after ~5000 iterations. The distribution of the magnitudes of the converged tap weights is plotted in **Figure 3(b)**, and it is found that the central tap weights take more dominant roles than the high-order tap weights [34, 66].

2.1.2. Static time-domain finite impulse response filter

Compared with the iteratively updated TD-LMS filter, the tap weights in STD-FIR filter have a relatively simple specification [34, 67–69], the tap weight in STD-FIR filter is given by the following equations:

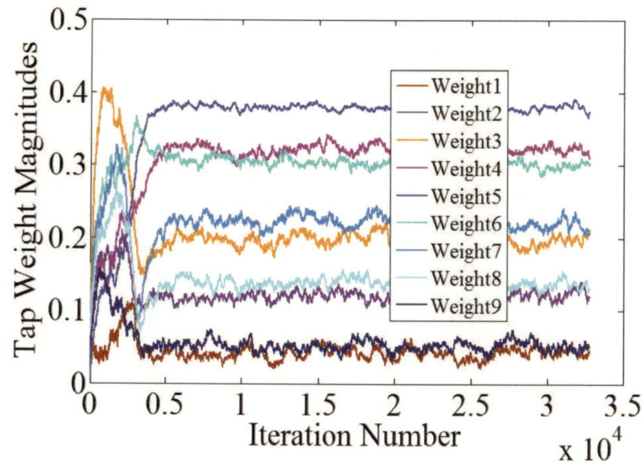

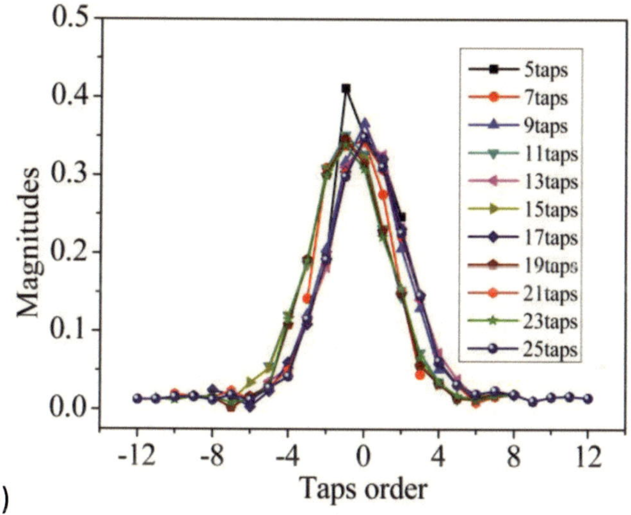

Figure 3. Taps weights of TD-LMS filter. (a) Tap weights magnitudes convergence and (b) converged tap weights magnitudes distribution.

$$a_k = \sqrt{\frac{jcT^2}{D\lambda^2 L}} \exp\left(-j\frac{\pi c T^2}{D\lambda^2 L}k^2\right) \quad -\left\lfloor\frac{N}{2}\right\rfloor \leq k \leq \left\lfloor\frac{N}{2}\right\rfloor \tag{4}$$

$$N^A = 2\times\left\lfloor\frac{|D|\lambda^2 L}{2cT^2}\right\rfloor + 1 \tag{5}$$

where D is the CD coefficient, λ is the carrier central wavelength, L is the length of fiber, T is the sampling period, N^A is the maximum number of taps, and $\lfloor x \rfloor$ means the nearest integer smaller than x.

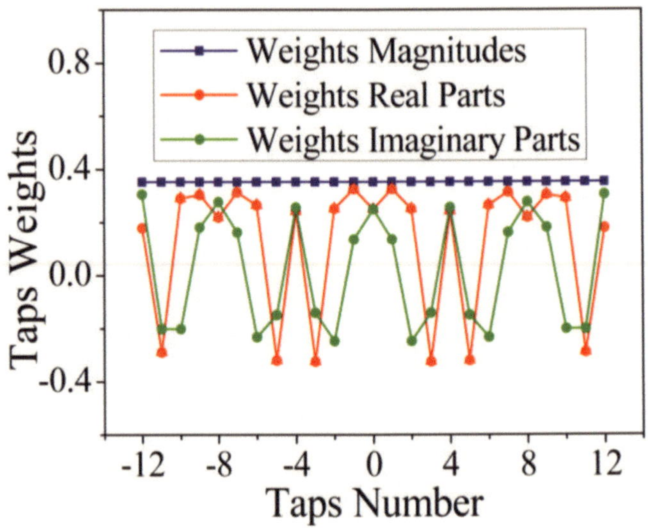

Figure 4. Tap weights of STD-FIR chromatic dispersion compensation filter.

For 20 km fiber with CD coefficient of $D = 16ps/(nm \cdot km)$, the distribution of the tap weights in the STD-FIR filter is shown in **Figure 4.**

2.1.3. Frequency domain equalizers

Since the complexity is very low for compensating large CD [34, 70], the most promising and popular chromatic dispersion compensation filters in coherent transmission systems are the frequency domain equalizers. The transfer function of the frequency domain equalizers is given by the following expression:

$$G_c(L, \omega) = \exp\left(\frac{-jD\lambda^2\omega^2 L}{4\pi c}\right) \tag{6}$$

where D is the chromatic dispersion coefficient, λ is the carrier central wavelength, ω is the angular frequency, L is the length of fiber, and c is the light speed in vacuum.

The frequency domain equalizers are generally implemented using the overlap-save (OLS) and the overlap-add (OLA) approaches based on the fast Fourier transform and the inverse fast Fourier transform (iFFT) convolution algorithms [71–73], as described in **Figure 5**.

2.2. Polarization mode dispersion equalization

Due to the random character of the polarization mode dispersion and the polarization rotation, the compensation of the PMD and the polarization rotation are generally realized by the adaptive algorithms such as the least-mean-square (LMS) and the constant modulus algorithm (CMA) filters.

http://dx.doi.org/10.5772/intechopen.68323

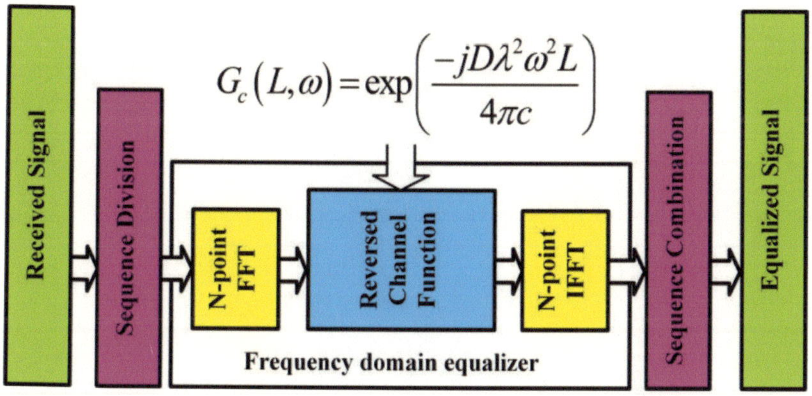

Figure 5. Schematic of frequency domain equalizer for chromatic dispersion compensation.

2.2.1. *LMS adaptive PMD equalization*

In the electrical domain, the impact of the PMD and the polarization fluctuation can be adaptively equalized using the decision-directed LMS (DD-LMS) filter [36, 63], of which the transfer function is given by:

$$\begin{bmatrix} x_{out}(n) \\ y_{out}(n) \end{bmatrix} = \begin{bmatrix} \vec{w}_{xx}^{H}(n) & \vec{w}_{xy}^{H}(n) \\ \vec{w}_{yx}^{H}(n) & \vec{w}_{yy}^{H}(n) \end{bmatrix} \cdot \begin{bmatrix} \vec{x}_{in}(n) \\ \vec{y}_{in}(n) \end{bmatrix} \tag{7}$$

$$\begin{cases} \vec{w}_{xx}(n+1) = \vec{w}_{xx}(n) + \mu_p \cdot \varepsilon_x(n) \cdot \vec{x}_{in}^{*}(n) \\ \vec{w}_{yx}(n+1) = \vec{w}_{yx}(n) + \mu_p \cdot \varepsilon_y(n) \cdot \vec{x}_{in}^{*}(n) \\ \vec{w}_{xy}(n+1) = \vec{w}_{xy}(n) + \mu_p \cdot \varepsilon_x(n) \cdot \vec{y}_{in}^{*}(n) \\ \vec{w}_{yy}(n+1) = \vec{w}_{yy}(n) + \mu_p \cdot \varepsilon_y(n) \cdot \vec{y}_{in}^{*}(n) \end{cases} \tag{8}$$

$$\begin{cases} \varepsilon_x(n) = d_x(n) - x_{out}(n) \\ \varepsilon_y(n) = d_y(n) - y_{out}(n) \end{cases} \tag{9}$$

where $\vec{x}_{in}(n)$ and $\vec{y}_{in}(n)$ are the vectors of the input signals, $x_{out}(n)$ and $y_{out}(n)$ are the equalized output signals, respectively, $\vec{w}_{xx}(n)$, $\vec{w}_{xy}(n)$, $\vec{w}_{yx}(n)$ and $\vec{w}_{yy}(n)$ are the complex tap weights vectors, $d_x(n)$ and $d_y(n)$ are the desired symbols, $\varepsilon_x(n)$ and $\varepsilon_y(n)$ are the estimation errors between the desired symbols and the output signals in the two polarizations, respectively, and μ_p is the step size in the DD-LMS algorithm.

2.2.2. *CMA adaptive PMD equalization*

The influence of the PMD and the polarization fluctuation can also be compensated employing the CMA adaptive filter [74, 75], of which the transfer function can be described as:

$$\begin{bmatrix} x_{out}(n) \\ y_{out}(n) \end{bmatrix} = \begin{bmatrix} \vec{v}^{H}_{xx}(n) & \vec{v}^{H}_{xy}(n) \\ \vec{v}^{H}_{yx}(n) & \vec{v}^{H}_{yy}(n) \end{bmatrix} \cdot \begin{bmatrix} \vec{x}_{in}(n) \\ \vec{y}_{in}(n) \end{bmatrix} \tag{10}$$

$$\begin{cases} \vec{v}_{xx}(n+1) = \vec{v}_{xx}(n) + \mu_q \cdot \eta_x(n) \cdot \vec{x}^{*}_{in}(n) \\ \vec{v}_{yx}(n+1) = \vec{v}_{yx}(n) + \mu_q \cdot \eta_y(n) \cdot \vec{x}^{*}_{in}(n) \\ \vec{v}_{xy}(n+1) = \vec{v}_{xy}(n) + \mu_q \cdot \eta_x(n) \cdot \vec{y}^{*}_{in}(n) \\ \vec{v}_{yy}(n+1) = \vec{v}_{yy}(n) + \mu_q \cdot \eta_y(n) \cdot \vec{y}^{*}_{in}(n) \end{cases} \tag{11}$$

$$\begin{cases} \eta_x(n) = 1 - |x_{out}(n)|^2 \\ \eta_y(n) = 1 - |y_{out}(n)|^2 \end{cases} \tag{12}$$

where $\vec{x}_{in}(n)$ and $\vec{y}_{in}(n)$ are the vectors of the input signals, $x_{out}(n)$ and $y_{out}(n)$ are the equalized output signals, respectively, $\vec{v}_{xx}(n)$, $\vec{v}_{xy}(n)$, $\vec{v}_{yx}(n)$, and $\vec{v}_{yy}(n)$ are the complex tap weights vectors, $\eta_x(n)$ and $\eta_y(n)$ are the estimation errors between the desired amplitude and the output signals in the two polarizations, respectively, and μ_q is the step size in the CMA algorithm.

It can be found that the CMA algorithm is based on the principle of minimizing the modulus variation of the output signal to update its weight vector.

2.3. Carrier phase estimation

In this section, the analyses on different carrier phase estimation algorithms, involving the one-tap normalized LMS, the differential phase estimation, the block-wise average (BWA), and the Viterbi-Viterbi (VV) methods in the coherent optical transmission systems, will be presented.

2.3.1. *The normalized LMS carrier phase estimation*

The one-tap normalized LMS filter can be employed effectively for carrier phase estimation [76–78], of which the tap weight is expressed as.

$$w_{NLMS}(n+1) = w_{NLMS}(n) + \frac{\mu_{NLMS}}{|x_{in}(n)|^2} x^{*}_{in}(n) e_{NLMS}(n) \tag{13}$$

$$e_{NLMS}(n) = d_{PE}(n) - w_{NLMS}(n) \cdot x_{in}(n) \tag{14}$$

where $w_{NLMS}(n)$ is the tap weight, $x_{in}(n)$ is the input signal, n is the symbol index, $d_{PE}(n)$ is the desired symbol, and $e_{NLMS}(n)$ is the carrier phase estimation error between the desired symbol and the output signal, and μ_{NLMS} is the step size in the one-tap normalized LMS filter.

It has been demonstrated that the one-tap normalized LMS carrier phase estimation behaves similar to the differential phase estimation [28, 53, 55, 76], of which the BER floor in the m-PSK coherent optical transmission systems can be approximately described by the following analytical expression:

$$BER_{floor}^{NLMS} \approx \frac{1}{\log_2 m} erfc\left(\frac{\pi}{m\sqrt{2}\sigma}\right) \tag{15}$$

where σ is the square root of the phase noise variance. The schematic of the one-tap normalized LMS carrier phase estimation is illustrated in **Figure 6**.

2.3.2. *Differential carrier phase estimation*

The differential signal demodulation can be also applied for carrier phase estimation in coherent transmission system [28, 53, 55], where the differentially encoded data can be recovered using the "delay and multiply" algorithm. Using differential carrier phase estimation, the encoded information can be recovered according to the phase difference between the two consecutive symbols, i.e., the decision variable $\Psi = x_n x_{n+1}^* \exp\{i\pi/m\}$, where x_n and x_{n+1} are the consecutive n-th and (n+1)-th received symbols. The BER floor of the differential carrier phase estimation can be evaluated using the principle of conditional probability. For the *m*-PSK coherent systems, the BER floor in differential phase estimation is expressed as the following equation [28, 53]:

$$BER_{floor}^{Differential} = \frac{1}{\log_2 m} erfc\left(\frac{\pi}{m\sqrt{2}\sigma}\right) \tag{16}$$

where σ is the square root of the phase noise variance. The schematic of the differential carrier phase estimation is described in **Figure 7**.

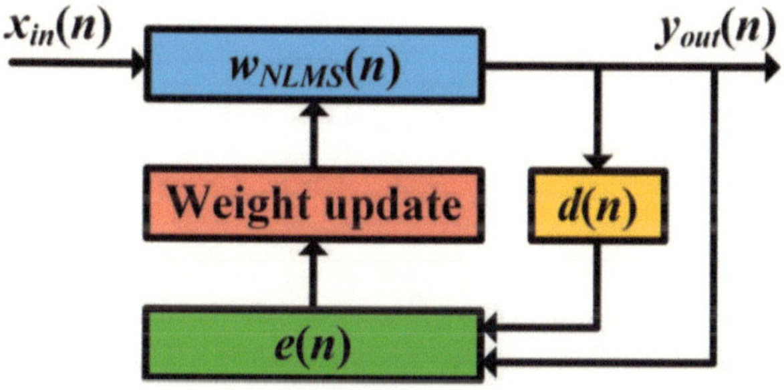

Figure 6. Schematic of one-tap normalized LMS carrier phase estimation.

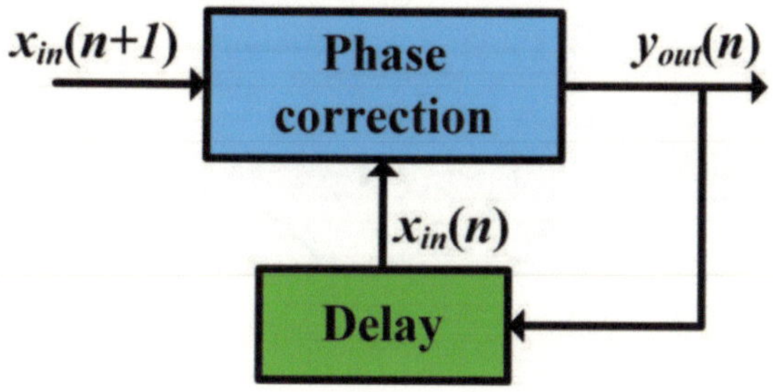

Figure 7. Schematic of differential carrier phase estimation.

2.3.3. The block-wise average carrier phase estimation

The block-wise average approach calculates the m-th power of the received symbols in each processing unit to remove the information of phase modulation, and the computed phase is summed and averaged over the entire process block, where the length of the process block is called block size. Then the averaged phase is divided by m, and the result leads to the phase estimate for the entire data block [79–81]. For the m-PSK coherent communication system, the estimated carrier phase in each process block using the block-wise average approach is given by the following expression:

$$\hat{\Phi}_{BWA}(n) = \frac{1}{m}\arg\left\{ \sum_{k=1+(M-1)\cdot N_b}^{M\cdot N_b} x^m(k) \right\} \tag{17}$$

$$M = \left\lceil \frac{n}{N_b} \right\rceil \tag{18}$$

where N_b is the block size in the BWA approach, and $\lceil x \rceil$ means the nearest integer larger than x.

The performance of the block-wise average carrier phase estimation method in the m-PSK coherent optical communication system can be derived based on the Taylor expansion of the estimated carrier phase error, and the BER floor in the block-wise average carrier phase estimation can be described using the following expression [52, 53, 55, 79]:

$$BER_{floor}^{BWA} \approx \frac{1}{N_b \cdot \log_2 m} \cdot \sum_{k=1}^{N_b} erfc\left(\frac{\pi}{m\sqrt{2}\sigma_{BWA,k}} \right) \tag{19}$$

$$\sigma_{BWA,k}^2 = \frac{\sigma^2}{6N_b^2} \cdot \left[2(k-1)^3 + 3(k-1)^2 + 2(N_b-k)^3 + 3(N_b-k)^2 + N_b - 1 \right] \tag{20}$$

where σ^2 represents the total phase noise variance in the coherent transmission system. The schematic of the block-wise average carrier phase estimation is shown in **Figure 8**.

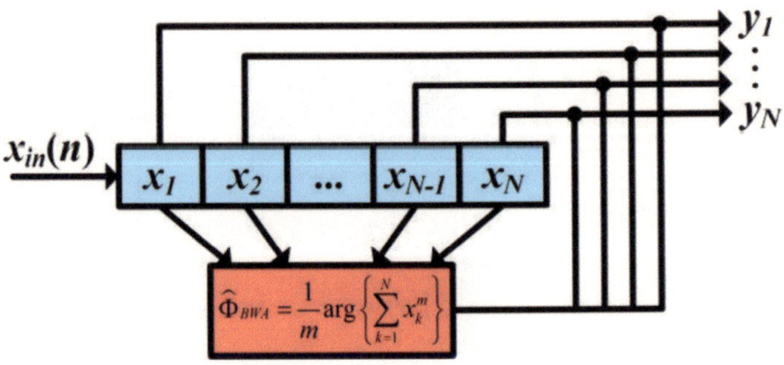

Figure 8. Schematic of block-wise average carrier phase estimation.

2.3.4. The Viterbi-Viterbi carrier phase estimation

The Viterbi-Viterbi carrier phase estimation approach also operates the symbols in each process block into the m-th power to remove the information of the phase modulation. The computed phase is also summed and averaged over the entire process block, where the length of the process block is also called block size. Then the averaged phase is divided by m as the estimated carrier phase. However, compared to the BWA approach, the estimated phase in the Viterbi-Viterbi carrier phase estimation approach is only applied in the phase recovery of the central symbol in each process block [55, 81–83]. The estimated carrier phase in the Viterbi-Viterbi approach in m-PSK optical communication systems is given by the following expression:

$$\hat{\Phi}_{VV}(n) = \frac{1}{m}\arg\left\{\sum_{k=-(N_v-1)/2}^{(N_v-1)/2} x^m(n+k)\right\},\ N_v = 1,3,5,7\ldots \tag{21}$$

where N_v is the block size in the Viterbi-Viterbi carrier phase estimation approach.

The performance of the Viterbi-Viterbi carrier phase estimation in the m-PSK coherent optical communication system can also be derived employing the Taylor expansion of the estimated carrier phase. The BER floor in the Viterbi-Viterbi carrier phase estimation for the m-PSK transmission system can be expressed as follows [52, 53, 55]:

$$BER_{floor}^{VV} \approx \frac{1}{\log_2 m} erfc\left(\frac{\pi}{m\sqrt{2}\sigma_{VV}}\right) \tag{22}$$

$$\sigma_{VV}^2 = \sigma^2 \cdot \frac{N_v^2 - 1}{12N_v} \tag{23}$$

where σ^2 represents the total phase noise variance in the coherent transmission system. The schematic of the Viterbi-Viterbi carrier phase estimation is illustrated in **Figure 9**.

According to Eqs. (20) and (23), it can be found that the phase estimate error in the Viterbi-Viterbi carrier phase estimation corresponds to the phase estimate error of the central symbol (the smallest error) in the block-wise average carrier phase estimation. Therefore, the Viterbi-

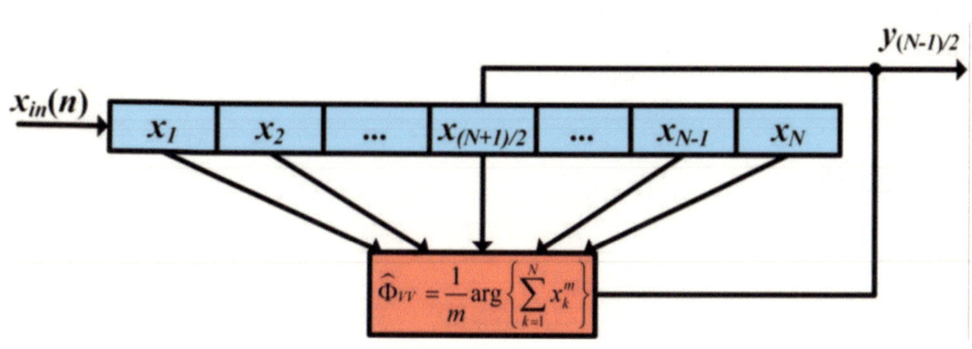

Figure 9. Schematic of Viterbi-Viterbi carrier phase estimation.

Viterbi approach will generally perform better than the block-wise average approach, in terms of the phase estimate error. However, it requires more computational complexity to update the process unit for the phase estimation of each symbol.

It is noted that the one-tap normalized LMS algorithm can also be employed for the m-QAM coherent transmission systems, while the block-wise average and the Viterbi-Viterbi methods cannot be easily used for the classical m-QAM coherent systems except the circular constellation m-QAM systems.

3. Conclusions

In this chapter, the digital signal processing techniques for compensating transmission impairments in optical communication systems including chromatic dispersion, polarization mode dispersion, and laser phase noise have been described and analyzed in detail. Chromatic dispersion can be compensated using the digital filters in both time domain and frequency domain. Polarization mode dispersion can be equalized adaptively using the least-mean-square method and the constant modulus algorithm. Phase noise from the laser sources can be estimated and compensated using the feed-forward and feed-back carrier phase recovery approaches.

Digital signal processing combined with coherent detection shows a very promising solution for long-haul high-capacity optical communication systems, which offers a great flexibility in the design, deployment, and operation of optical communication networks. Fiber nonlinearities, including self-phase modulation, cross-phase modulation, and four-wave mixing, can be mitigated using single-channel and multichannel digital back-propagation in the electrical domain, which will be discussed in future work.

Acknowledgements

This work is supported in part by UK Engineering and Physical Sciences Research Council (project UNLOC EP/J017582/1), in part by European Commission Research Council FP7-PEOPLE-2012-IAPP (project GRIFFON, No. 324391), in part by European Commission Research Council FP7-PEOPLE-2013-ITN (project ICONE, No. 608099), and in part by Swedish Research Council Vetenskapsradet (No. 0379801).

Author details

Tianhua Xu

Address all correspondence to: tianhua.xu@ucl.ac.uk

Department of Electronic and Electrical Engineering, University College London, London, United Kingdom

References

[1] Agrawal GP. Fiber-Optic Communication Systems. 4th ed. John Wiley & Sons, Inc, New Jersey, USA.; 2010. ISBN: 978-0-470-50511-3

[2] Kaminow I, Li T, Willner AE. Optical Fiber Telecommunications VB: System and Networks. 5th ed. Oxford: Academic Press; 2010. ISBN: 978-0-12-374172-1

[3] Semrau D. et al. Achievable information rates estimates in optically amplified transmission systems using nonlinearity compensation and probabilistic shaping. Optics Letters. 2017;**41**(21):121-124. DOI: 10.1364/OL.42.000121

[4] Li Y. et al. Dynamic dispersion compensation in a 40 Gb/s single-channeled optical fiber communication system, Acta Optica Sinica. 2007;**27**:1161-1165. ISSN: 0253-2239.2007.07.004

[5] IEEE, Xu T. et al. Overcoming fibre nonlinearities to enhance the achievable transmission rates in optical communication systems. In: Asia Communications and Photonics Conference, Workshop 3, Hong Kong; 2015. (Invited Talk)

[6] IEEE, Liga G. et al. Ultra-wideband nonlinearity compensation performance in the presence of PMD. In: European Conference on Optical Communication, Düsseldorf , Germany; 2016. 794-796. ISBN: 978-3-8007-4274-5

[7] KTH, Xu T. DSP Based Chromatic Dispersion Equalization and Carrier Phase Estimation in High Speed Coherent Optical Transmission Systems (KTH Ph.D. thesis), Stockholm, Sweden; 2012. ISBN 978-91-7501-346-6

[8] Sergeyev S. et al., All-optical polarisation control in fibre Raman amplifiers. International Scientific and Technical Conference on Quantum Electronics. 2013;**81**. ISBN: 978-985-553-157-0

[9] Jacobsen G. et al. Phase noise influence in coherent optical DnPSK systems with DSP based dispersion compensation. Journal of Optical Communications. 2014;**35**(1):57–61. DOI: 10.1515/joc-2013-0065

[10] Xu T. et al. Field trial over 820 km installed SSMF and its potential Terabit/s superchannel application with up to 57.5-Gbaud DP-QPSK transmission. Optics Communications. 2015;**353**:133-138. DOI: 10.1016/j.optcom.2015.05.029

[11] Ip E, Lau APT, Barros DJF, Kahn JM. Coherent detection in optical fiber systems. Optics Express. 2008;**16**(2):753-791. DOI: 10.1364/OE.16.000753

[12] Li G. Recent advances in coherent optical communication. Advances in Optics and Photonics. 2009;**1**(2):279-307. DOI: 10.1364/AOP.1.000279

[13] G. Jacobsen, et al., Phase noise influence in optical OFDM systems employing RF pilot tone for phase noise cancellation. Journal of Optical Communications. 2011;**32**(2):141-145. DOI: 10.1515/joc.2011.017

[14] IEEE, Maher R. et al. Digital pulse shaping to mitigate linear crosstalk in Nyquist spaced 16QAM WDM transmission systems. In: OptoElectronics and Communication Conference, MO2B2, Melbourne, Australia; 2014. ISBN: 978-1-922107-21-3

[15] IEEE, Xu T. et al. Phase noise mitigation in coherent transmission system using a pilot carrier. In: Asia Communications and Photonics Conference, Proceedings SPIE. Vol. 8309, 8309Z, Shanghai, China; 2011. DOI: 10.1117/12.904038

[16] IEEE, Liga G. et al. Digital back-propagation for high spectral-efficiency Terabit/s superchannels. In: Optical Fiber Communication Conference, San Francisco, California, USA; 2014, W2A.23. DOI: 10.1364/OFC.2014.W2A.23

[17] Savory SJ, Digital equalization in coherent optical transmission systems. In: Enabling Technologies for High Spectral-Efficiency Coherent Optical Communication Networks. John Wiley & Sons, Inc., New Jersey, USA; 2016. p. 311-332. DOI: 10.1002/9781119078289.ch8

[18] Xu T. et al. Digital adaptive carrier phase estimation in multi-level phase shift keying coherent optical communication systems. International Conference on Information Science and Control Engineering. 2016;1293-1297. DOI:10.1109/ICISCE.2016.276

[19] IEEE, Kikuchi K. Coherent transmission systems. In: European Conference on Optical Communication, Brussels, Belgium; 2008, Th.2.A.1. DOI: 10.1109/ECOC.2008.4729551

[20] IEEE, Shevchenko NA. et al. Achievable information rates estimation for 100-nm Raman-amplified optical transmission system. In: European Conference on Optical Communication, Düsseldorf , Germany; 2016. pp. 878-880. ISBN: 978-3-8007-4274-5

[21] IEEE, Maher R. et al. Linear and nonlinear impairment mitigation in a Nyquist spaced DP-16QAM WDM transmission system with full-field DBP. In: European Conference on Optical Communication, Cannes, France; 2014, P.5.10. DOI: 10.1109/ECOC.2014.6963971

[22] T. Xu, et al., Analytical estimation in differential optical transmission systems influenced by equalization enhanced phase noise. In: Progress in Electromagnetics Research Symposium; 2016. pp. 4844-4848. DOI: 10.1109/PIERS.2016.7735770

[23] Mynbaev DK, Scheiner LL. Fiber-Optic Communication Technology. Prentice Hall, New Jersey, USA; 2000. p. 1-68. ISBN: 978-0139620690

[24] Okoshi T, Kikuchi K. Coherent Optical Fiber Communications. Kluwer Academic Publishers, Tokyo, Japan; 1988. p. 1-32. ISBN: 978-90-277-2677-3

[25] Kikuchi K. Coherent optical communications - history, state-of-the-art technologies, and challenges for the future. In: Opto-Electronics and Communications Conference; 2008. pp. 1-4. DOI: 10.1109/OECCACOFT.2008.4610574

[26] Kikuchi K. History of coherent optical communications and challenges for the future. IEEE Summer Topical Meetings, TuC1.1; 2008. DOI: 10.1109/LEOSST.2008.4590512

[27] Xu T. et al. Modulation format dependence of digital nonlinearity compensation performance in optical fibre communication systems. Optics Express, 20147;**25**(4):3311-3326. DOI: 10.1364/OE.25.003311

[28] Xu T. et al. Analytical BER performance in differential n-PSK coherent transmission system influenced by equalization enhanced phase noise. Optics Communications. 2015;**334**:222-227. DOI: 10.1016/j.optcom.2014.07.094

[29] Bayvel P. et al. Maximising the optical network capacity. Philosophical Transactions of the Royal Society A. 2016;**374**(2062):20140440. DOI: 10.1098/rsta.2014.0440

[30] Jacobsen G. et al. Error-rate floors in differential n-level phase-shift-keying coherent receivers employing electronic dispersion equalization. Journal of Optical Communications. 2011;**32**(3):191-193. DOI: 10.1515/JOC.2011.031

[31] Kahn J, Ho KP. Spectral efficiency limits and modulation/detection techniques for DWDM systems. Journal of Selected Topics in Quantum Electronics. 2004;**10**(2):259-272. DOI: 10.1109/JSTQE.2004.826575

[32] IEEE, Griffin RA. et al. 10 Gb/s optical differential quadrature phase shift key (DQPSK) transmission using GaAs/AlGaAs integration. In: Optical Fiber Communication Conference, WX6, Anaheim, California, USA; 2002. DOI: 10.1109/OFC.2002.1036787

[33] Tsukamoto S. et al. Coherent demodulation of 40-Gbit/s polarization-multiplexed QPSK signals with 16-GHz spacing after 200-km transmission. In: Optical Fiber Communication Conference, PDP29, 2005. DOI: 10.1109/OFC.2005.193207

[34] Xu T. et al. Chromatic dispersion compensation in coherent transmission system using digital filters. Optics Express. 2010;**18**(15):16243-16257. DOI: 10.1364/OE.18.016243

[35] Jacobsen G. et al. Phase noise influence in coherent optical OFDM systems with RF pilot tone: IFFT multiplexing and FFT demodulation. Journal of Optical Communications. 2012;**33**(3):217-226. DOI: 10.1515/joc-2012-0038

[36] Ip E, Kahn JM. Digital equalization of chromatic dispersion and polarization mode dispersion. Journal of Lightwave Technology. 2007;**25**(8):2033-2043. DOI: 10.1109/JLT.2007.900889

[37] Maher R. et al. Spectrally shaped DP-16QAM super-channel transmission with multi-channel digital back propagation. Scientific Reports. 2015;**5**:08214. DOI: 10.1038/srep08214

[38] Xu T. et al. Normalized LMS digital filter for chromatic dispersion equalization in 112-Gbit/s PDM-QPSK coherent optical transmission system. Optics Communications. 2010;**283**:963-967. DOI: 10.1016/j.optcom.2009.11.011

[39] Jacobsen G. et al. Phase noise influence in long-range coherent optical OFDM systems with delay detection: IFFT multiplexing and FFT demodulation. Journal of Optical Communications. 2012;**33**(4):289-295. DOI: 10.1515/joc-2012-0047

[40] IEEE, Liga G, Czegledi CB, Xu T. PMD and wideband nonlinearity compensation: next bottleneck or fundamental limitation? In: European Conference on Optical Communication (ECOC), Workshop WS06, Düsseldorf, Germany; 2016. (Invited Talk)

[41] Jacobsen G. et al. Receiver implemented RF pilot tone phase noise mitigation in coherent optical nPSK and nQAM systems. Optics Express. 2011;**19** (15):14487-14494. DOI: 10.1364/OE.19.014487

[42] Xu T. et al. Analysis of chromatic dispersion compensation and carrier phase recovery in long-haul optical transmission system influenced by equalization enhanced phase noise. Optik, to appear, 2017.

[43] PIERS, Jacobsen G. et al. Capacity constraints for phase noise influenced coherent optical DnPSK systems. In: Progress in Electromagnetics Research Symposium, Guangzhou, China; 2014. 140319195903 (Invited Talk)

[44] Xu T. et al. Equalization enhanced phase noise in Nyquist-spaced superchannel transmission systems using multi-channel digital back-propagation. Scientific Reports. 2015;**5**:13990. DOI: 10.1038/srep13990

[45] SPIE, Yoshida T, Sugihara T, Uto K. DSP-based optical modulation technique for long-haul transmission. Next-Generation Optical Communication: Components, Sub-Systems, and Systems IV Proceedings SPIE; Vol. 9389, 93890K, San Francisco, California, USA; 2015. DOI: 10.1117/12.2078042

[46] IEEE, Xu T. et al. Quasi real-time 230-Gbit/s coherent transmission field trial over 820 km SSMF using 57.5-Gbaud dual-polarization QPSK. In: Asia Communications and Photonics Conference, Beijing, China; 2013. AF1F.3. DOI: 10.1364/ACPC.2013.AF1F.3

[47] Ip E, Kahn JM. Feedforward carrier recovery for coherent optical communications. Journal of Lightwave Technology. 2007;**25**(9):2675-2692. DOI: 10.1109/JLT.2007.902118

[48] Haunstein HF. et al. Principles for electronic equalization of polarization-mode dispersion. Journal of Lightwave Technology. 2004;**22**(4):1169-1182. DOI: 10.1109/JLT.2004.825333

[49] IEEE, Xu T. et al. Mitigation of EEPN in long-haul n-PSK coherent transmission system using modified optical pilot carrier. In: Asia Communications and Photonics Conference, Beijing, China; 2013. AF3E.1. DOI: 10.1364/ACPC.2013.AF3E.1

[50] Jacobsen G. et al. Study of EEPN mitigation using modified RF pilot and Viterbi-Viterbi based phase noise compensation. Optics Express. 2013;**21**(10):12351-12362. DOI: 10.1364/OE.21.012351

[51] Taylor MG. Phase estimation methods for optical coherent detection using digital signal processing. Journal of Lightwave Technology. 2009;**17**(7):901-914. DOI: 10.1109/JLT.2008.927778

[52] Xu T. et al. Analytical investigations on carrier phase recovery in dispersion-unmanaged n-PSK coherent optical communication systems. Photonics. 2016;**3**(4):51. DOI: 10.3390/photonics3040051

[53] Xu T. et al. Comparative study on carrier phase estimation methods in dispersion-unmanaged optical transmission systems. In: Advanced Information Technology, Electronic and Automation Control Conference, to appear, 2017.

[54] Fatadin I, Ives D, Savory SJ. Differential carrier phase recovery for QPSK optical coherent systems with integrated tunable lasers. Optics Express. 2013;**21**(8):10166-10171. DOI: 10.1364/OE.21.010166

[55] Xu T. et al. Carrier phase estimation methods in coherent transmission systems influenced by equalization enhanced phase noise. Optics Communications. 2013;**293**:54-60. DOI: 10.1016/j.optcom.2012.11.090

[56] Jacobsen G. et al. EEPN and CD study for coherent optical nPSK and nQAM systems with RF pilot based phase noise compensation. Optics Express. 2012;**20**(8):8862-8870. DOI: 10.1364/OE.20.008862

[57] IEEE, Ellis A. et al. The impact of phase conjugation on the nonlinear-Shannon limit: the difference between optical and electrical phase conjugation. In: IEEE Summer Topical Conference, Nassau, Bahamas; 2015. pp. 209-210. DOI: 10.1109/PHOSST.2015.7248271

[58] Jacobsen G. Influence of pre- and post-compensation of chromatic dispersion on equalization enhanced phase noise in coherent multilevel systems. Journal of Optical Communications. 2011;**32**(4):257-261. DOI: 10.1515/JOC.2011.053

[59] Liga G. et al. On the performance of multichannel digital backpropagation in high-capacity long-haul optical transmission. Optics Express. 2014;**22**(24):30053-30062. DOI: 10.1364/OE.22.030053

[60] IEEE, Xu T. et al. Receiver-based strategies for mitigating nonlinear distortion in high-speed optical communication systems. In: Tyrrhenian International Workshop on Digital Communications, IEEE Photonics in Switching, Florence, Italy; 2015. P2.2 (Invited Talk)

[61] IEEE, Killey RI. et al. Experimental characterisation of digital Nyquist pulse-shaped dual-polarisation 16QAM WDM transmission and comparison with the Gaussian noise model of nonlinear propagation. In: International Conference on Transparent Optical Networks, Graz, Austria; 2014. Tu.D1.3. DOI: 10.1109/ICTON.2014.6876439

[62] IEEE, Le S. et al. Optical and digital phase conjugation techniques for fiber nonlinearity compensation. In: OptoElectronics and Communication Conference, Shanghai, China; 2015. pp. 1-3. DOI: 10.1109/OECC.2015.7340113

[63] Haykin S. Adaptive Filter Theory. 5th ed. Prentice Hall, New Jersey, USA; 2013. p. 231-341. ISBN: 978-0132671453

[64] IEEE, Xu T. et al. Digital compensation of chromatic dispersion in 112-Gbit/s PDM-QPSK system. In: Asia Communications and Photonics Conference Proceedings; Vol. 7632, 763202, Shanghai, China; 2009. DOI: 10.1364/ACP.2009.TuE2

[65] SPIE, Xu T. et al. Variable-step-size LMS adaptive filter for digital chromatic dispersion compensation in PDM-QPSK coherent transmission system, In: International Conference on Optical Instruments and Technology, Shanghai, China; 2009. p. 7506, 75062I. DOI: 10.1117/12.837834

[66] www.diva-portal.org

[67] IEEE, Xu T. et al. Influence of digital dispersion equalization on phase noise enhancement in coherent optical system. In: Asia Communications and Photonics Conference, Guangzhou, China; 2012. AS1C.3. DOI: 10.1364/ACPC.2012.AS1C.3

[68] Savory SJ. Digital filters for coherent optical receivers. Optics Express. 2008;**16**(2):804-817. DOI: 10.1364/OE.16.000804

[69] IEEE, Xu T. et al. Digital chromatic dispersion compensation in coherent transmission system using a time-domain filter. In: Asia Communications and Photonics Conference, Shanghai, China; 2010. pp. 132-133. DOI: 10.1109/ACP.2010.5682798

[70] Kudo R. et al. Coherent optical single carrier transmission using overlap frequency domain equalization for long-haul optical systems. Journal of Lightwave Technology. 2009;**27**(16):3721-3728. DOI: 10.1109/JLT.2009.2024091

[71] Benvenuto N, Cherubini G. Algorithms for Communications Systems and Their Applications. New York: John Wiley & Sons, Inc.; 2004. ISBN: 978-0-470-84389-5

[72] IEEE, Lowe D, Huang X. Adaptive overlap-add equalization for MB-OFDM ultra-wideband. In: International Symposium on Communications and Information Technologies, Bangkok, Thailand; 2006. 644-648. DOI: 10.1109/ISCIT.2006.339826

[73] Xu T. et al. Frequency-domain chromatic dispersion equalization using overlap-add methods in coherent optical system. Journal of Optical Communications. 2011;**32**(2):131-135. DOI: 10.1515/joc.2011.022

[74] Kwon OW, Un CK, Lee JC. Performance of constant modulus adaptive digital filters for interference cancellation. Signal Processing. 1992;**26**(2):185-196. DOI: 10.1016/0165-1684(92) 90129-K

[75] Benesty J, Duhamel P. Fast constant modulus adaptive algorithm. IEE Radar and Signal Processing. 1991;**138**(4):379-387. DOI: 10.1049/ip-f-2.1991.0049

[76] Xu T. et al. Analytical estimation of phase noise influence in coherent transmission system with digital dispersion equalization. Optics Express. 2011;**19**(8):7756-7768. DOI: 10.1364/OE.19.007756

[77] Mori Y. et al. Unrepeated 200-km transmission of 40-Gbit/s 16-QAM signals using digital coherent receiver. Optics Express. 2009;**17**(3):1435-1441. DOI: 10.1364/OE.17.001435

[78] SPIE, Xu T. et al. Close-form expression of one-tap normalized LMS carrier phase recovery in optical communication systems. International Conference on Wireless and Optical Communications; Vol. 9902, 990203, Beijing, China; 2016. DOI: 10.1117/12.2261932

[79] Goldfarb G, Li G. BER estimation of QPSK homodyne detection with carrier phase estimation using digital signal. Optics Express. 2006;**14**(18):8043-8053. DOI: 10.1364/OE.14.008043

[80] Ly-Gagnon DS. et al. Coherent detection of optical quadrature phase-shift keying signals with carrier phase estimation., Journal of Lightwave Technology. 2006;**24**(1):12-21. DOI: 10.1109/JLT.2005.860477

[81] www.mdpi.com

[82] Viterbi AJ, Viterbi AM. Nonlinear estimation of PSK-modulated carrier phase with application to burst digital transmission. IEEE Transactions on Information Theory. 1983;**29**(4):543-551. DOI: 10.1109/TIT.1983.1056713

[83] IEEE, Xu T. et al. Analysis of carrier phase extraction methods in 112-Gbit/s NRZ-PDM-QPSK coherent transmission system. In: Asia Communications and Photonics Conference, Guangzhou, China; 2012, AS1C.2. DOI: 10.1364/ACPC.2012.AS1C.2

Chapter 16

Physical-Layer Encryption Using Digital Chaos for Secure OFDM Transmission

Xuelin Yang, Adnan A.E. Hajomer and Weisheng Hu

Additional information is available at the end of the chapter

http://dx.doi.org/10.5772/intechopen.68238

Abstract

Due to the broadcasting nature of passive optical network (PON), data security is challenging. For the transmission of orthogonal frequency division multiplexing (OFDM) signals, the high peak-to-average power ratio (PAPR) is considered as one of the major drawbacks. This chapter reviews the digital chaos-based secure OFDM data encryption schemes, where the transmission performance is improved via PAPR reduction. The digital chaos is incorporated into the signal scrambling approaches: selective mapping (SLM), partial transmit sequence (PTS); and precoding approaches: discrete Fourier transform (DFT) and Walsh-Hadamard transform (WHT) for PAPR reduction. Multi-fold data encryption is achieved with a huge key space provided by digital chaos, to enhance the physical-layer security for OFDM-PON, while the pseudo-random properties of digital chaos are applied for PAPR reduction, which consequently improves the transmission performance. The evidences of these encryption approaches are presented in terms of theories, simulations, as well as experimental demonstrations. The chaotic data encryption schemes could be promising candidates for next-generation OFDM-PON.

Keywords: orthogonal frequency division multiplexing (OFDM), peak-to-average power ratio (PAPR), digital chaos, passive optical network (PON)

1. Introduction

Over the last decades, passive optical network (PON) has been playing a vital role in data traffic explosion driven by broadband services such as high-definition television (HDTV), cloud computing, 3D television, video on demand (VoD) [1, 2], and so on, because it offers several potential benefits such as high capacity, low cost, and energy efficiency. In fact, PON is a broadcasting structure that extends for ~20–100 km from optical line terminal (OLT) to optical network units (ONUs), in which no active component (such as Erbium-doped fiber amplifier, EDFA) is employed. The broadcasting nature in the downstream data traffic and the huge number of subscribers in PON make the data more susceptible to be eavesdropped or attacked by illegal ONUs. For instance, during the ranging process, the OLT has to broadcast the serial number and ID information of the ranged ONUs; a malicious user could make use of this information for spoofing.

Comparing with the encryption in higher layers, for instance, media access control (MAC) layer, physical-layer encryption can protect the data as well as the control and header information. If the physical-layer data encryption is implemented, this type of spoofing can be avoided since the header and ID information are all encrypted within the physical-layer; thus, it becomes desirable for security enhancement in PON. For security reasons, churning procedure of scrambling the data for downstream connection has been defined in ITU-T G.983.1 standard (Section 8.3.5.6 [3]), which is based on a key sent from ONU to OLT through a secure channel with a defined protocol; however, the key is vulnerable to be broken due to its limited short key length.

Chaos communication has been proposed to provide a huge key space for data security enhancement attributed to its unpredictable nature of randomness, noise-like nature, and broadband. However, the implementation of chaotic optical communication (i.e., fast changing of chaotic optical carriers) requires identical devices with identical parameters for transmitter and receiver, which is quite restricted from real implementation. On the other hand, digital chaos has attracted notable attention recently as a flexible alternative to avoid the implementation difficulty for device-based optical chaos [4–10]. Because it offers very appealing properties from the perspective of data encryption such as ergodicity, pseudo-randomness, and high sensitivity to the initial values, digital chaos provides a huge key space for security applications. Moreover, due to flexible digital signal processing (DSP) in electric domain, digital chaos is easier to be applied.

Optical orthogonal frequency division multiplexing (OFDM) is regarded as a promising modulation technique for next-generation PON, owing to the advantages in high spectrum efficiency, cost-effectiveness, and tolerance to fiber dispersion. High-speed data rate of OFDM signals is achieved by parallel transmission of partially overlapped spectra, lower rate frequency domain tributaries [11]. Moreover, the generation, modulation, and demodulation of OFDM signals have to be performed using DSP in electric domain, therefore it provides a natural physical-layer environment, where digital chaos can be incorporated into OFDM data encryption during transmission. However, OFDM modulation often leads to high peak-to-average power ratio (PAPR), which is one of the most detrimental factors in OFDM signal transmission, as it causes power saturation and nonlinear distortion at the optical receiver while degrading the transmission performance. The pseudo-random properties of digital chaos are helpful for PAPR reduction of OFDM signals, which consequently improves the transmission performance.

In this chapter, OFDM data encryption schemes are reviewed in detail for physical-layer security enhancement based on digital chaos during transmission, while jointly the transmission performances are significantly improved because of the effective reduction in PAPR of OFDM signals. The rest of the chapter is organized as follows. In Section 2, the fundamental theory of PAPR reduction of OFDM signals is shown. The properties of digital chaos are presented in Section 3. In Section 4, the encryption schemes are illustrated in details. Conclusions are given in Section 5.

http://dx.doi.org/10.5772/intechopen.68238

2. PAPR of OFDM signals

OFDM modulation is the superposition of many independent signals modulated onto individual subcarriers with equal-spaced bandwidth. **Figure 1** shows the overlapping of the subcarriers in frequency domain. High PAPR could be inevitable especially when a large number of subcarriers are in phase. As a result, the optical receiver with a wide linear range is required to accommodate a large dynamic range of PAPR [12].

If a block of N symbols is denoted as the vector $X = [X_0, X_1, \ldots, X_{N-1}]$ for OFDM signals, the vector X is oversampling by g (i.e., $g(N-1)$ zero-padding, ZP), where g is an integer greater than or equal one. Therefore, the complex envelop of OFDM signals is

$$x[n] = \frac{1}{\sqrt{N}} \sum_{i=0}^{Ng-1} X_i e^{\frac{2\pi in}{Ng}}, 0 \leq n \leq Ng - 1 \tag{1}$$

By definition, the PAPR of OFDM signals is

$$PAPR = 10 \log(\max(|x[n]|^2)/E(|x[n]|^2)) \tag{2}$$

From Eqs. (1) and (2), the oversampling factor must be greater than one for sufficient accuracy of PAPR calculation [13]. To evaluate the PAPR performance, the complementary cumulative distribution function (CCDF) is commonly simulated, which is defined as the PAPR probability exceeding a given threshold for a certain OFDM data block. Based on the central limit theorem, the real and imaginary parts of the complex OFDM signals after inverse fast Fourier transformation (IFFT) have Gaussian distribution in the case of sufficient large number of subcarriers, thus the amplitudes follow the Rayleigh distribution. For instance, if the CDF of the amplitude of a signal sample is given by

$$F(z) = 1 - e^{-z} \tag{3}$$

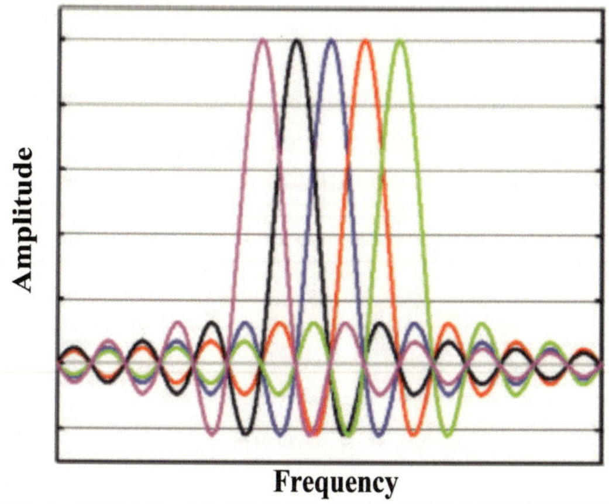

Figure 1. Spectrum of equal-spaced subcarriers in OFDM signals.

and N is a large number of samples, the CCDF of PAPR of the signal is [14]

$$P(PAPR > z) = 1 - P(PAPR \leq z) = 1 - F(z)^N = 1 - (1 - e^{-z})^N \tag{4}$$

3. Characteristics of digital chaos

Digital chaos has recently attracted numerous applications in OFDM-PONs [4–9], especially for data encryption, which is mainly due to the chaotic characteristics including pseudo-randomness, ergodicity, and high sensitivity to the initial values. Secure optical OFDM transmission is achieved by digital chaos, in which a huge key space is generated and predetermined by chaotic equations. Since the initial values and the other control parameters are utilized as the secure keys between OLT and ONUs, it provides a huge key space, which guarantees the physical-layer confidentiality. At ONUs, the same chaotic sequences are generated using the same keys for data recovering after reception.

The fundamental properties of digital chaos can be described, for example, via a 4-dimensional (4D) hyper chaos [15]

$$\begin{cases} \dot{x} = a(-x + y) + yzu \\ \dot{y} = b(x + y) - xzu \\ \dot{z} = cy - u + dxyu \\ \dot{u} = -eu + xyz \end{cases} \tag{5}$$

where a, b, c, d, and e are constant parameters. When $a = 35$, $b = 10$, $c = 80$, $d = 0.5$, and $e = 10$, these appropriate initial values bring the system into chaotic zones. Eq. (5) can be solved by Runge-Kutta method with a time step of $h = 0.001$. The solutions of Eq. (5) output the attractor diagrams of the 4D chaos, as plotted in **Figure 2**, where excellent chaotic behaviors in terms of pseudo-randomness and phase dynamics are illustrated.

In digital chaos, the chaotic state is very sensitive with respect to the initial values, thus even a tiny change or modification of the original initial values will let it enter into another different chaotic state. In **Figure 3(a)**, the variation curve is illustrated for the digital chaotic sequence $\{y_i\}$, under a slight change (1×10^{-15}) in the initial value of y_0. The auto- and cross-correlation functions $R_{ac}(\tau)$ and $R_{cc}(\tau)$ are plotted in **Figure 3(b) and (c)** respectively, which reveal the high sensitivity associated with the chaotic initial values, where τ is the time lag. The good quality of randomness observed in **Figure 3** is essential to guarantee high-level security reliability for data encryption.

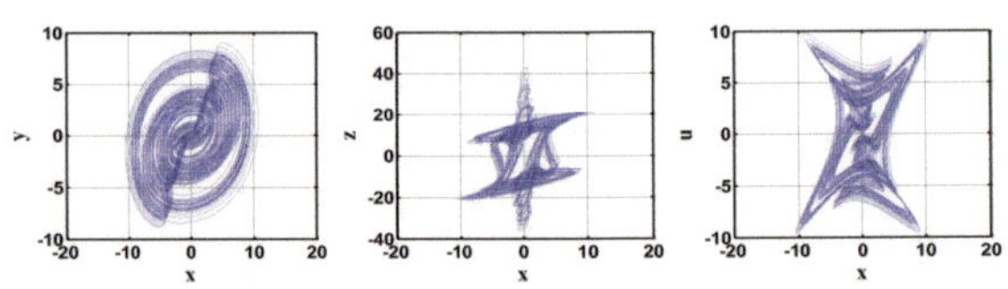

Figure 2. Chaotic attractor diagrams in phase planes of (x, y), (x, z), and (x, u).

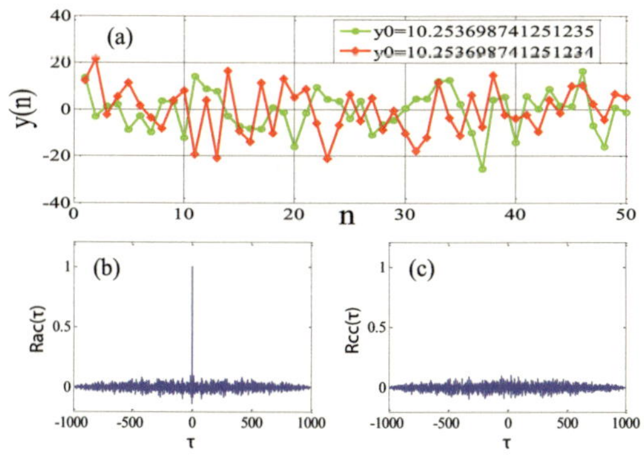

Figure 3. (a) Chaotic sequence of $\{y_i\}$ under tiny change of the initial values; (b) autocorrelation for y_0 =1.428121243912452; (c) cross-correlation for y_0 =1.428121243912452 and y_0 =1.428121243912453.

For a conservative estimate, a tiny change (~1×10^{-15}) of the initial values leads to a totally different chaotic state, as shown in **Figure 3**, therefore, the key space of 4D digital hyper chaos is ~10^{60} ($10^{15}\times10^{15}\times10^{15}\times10^{15}$).

Furthermore, the iteration times of chaotic differential equations and even the equations themselves can be served as the additional secure keys as well, so the actual key space will be >>10^{60}. Currently, the fastest computing speed is about $2.5\times10^{13}s^{-1}$, thus it will take ~1.3×10^{39} years to work out the possible initial keys of the 4D chaos via brutal-force trials [16]. Consequently, the chaotic data encryption provides a huge key space, which is large enough to resist exhaustive attacks.

4. Chaotic OFDM encryption with PAPR reduction

A variety of chaotic encryption schemes has been proposed for the physical-layer security in OFDM-PON [4–9], in which digital chaos has been incorporated into OFDM signal generation or modulation to enhance data security during transmission. However, in the existing chaotic secure schemes, the encryption is generally achieved without considering the improvement of transmission performance through the effective PAPR reduction in OFDM signals.

This section presents some systematic novel approaches to enhance the physical-layer security and jointly improve the OFDM transmission performances. The key idea behind these schemes is the combination of chaos-based encryption along with PAPR reduction; these approaches mainly belong to two categories of PAPR reduction: signal scrambling and matrix precoding, such as

- Chaotic signal scrambling: Chaotic partial transmit sequence (PTS) and chaotic selective mapping (SLM).
- Chaotic precoding: Discrete Fourier transform (DFT) and Walsh-Hadamard transform (WHT).

4.1. Chaotic signal scrambling

Signal scrambling is one of the traditional approaches in PAPR reduction. Chaotic sequences can be employed to scramble an input data block of OFDM symbols randomly, and then one of the scrambled OFDM signals with minimum PAPR is selected for final transmission. With the use of chaotic sequences in OFDM symbol scrambling, the physical-layer security is enhanced since the initial keys are required for correct recovery of the original OFDM data via correct counter-scrambling. At the same time, the transmission performance is improved simultaneously, owing to the effective reduction in PAPR via signal scrambling.

4.1.1. Chaotic partial transmit sequence (PTS)

The block diagram of OFDM data encryption using chaotic PTS is plotted in **Figure 4** [17]. After serial-to-parallel (S/P) conversion, a pseudo-random binary sequence (PRBS) is mapped onto QAM subcarriers, and then is sent for data encryption, which is applied via the following threefold encryption: chaotic training sequence (TS) insertion, chaotic random partition generation, and chaotic phase weighing factors generation.

Figure 5(a)–(c) shows the different partition examples in PTS scheme. Compared with the adjacent and interleaved partitions, the random partition provides the weakest cross-correlation among the sub-blocks, which leads to the largest PAPR reduction after selecting the optimum phase weighing factors. The PAPR can be minimized after searching for the optimized OFDM symbols for chaotic (random) signal scrambling.

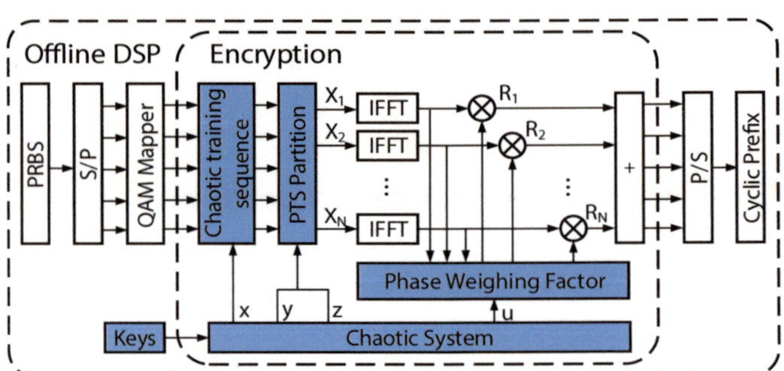

Figure 4. Schematic diagram of OFDM signal encryption using chaotic PTS.

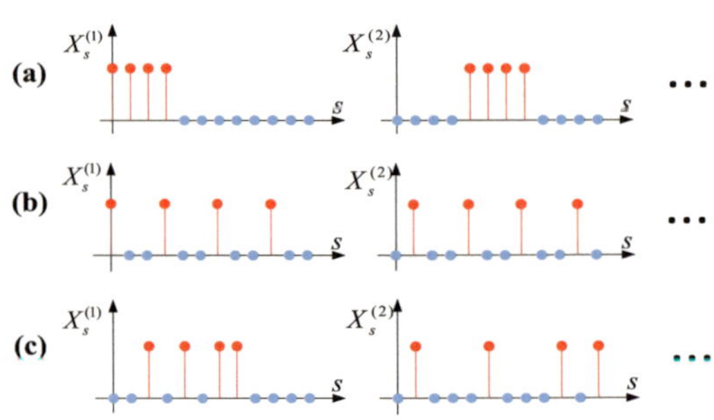

Figure 5. Partition design in PTS, for (a) adjacent; (b) interleaved; (c) random partitions.

http://dx.doi.org/10.5772/intechopen.68238

To generate digitized chaotic sequences for OFDM signal encryption, the chaotic sequences obtained in Eq. (5) are processed using [18]

$$D_{x,i} = \text{mod}(Extract(x_i, m, n, p), M) \tag{6}$$

where $Extract(x_i, m, n, p)$ outputs an integer, which is obtained by the mth, nth, and pth digits in the decimal part of x_i, mod(a, b) outputs the remainder of a divided by b, M is the maximum digital value in the digital sequence, which is 256 in all of our schemes [4]. Similarly, the other sequences $\{y_i\}$, $\{z_i\}$, and $\{u_i\}$ can be digitalized into $\{D_{y,i}\}$, $\{D_{z,i}\}$, and $\{D_{u,i}\}$. The details of chaotic PTS are given as follows.

The first sequence $\{D_{x,i}\}$ is applied to generate the chaotic TS for time synchronization of OFDM symbols. Since the illegal ONU does not have the information of chaotic TS, it has to try $(N_0+N)/N_0$ times on average to realize correct synchronization and demodulation, where N_0 is the length of cyclic prefix (CP). Moreover, due to the pseudo-random characteristic of digital chaos, it produces an autocorrelation similar to the δ function, which is advantageous for accurate OFDM symbol synchronization, as shown in **Figure 6(a)**. A legitimate ONU can correctly demodulate the original data; however, with a slight change in the initial keys, the peak does not appear, so that the data cannot be fully recovered, as illustrated in **Figure 6(b)**.

In chaotic (random) PTS, the chaotic sequences $\{D_{y,i}\}$, $\{D_{z,i}\}$, and $\{D_{u,i}\}$ can be employed to generate the partition information and the phase weighing factors. First, the digitized chaotic sequence $\{D_{y,i}\}$ is applied to generate the binary sequence $\{A_1\}$ using $\{\text{mod}(D_{y,i}, 2)\}$. Similarly, the binary sequence can be generated by $\{D_{z,i}\}$. After bitwise logical operation of these two binary sequences, as shown in **Figure 7**, the chaotic random partition information $\{P^{(l)}\}$ is formed. The sub-blocks are then generated using $X^{(l)} = X \cdot P^{(l)}$, $l = 1, 2, \cdots, L$. where L denotes the total number of sub-blocks.

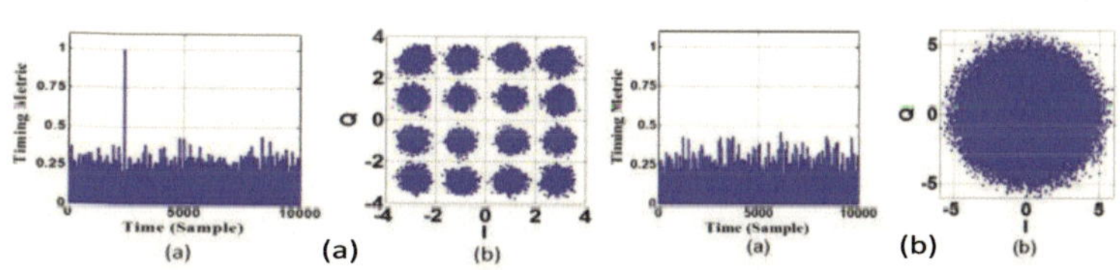

Figure 6. Effect of OFDM symbol time synchronization with (a) the right key; (b) a wrong key.

	← N →															
A_1	1	0	1	0	1	0	0	0	1	0	0	1	···	1	1	1	0
A_2	1	1	0	1	0	0	1	1	1	0	1	0	···	0	0	0	1
$P^{(1)} = A_1A_2$	1	0	0	0	0	0	0	0	1	0	0	0	···	0	0	0	0
$P^{(2)} = A_1\bar{A}_2$	0	0	1	0	1	0	0	0	0	0	0	1	···	1	1	1	0
$P^{(3)} = \bar{A}_1A_2$	0	1	0	1	0	0	1	1	0	0	1	0	···	0	0	0	1
$P^{(4)} = \bar{A}_1\bar{A}_2$	0	0	0	0	0	1	0	0	0	1	0	0	···	0	0	0	0

Figure 7. Generation of partition information (four sub-blocks) in chaotic PTS.

Second, the digitalized chaotic sequence $\{D_{u,i}\}$ is applied to generate the phase weighing factors R_l in PTS,

$$R_l = \exp\left(j \cdot 2\pi \cdot D_{u,i}/M\right) \quad (i = 1, 2, \ldots, K) \tag{7}$$

where K is the total number of phase weighing factors. After multiplying the phase weighing factors and transforming from frequency into time-domain, the corresponding time-domain vector becomes

$$x = \sum_{l=1}^{L} R_i x^{(l)} \tag{8}$$

$$x_n^{(l)} = \frac{1}{\sqrt{N}} \sum_{s=0}^{N-1} X_s^{(l)} \exp\left(j \frac{2\pi sn}{N}\right) \tag{9}$$

where $x^{(l)}$ is the IFFT of $X^{(l)}$, s and n denote the indices of OFDM subcarrier and symbol, respectively.

In PTS, the combinations of sub-blocks and phase weighing factors are calculated to find the minimum PAPR, thus the final PAPR is

$$PAPR_{\mathrm{PTS}} = \min\left(PAPR\left(\sum_{l=1}^{L} R_i x^{(l)}\right)\right) \tag{10}$$

The final OFDM vector becomes

$$x_{pts} = \sum_{l=1}^{L} R_{l,\mathrm{opt}} x^{(l)} \tag{11}$$

where $R_{l,opt}$ is the optimized phase weighing factors. At ONUs, the decryption and demodulation procedures are the inverse operations that are implemented at OLT.

Since the initial values in digital chaos serve as the secure keys to recover the original OFDM signals, only legal ONUs can generate the same chaotic scrambling information with the correct secure keys that applied at OLT. By introducing digital chaos into signal scrambling, not only the security confidentiality of OFDM signal transmission is greatly enhanced, but also a performance-improved transmission can be expected with a lower PAPR.

As for the computational complexity, it is the same for the chaotic PTS and conventional random PTS. If compared with the case of OFDM transmission without PTS, the computational complexity of chaotic PTS scheme will be L^k times higher, which is inevitable for PAPR optimization. In our case, assuming four sub-blocks and two phase weighing factors, the computational complexity is 16 times higher for PAPR optimization; however, it guarantees both higher security and lower PAPR. Since the number of possible PTS partition states is 4^N ($4^{256} \approx 10^{154}$ in our case), it is hardly to be broken through brutal-force attacks. However, chaotic PTS still requires sideband information to transmit the partition and the phase factor information, which reduces the overall spectral efficiency in transmission.

4.1.2. Chaotic selected mapping (SLM)

As shown in **Figure 8**, chaotic SLM is another alternative data encryption scheme based on signal scrambling using digital chaotic sequences. In chaotic SLM, the digitized chaotic sequences can be obtained by processing of the chaotic sequences from 2D Henon chaos [19],

$$\begin{cases} x_{i+1} = 1 - \eta x_i^2 + y_i \\ y_{i+1} = \psi x_i \end{cases} \tag{12}$$

The input OFDM sequence is denoted as $X = [X_0, X_1, \ldots, X_{N-1}]$ in frequency domain, where X_k represents the complex data of the kth subcarrier, and N represents the total number of subcarriers. First, to implement the digital chaos into SLM scheme, the total number of the chaotic phase sequences is set to be V, and all of the phase sequences $P_v = [P_{v,0}, P_{v,1}, \ldots, P_{v,N-1}]$ $(1 \le v \le V)$ are obtained from the chaotic sequence $\{D_{y,i}\}$, where $P_{v,k} = exp(j \cdot 2\pi \cdot D_{y,k}/M)$, $(0 \le k \le N-1)$. By component-wise vector multiplication of the input OFDM sequence X with the phase sequence $\{P_v\}$, the input OFDM signals are encrypted fully via the chaotic sequences obtained in Eq. (12). The encrypted OFDM sequence $\{Y_v\}$ is

$$Y_v = X \otimes P_v = [X_0 P_{v,0}, X_1 P_{v,1}, \cdots, X_{N-1} P_{v,N-1}] \tag{13}$$

where ⊗ denotes the component-wise vector multiplication. Second, the IFFT is performed for all of the encrypted OFDM sequences $\{Y_v\}$

$$x_v = \text{IFFT}(Y_v) \tag{14}$$

Finally, the OFDM signal sequence $\{x_v\}$ with minimum PAPR is chosen for OFDM transmission. It should be mentioned that in terms of computational complexity, better PAPR reduction can be expected with a larger V; however, the complexity will also increase due to the increased times of IFFT.

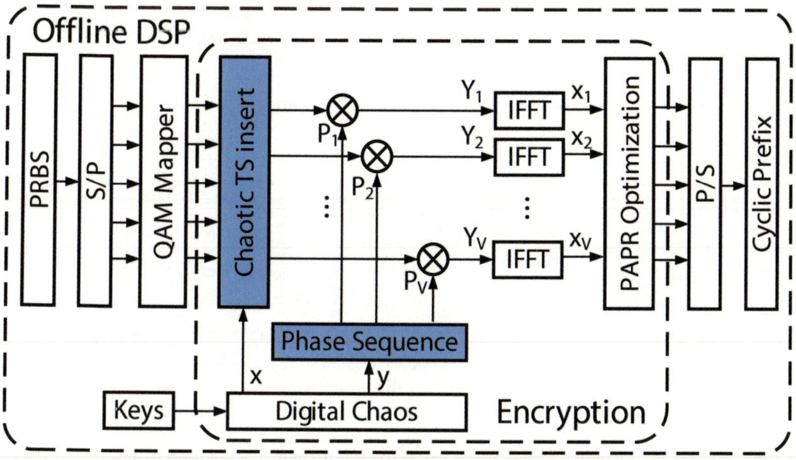

Figure 8. Schematic diagram of OFDM signal encryption using chaotic SLM.

4.1.3. Experiment results of chaotic PTS and SLM

The experimental setup of encrypted OFDM-PON transmission using chaotic PTS/SLM is shown in **Figure 9**. The input data stream was mapped onto 129 OFDM subcarriers at OLT, where 64 subcarriers were in the format of 16-QAM, and one subcarrier was unfilled with DC. The other 64 subcarriers had to be set as the complex conjugate of the aforementioned 64 subcarriers, where Hermitian symmetry had to be satisfied to realize intensity modulation and direct detection (IM/DD) after IFFT. After TS insertion for OFDM symbol time synchronization, chaotic PTS/SLM optimization and parallel-to-serial (P/S) conversion, a CP of 1/8 length of OFDM symbol was added into each OFDM symbol.

The above processing steps were performed offline using Matlab programs. The encrypted signal was then loaded into an arbitrary waveform generator (AWG), which had a sample rate of 10 Gs/s of electrical OFDM signal generation. A continuous-wave (CW) laser ($\lambda = 1550.3$ nm) was applied as the optical carrier. After that, the electrical signal in AWG was changed into optical by an optical Mach-Zehnder modulator (MZM). Thus, the net data rate applied was actually at 8.9 Gb/s (10 Gs/s × 4 × 64/256 × 8/9), with the electrical bandwidth of 2.5 GHz (10 Gs/s × 65/256). At ONU, the received data were recorded with a 20 Gs/s real-time oscilloscope after direct detection via a 10-GHz photodiode (PD). The waveform, electric and optical spectra of OFDM signals are measured and plotted in **Figure 10**.

The CCDFs of encrypted OFDM signals are plotted in **Figure 11**, where 10,000 OFDM symbols with 16-QAM modulation format were employed. From **Figure 11** it can be noted that, by employing chaotic signal scrambling, significant PAPR reduction is achieved, with respect to the conventional OFDM transmission without any signal scrambling. The total number of sub-blocks and phase factors applied in PTS was $L = 4$ and $K = 2$ respectively, and the phase factor in SLM was $V = 4$.

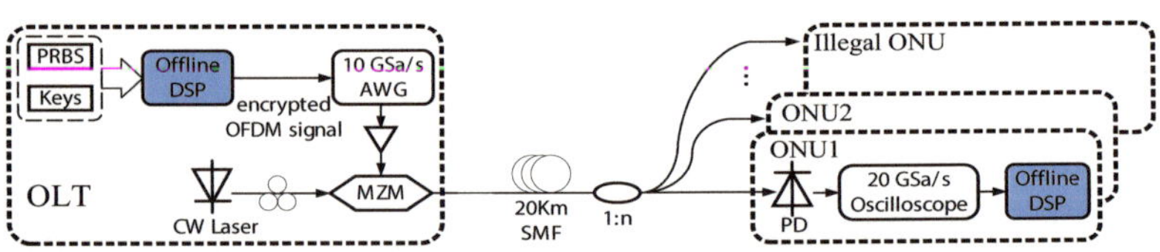

Figure 9. Experimental setup of encrypted OFDM transmission using chaotic PTS/SLM.

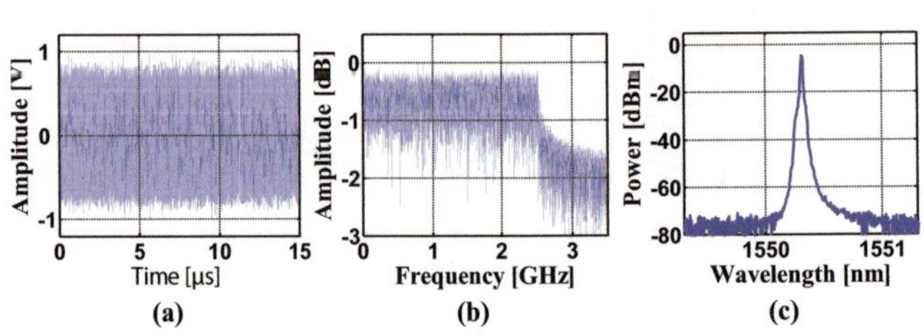

Figure 10. Encrypted OFDM signals. (a) Waveform; (b) electric spectrum; (c) optical spectrum.

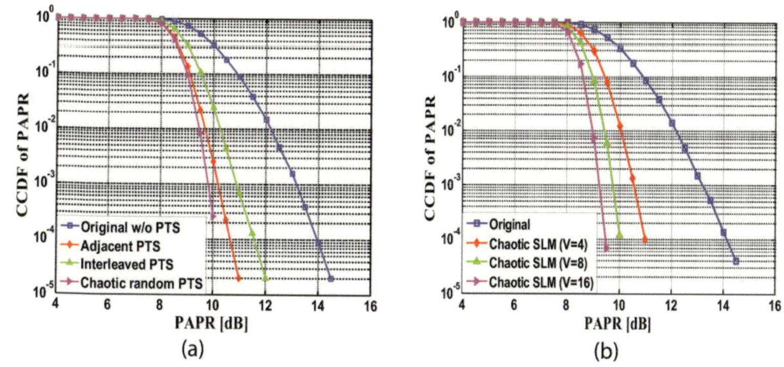

Figure 11. CCDFs of PAPR in chaotic (a) PTS; (b) SLM.

The bit error ratio (BER) measurements of secure OFDM transmission using chaotic PTS/SLM are plotted in **Figure 12**. The OFDM signals can be correctly decrypted after transmission over 20 km standard single-mode fiber (SSMF) for legitimate ONUs with the correct keys, while any illegal ONU cannot decrypt OFDM signals with a wrong key, which is even a tiny discrepancy of 1×10^{-15} away from the correct key. The corresponding constellations are plotted as the insets in **Figure 12**. If compared with the conventional OFDM transmission, the BER performance was improved ~1 dB (BER@10^{-3}) for encrypted OFDM signals, if compared with back-to-back (B2B) signals, which can mainly be attributed to the effective reduction of PAPR through the signal scrambling schemes, either chaotic PTS or SLM.

4.2. Chaotic precoding of OFDM signals

Matrix precoding is a set of typical approaches for PAPR reduction, which provides unique properties such as lower computational complexity and higher spectral efficiency (without requirement of sideband), if compared with the signal scrambling schemes. Actually, PAPR

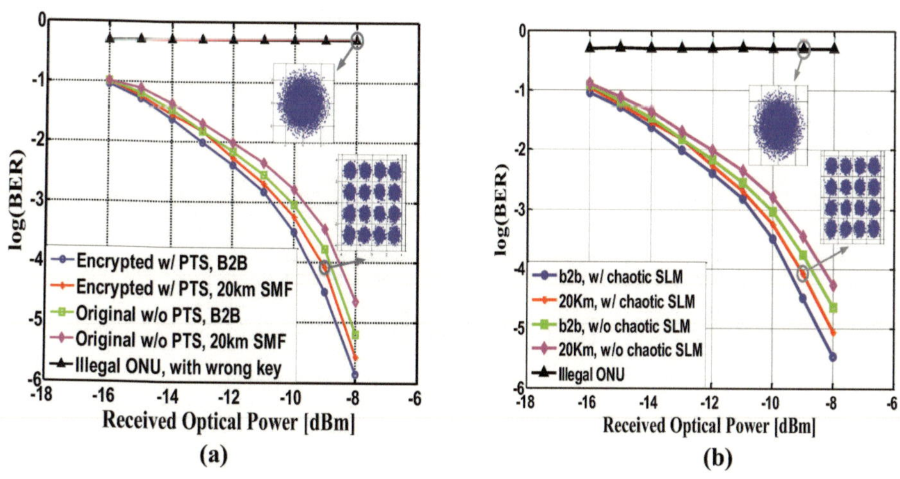

Figure 12. BERs of encrypted OFDM transmission using chaotic (a) PTS; (b) SLM.

reduction is achieved through the proper selection and distribution of power between OFDM blocks. In this section, the digital chaos will be employed into the precoding schemes, aimed to construct or reconfigure the standard precoding matrices into alternative chaotic ones, so that the security can be enhanced during transmission via the introduction of digital chaos into the procedure of OFDM signal precoding, and jointly the PAPR can be reduced.

4.2.1. Chaotic discrete Fourier transform spread OFDM (DFT-S-OFDM)

As depicted in **Figure 13**, the DFT-S-OFDM data encryption is implemented multi-fold, such as chaotic TS insertion, chaotic DFT matrix generation, and chaotic subcarrier allocation, all of which are predetermined by digital chaotic sequences [20]. The details of encryption are given as follows.

In the first fold of encryption, one chaotic sequence is used to generate TS, the same way as described in the previous chaotic signal scrambling schemes. The second fold of encryption is to generate the chaotic reconfigurable DFT precoding matrix. Assuming $\mathbf{F}$ is the standard $M \times M$ DFT matrix, the matrix element is given by

$$\mathbf{F}_{\alpha,\beta} = \frac{1}{\sqrt{M}} e^{-j2\pi(\alpha-1)(\beta-1)/M},\ 0 \le \alpha, \beta \le M-1 \tag{15}$$

where α, β are the indexes of row and column, respectively. From Eq. (15), a reconfigurable matrix can be generalized

$$\mathbf{F}'_{\alpha,\beta} = \frac{1}{\sqrt{M}} e^{-j2\pi(\alpha-m)(\beta-n)/M},\ 0 \le \alpha, \beta \le M-1 \tag{16}$$

where m, n are the real constants, which can be predetermined using chaotic sequences obtained in Eq. (5) as

$$m_{y,i} = \cdots + 10n_1 + n_2 + 0.1n_3 + 0.01n_4 + \cdots \tag{17}$$

where n_j donates the n_jth digit in the decimal part of y. Similarly, $\{n_{z,i}\}$ can be generated from $\{z_i\}$ in the same way as $\{m_{y,i}\}$.

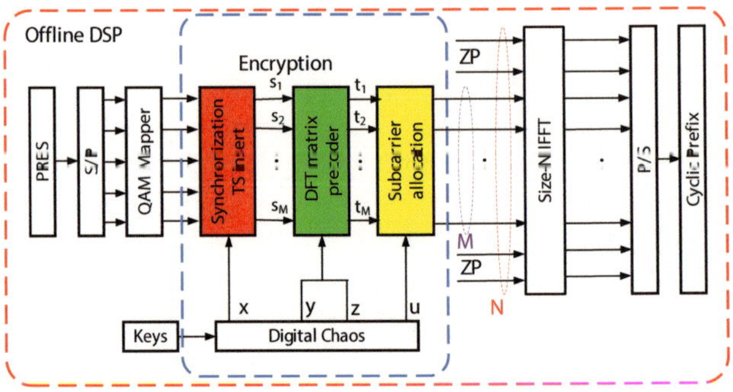

Figure 13. Schematic diagram of chaotic encrypted DFT-S-OFDM signal transmission.

Mathematically, the same basic features in the original standard **F** are kept for the reconfigurable DFT matrix **F′**. For example, the inverse matrix of **F′** is equal to its conjugate matrix. Moreover, all of the elements in **F′** are in geometric sequence with the same common ratio of $e^{-j2\pi/M}$ for the same row or column indexes.

To verify the capability of PAPR reduction of OFDM signals by the new DFT matrix **F′**, the upper bound of the peak factor should be considered, which is the square root of PAPR and depends on the aperiodic autocorrelation functions of the input symbols.

$$\gamma \le 1 + \frac{2}{N}\sum_{n=1}^{N}|\rho_n| \tag{18}$$

$$\rho_n = \sum_{l=1}^{N-l} a_l a_{l+n}, \quad n = 1,\ 2, \ldots N-1 \tag{19}$$

where ρ_n are the aperiodic autocorrelation coefficients, a_n are the input symbols (QAM, QPSK, etc.), and N is the total number of subcarriers. Eq. (18) shows that, if the autocorrelation coefficients in the corresponding input symbols are small, the peak factor of OFDM signals will be small.

Assuming the kth OFDM symbol is $s^{(k)} \triangleq [s_1^{(k)}, s_2^{(k)}, \cdots, s_M^{(k)}]^T$, where T denotes the transpose of matrix, the DFT precoded OFDM signal becomes

$$\boldsymbol{t}^{(k)} \triangleq [t_1^{(k)}, t_2^{(k)}, \cdots, t_M^{(k)}]^T = \mathbf{F}'\boldsymbol{s}^{(k)} \tag{20}$$

As plotted in **Figure 14(a)**, the same non-periodic autocorrelation functions are obtained for the OFDM symbol $\boldsymbol{t}^{(k)}$, either precoded with DFT matrix **F** or **F′**. From **Figure 14(a)**, it should be noted that different shapes of sidelobe appear in the autocorrelation functions, while the sidelobe shapes of OFDM signals with DFT precoding are lower than those without DFT precoding, when the same modulation format is applied. This verifies that, effective PAPR reduction of OFDM signals is achieved by DFT precoding with the reconfigurable matrix **F′**, the same effect as the standard DFT matrix **F**.

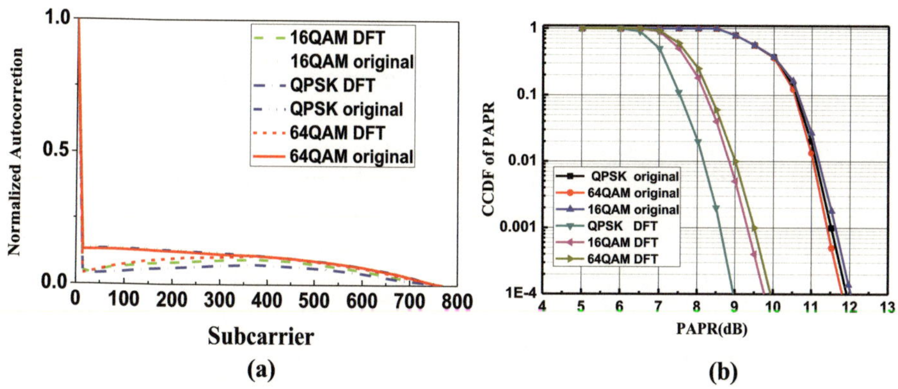

Figure 14. PAPR reduction using chaotic DFT, (a) normalized aperiodic autocorrelation function; (b) CCDFs of PAPR.

To directly evaluate the PAPR reduction, the CCDF curves is plotted in **Figure 14(b)**, for the cases of OFDM signals with and without the reconfigurable DFT precoding, where 100,000 OFDM symbols are applied in the formats of QPSK, 16-QAM, and 64-QAM, while Nyquist sampling rate is adopted. It can be observed that, if compared with the cases without DFT, the OFDM signals precoding with the reconfigurable matrix DFT can reduce the PAPR, regardless of the modulation formats, while the differences at the probability of 10^{-4} are about 3, 2.3, and 1.9 dB respectively for QPSK, 16-QAM, and 64-QAM. The CCDF curves in **Figure 14(a)** prove that, the new reconfigurable DFT does significantly reduce the PAPR of OFDM signals.

Finally, the fourth chaotic sequence $\{u_i\}$ is used to generate the scrambling criterion for subcarrier allocation before ZP. After an array $\{w_i\}$ (i=1,...,M) is obtained from $\{u_i\}$, where M is the total number of subcarriers (as shown in **Figure 13**), the elements in $\{w_i\}$ are swapped into $\{w_i'\}$ in the new order with respect to their values, for instance, from the lowest value to the highest. Then the new index order of subcarriers is rearranged as that of $\{w_i'\}$, details as shown in **Figure 15**. For example, assuming an original chaotic array of $\{0.8, 0.3, 0.5, 1.2, 1.1\}$, it is swapped from the small value to the big one, and then the new array $\{0.3, 0.5, 0.8, 1.1, 1.2\}$ can be obtained. In the original OFDM symbol, the first subcarrier is changed into the third position of the new OFDM symbol; similarly, the second is assigned to the first position, and so on. Following this scrambling criterion, a key space of $M!$ is created for OFDM data encryption.

The setup of IM/DD OFDM transmission experiment was the same as applied in the previous signal scrambling schemes; however, with the FFT points of 1024 and the number of subcarriers of 384, consequently the data rate was 13.3 Gb/s (10Gs/s$\times$4$\times$384/1024$\times$8/9). The BER measurements were shown in **Figure 16**, for the encrypted 16-QAM OFDM signals. The encrypted signals with DFT-S-OFDM precoding were fully recovered by the legal ONUs. Due to the effective PAPR reduction via DFT precoding, the BER performance was improved ~2 dB (BER@10^{-3}) for the encrypted DFT-S-OFDM signals, if compared with the cases of original OFDM transmission without DFT precoding. The power penalty was only 0.9 dB (at BER@10^{-3}) between the case of B2B and transmission after 20 km SSMF. Moreover, for any illegal ONU with a wrong key, even a very small deviation of 1×10^{-15}, it was not possible to decrypt the original OFDM signals. The corresponding constellations for correct and incorrect chaotic keys are plotted as insets in **Figure 16**, which verified the security feasibility of chaotic DFT-S-OFDM signal transmission.

To quantitatively evaluate the performance dependence on the mismatches of row or column parameters $\Delta m, \Delta n$ in the reconfigurable DFT matrix $\mathbf{F}'$, the BERs are shown with respect to the parameters $\Delta m, \Delta n$ in **Figure 17**, where the modulation format is 16-QAM, and the received optical power is -9 dBm. In **Figure 17(a)**, when $\Delta m > 0.01$, the BER is significantly deteriorated

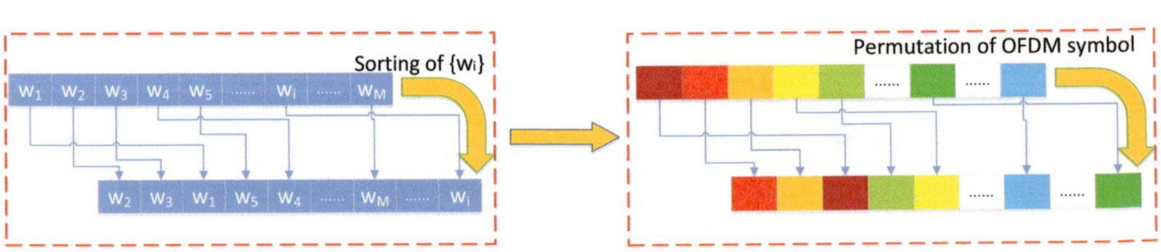

Figure 15. Subcarrier allocation in OFDM symbol using digital chaotic sequences.

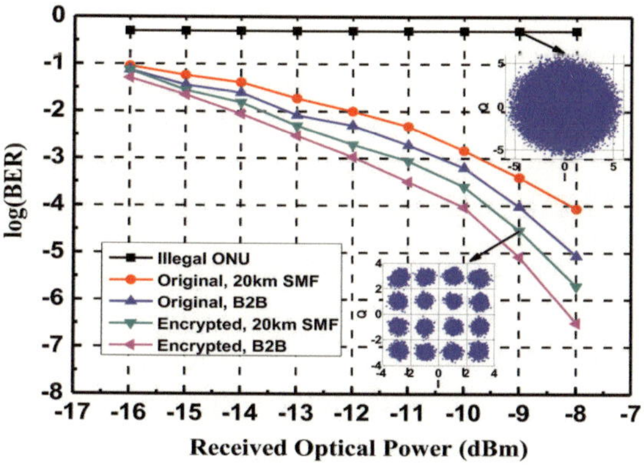

Figure 16. BERs of chaotic encrypted DFT-S-OFDM transmission.

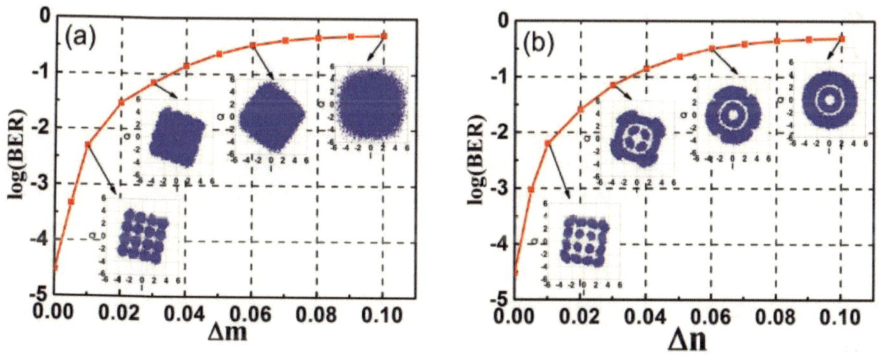

Figure 17. BERs versus the index mismatches in (a) Δm (b) Δn of chaotic DFT matrix.

(BER>10^{-3}), thus the data cannot be correctly received, even after using forward error correction approaches. It can also be noted that, if Δm is increased further, deterioration becomes more serious, as shown in the corresponding BERs and constellations. Similarly, the BERs variation with respect to the column parameter mismatch in the reconfigurable DFT matrix in $\mathbf{F}'$ is plotted in **Figure 17(b)**. From **Figure 17(b)**, the original data can be only recovered when $\Delta n < 0.01$. The results also verify that, the rotation angles in the corresponding constellations become larger with the increase of Δn. The difference of BERs and constellations with respect to $\Delta m, \Delta n$ again confirms that it is helpful to set two independent values of m, n in the new reconfigurable DFT matrix in $\mathbf{F}'$, for the purpose of key space expansion in encryption scheme using chaotic DFT precoding.

For the proposed multi-fold DFT-S-OFDM data encryption, the total key space can be given as follows. First, the parameters in DFT matrix (both of row and column) can be varied from 0 to 384 with the sensitivity of 0.01, which generates a key space of ~$1.5\times10^9(38,400\times38,400)$ for each OFDM symbol. Second, the subcarrier allocation generates a key space of 384!. Finally, the chaotic TS for OFDM symbol time synchronization enlarges the key space by a factor of ~10.

4.2.2. Chaotic Walsh-Hadamard transform (WHT)

Chaotic WHT is an alternative encryption technique using signal precoding, which provides lower computational complexity if compared with DFT precoding. The encryption scheme based on a reconfigurable chaotic WHT precoding matrix is shown in **Figure 18** [21], where the modified version of reconfigurable WHT matrix is also determined by digital chaos.

Theoretically, it can be proven that the permutation on the standard WHT matrix does not degrade the precoding properties. Since the WHT matrix is an orthogonal matrix whose elements are constrained from the set $\{-1, 1\}$,

$$\mathbf{H}\mathbf{H}^{\mathrm{T}} = \mathbf{H}^{\mathrm{T}}\mathbf{H} = n\mathbf{I} \tag{21}$$

Assuming $\mathbf{Q}_{n\times n}$ and $\mathbf{R}_{n\times n}$ are the row and column permutation matrices on WHT matrix $\mathbf{H}_{n\times n}$ respectively, then the chaotic row/column permutated WHT matrix becomes $\mathbf{P}_{n\times n}$.

$$\mathbf{P} = \mathbf{QHR}, \quad \mathbf{P}^{\mathrm{T}} = \mathbf{R}^{\mathrm{T}}\mathbf{H}^{\mathrm{T}}\mathbf{Q}^{\mathrm{T}} \tag{22}$$

$$\mathbf{P}\mathbf{P}^{\mathrm{T}} = \mathbf{QH}(\mathbf{R}\mathbf{R}^{\mathrm{T}})\mathbf{H}^{\mathrm{T}}\mathbf{Q}^{\mathrm{T}} = \mathbf{QHI}\mathbf{H}^{\mathrm{T}}\mathbf{Q}^{\mathrm{T}} = \mathbf{Q}n\mathbf{I}\mathbf{Q}^{\mathrm{T}} = n\mathbf{I} \tag{23}$$

To specify the PAPR reduction capability of chaotic WHT matrix, the same analysis based on the autocorrelation coefficient can be applied as for chaotic DFT.

As plotted in **Figure 19(a)**, the original 16-QAM OFDM symbols are applied and compared with the standard and chaotic WHT. The autocorrelation function is reduced in sidelobe if either chaotic or standard WHT is employed. An enlarged sidelobe for a specific zone is inserted in **Figure 19(a)** for clear view. Moreover, the sidelobe shape is exactly the same for both the standard and chaotic WHT precoders.

Furthermore, the CCDF curves are calculated for the cases of chaotic WHT, standard WHT, and un-precoded original OFDM signals, as shown in **Figure 19(b)**. The PAPR is effectively reduced by ~1 dB by either the standard or chaotic WHT precoder, compared to the un-precoded OFDM signals. In addition, the CCDFs are exactly the same for the standard and chaotic WHT, which again verifies the above theoretical analysis.

Therefore, row/column permutated WHT can be still applied as a precoder not only to reduce the PAPR but also to enhance the physical-layer security of OFDM data, since the correct

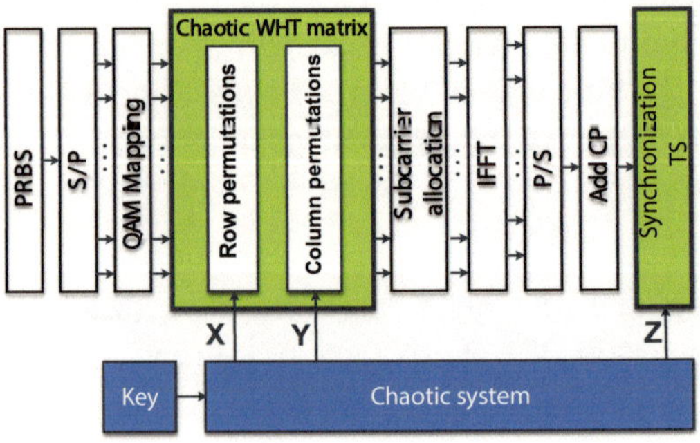

Figure 18. Schematic diagram of OFDM encryption precoded with chaotic WHT.

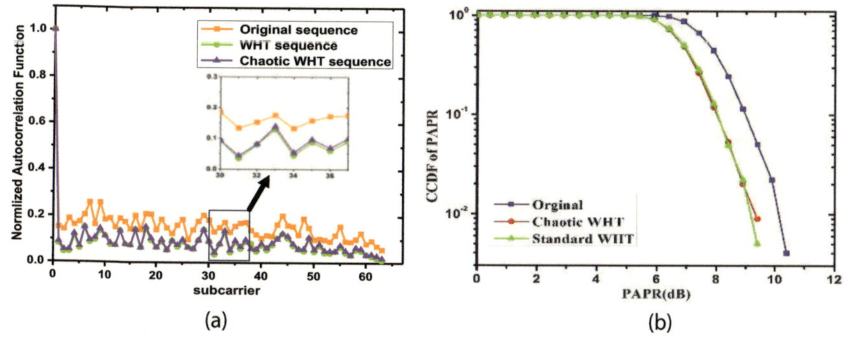

Figure 19. PAPR reduction preceded by the standard/chaotic WHT, (a) normalized aperiodic autocorrelation function; (b) CCDF of PAPR.

original OFDM data can be recovered only after obtaining the correct row/column sequences to reconstruct the chaotic WHT matrix through legal ONUs.

In the proposed chaotic WHT scheme, a 3D hyper Chen's chaos is implemented to generate the digital chaotic sequences for data encryption [22]:

$$\begin{cases} \dot{x} = -ax + y \\ \dot{y} = (c-a)x + cy + xz \\ \dot{z} = bz + xy \end{cases} \tag{24}$$

where a, b, and c are constants. Generally, with a=35, b=3, $c\in[20, 28.4]$, it enters into chaotic zones. To generate the permutation vectors D_x, D_y and D_z, each of the chaotic sequences $\{x\}$, $\{y\}$, and $\{z\}$ is post-processed as follows.

First, we assume that the length of chaotic sequences $\{x\}$ and $\{y\}$ is M, where M is the order of WHT matrix, and then

$$D_x = sort\{x\}, \quad D_y = sort\{y\}, \quad D_z = sort\{z\} \tag{25}$$

where *sort* function returns an index vector, according to the ascending order of the values in the chaotic sequences.

The procedure of the proposed WHT chaotic encryption is given as follows. The two permutation vectors D_x and $D_{y,}$ are used to permute the row and column indexes of the standard WHT matrix, respectively. Assuming the standard WHT matrix $\mathbf{H}^{(0)}_{M\times M}$ is

$$\mathbf{H}^{(0)}_{M\times M} = [\boldsymbol{r}_1^T, \boldsymbol{r}_2^T, \cdots, \boldsymbol{r}_M^T] \tag{26}$$

where $r_1, r_{2,\ldots,} r_M$ are the row vectors. After row permutation, it becomes

$$\mathbf{H}^{(1)}_{M\times M} = [\boldsymbol{r}_{Dx(1)}^T, \boldsymbol{r_{Dx}}^T{}_{(2)}, \cdots \boldsymbol{r_{Dx}}^T{}_{(M)}] = [c_1, c_2 \cdots c_M] \tag{27}$$

Similarly, the columns are permutated via vector D_y, and the corresponding chaotic WHT matrix after both row and column permutations becomes

$$\mathbf{H}^{(2)}_{M\times M} = [c_{D_y(1)}, c_{D_y(2)}, \cdots, c_{D_y(M)}] = [v_1^T, v_2^T, \cdots, v_M^T] \tag{28}$$

The chaotic sequence D_z is used to generate the chaotic TS for OFDM symbol synchronization, which is composed of random {−1, 1} and defined as [23]

$$\text{TS} = \{(\text{mod}(D_{z,i}, 2) - 0.5)\times 2\} \tag{29}$$

Since the hyper-chaotic sequences have high sensitivity to the initial values ($\sim 1\times 10^{-15}$) and present exceptional random behavior, the initial values are served as the security key between OLT and ONUs. A secure channel has to be employed to transfer the keys between OLT and ONUs, for instance, using quantum key distribution (QKD), which is out of the scope of this chapter and will not be given in detail here. **Figure 20** shows the sensitivity of the row/column permutation indexes of chaotic WHT matrix with respect to a tiny change in the initial conditions.

Figure 21 shows the experimental setup of the encrypted transmission of OFDM signals using chaotic WHT precoding. The IM/DD was also applied, and the details were described in Section 4.1. At OLT, there were 256 points in IFFT, among which 64 subcarriers were used for OFDM data. The effective data rate was 8.9 Gb/s (10 Gs/s×4×64/256/(1+1/8)). The signal waveform, electric and optical spectra are plotted as insets in **Figure 21**.

The transmission performance was evaluated for OFDM signals precoded by chaotic WHT via BER measurements of 128 OFDM symbols, for the cases of B2B and transmission after 20 km

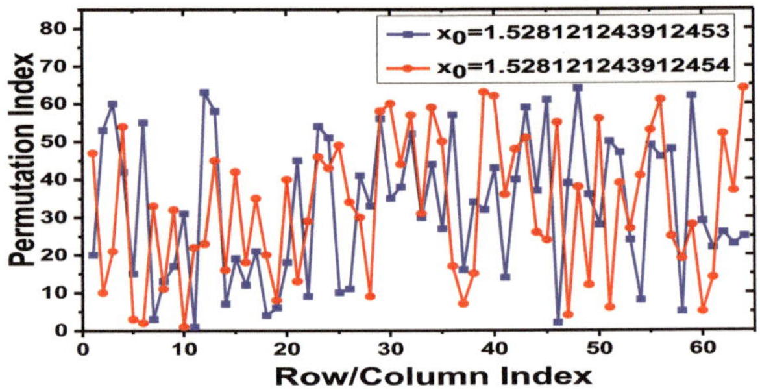

Figure 20. Chaotic permutation index versus a slight change in the initial values.

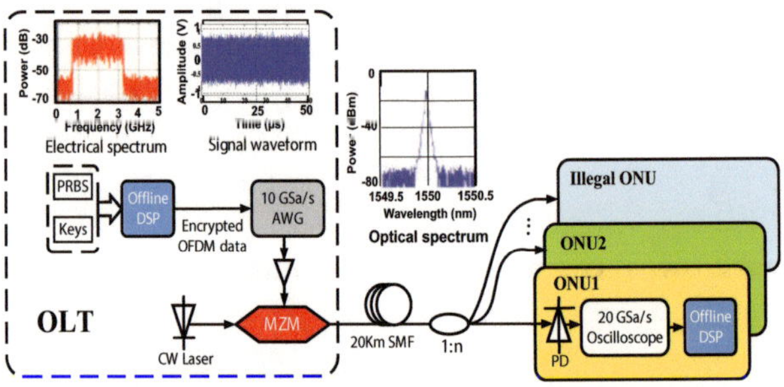

Figure 21. Experimental setup of encrypted optical OFDM transmission using chaotic WHT.

http://dx.doi.org/10.5772/intechopen.68238

SSMF, as shown in **Figure 22**. The BER was improved by ~1 dB (BER@10^{-3}) in receiver sensitivity for chaotic WHT, compared with the un-precoded OFDM signals. This improvement can be mainly attributed to the precoding gain from chaotic WHT, since BER was improved also for the case of B2B, which verified the effective PAPR reduction in OFDM signals through chaotic WHT precoding. For transmission after 20 km SSMF, the effect of fiber dispersion is negligible due to the narrow spectrum of OFDM signals.

The robustness of the multi-fold data encryption is evaluated via the total key space created by chaotic WHT, since only legal ONUs can generate the correct chaotic WHT matrix to recover the original OFDM data. Assuming that the dimension of chaotic WHT matrix is 64×64, it creates a total key space of $64!\times64!$ (~10^{178}). In addition, the chaotic TS for OFDM symbol synchronization further increases the key space by a factor of ~10. As a result, a total key space of ~10^{179} is achieved in chaotic WHT encryption scheme.

The computational complexities the chaotic encryption schemes with PAPR reduction are listed in **Table 1**. The computational complexity is reduced significantly in WHT, since it does not require complex multiplication if compared with DFT, SLM, and PTS, because the elements in WHT matrix are constrained to the value set of {1,−1}. Meanwhile, WHT has the same

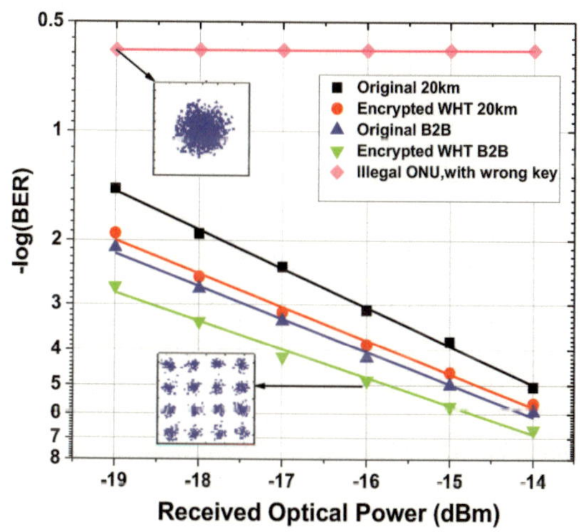

Figure 22. BER measurements of the OFDM transmission with and without chaotic WHT.

	WHT	DFT	SLM	PTS
Multiplication	None	N^2	$B\frac{gN}{2}\log_2 gN$	$V\frac{gN}{2}\log_2 gN$
Addition	$N(N-1)$	$N(N-1)$	$BgN\log_2 gN$	$VgN\log_2 gN+(V-1)gNW^V$
Sideband	None	None	Needed	Needed

B is the number of candidate signals in SLM, *V* is the number of sub-blocks, and *W* is the number of phase factors.

Table 1. Computation complexity and sideband information of chaotic encryption schemes with PAPR reduction.

addition complexity as DFT. In addition, if compared with PTL or SLM schemes, simpler optical OFDM transmission structure and higher spectral efficiency can be expected for the WHT and DFT precoding schemes, since no additional sideband information needs to be transmitted.

5. Conclusions

We have reviewed a serial of novel data encryption schemes for physical-layer security enhancement for optical OFDM transmission. By incorporating multi-fold digital chaos-based data encryption in various DSP procedures in OFDM signal generation, modulation as well as OFDM symbol synchronization, data security is greatly enhanced with a huge key space. Moreover, the OFDM transmission performance is significantly improved because the pseudo-random properties of digital chaos are employed for effective PAPR reduction. Moreover, for the chaotic precoding schemes, the reconfigurable chaotic precoding matrices are predetermined by digital chaos; therefore, the sideband information is no longer necessary, which also improves the spectral efficiency in transmission. The proposed chaotic data encryption schemes have been successfully demonstrated by ~10 Gb/s OFDM transmission experiments, which verifies that these schemes could be promising candidates for next-generation secure OFDM-PON.

Acknowledgements

This work was supported in part by International S&T Cooperation Program of China, 2016YFE0104500, Natural Science Fund of China under Grants 61571291, 61431009, and 61221001.

Author details

Xuelin Yang*, Adnan A.E. Hajomer and Weisheng Hu

*Address all correspondence to: x.yang@sjtu.edu.cn

State Key Laboratory of Advanced Optical Communication Systems and Networks, Shanghai Jiao Tong University, Shanghai, China

References

[1] Fang Y, Yu J, Zhang J, Chi N, Xiao J, Chang G. Ultrahigh capacity access network architecture for mobile data backhaul using integrated W-band wireless and free-space optical links with OAM multiplexing. Optics Letters. 2014;**39**(14):4168–4171. DOI: 10.1364/OL.39.004168

[2] Reis J, et al. Terabit+ (192 × 10Gb/s) Nyquist shaped UDWDM coherent PON with upstream and downstream over a 12.8 nm band. Journal of Lightwave Technology. 2014;**32**(4):729–735. DOI: 10.1109/JLT.2013.2283017

[3] ITU-T Recommendation G.983.1, Broadband Optical Access Systems Based on Passive Optical Networks, International Telecommunication Union-Telecommunication Sector (ITU-T); 2005

[4] Zhang L, Xin X, Liu B, Yin X. Physical secure enhancement in optical OFDMA-PON based on two-dimensional scrambling. Optics Express. 2012;**20**(26):B32–B37. DOI: 10.1364/OE.20.000B32

[5] Liu B, Zhang L, Xin X, Wang Y. Physical-layer security in OFDM-PON based on dimension-transformed chaotic permutation. IEEE Photonics Technology Letters. 2014;**26**(2):127–130. DOI: 10.1109/LPT.2013.2290041

[6] Zhang L, Xin X, Liu B, Wang Y. Secure OFDM-PON based on chaos scrambling. IEEE Photonics Technology Letters. 2011;**23**(14):998–1000. DOI: 10.1109/LPT.2011.2149512

[7] Liu B, et al. Piecewise chaotic permutation method for physical-layer security in OFDM-PON. IEEE Photonics Technology Letters. 2016;**28**(21):2359–2362. DOI: 10.1109/LPT.2016.2594042

[8] Zhang W, et al. Joint PAPR reduction and physical-layer security enhancement in OFDMA-PON. IEEE Photonics Technology Letters. 2016;**28**(9):998–1001. DOI: 10.1109/LPT.2016.2522965

[9] Zhang W, Zhang CF, Jin W, Chen C, Jiang N, Qiu K. Chaos coding-based QAM IQ-encryption for improved security in OFDMA-PON. IEEE Photonics Technology Letters. 2014;**26**(19):1964–1967. DOI: 10.1109/LPT.2014.2343616

[10] Yang X, Hu X, Shen Z, He H, Hu W, Bai C. Physical layer signal encryption using digital chaos in OFDM-PON. In: 10th International Conference on Information Communications and Signal Processing (ICICS); November 2015. DOI: 10.1109/ICICS. 2015.7459872

[11] Cvijetic N. OFDM for next generation optical access networks. Journal of Lightwave Technology. 2012;**30**(4):384–398. DOI: 10.1109/JLT.2011.2166375

[12] Popoola WO, et al. Pilot-assisted PAPR reduction technique for optical OFDM communication systems. Journal of Lightwave Technology. 2014;**32**(7):1374–1382. DOI: 10.1109/JLT.2014.2304493

[13] Jayalath ADS, Tellambura C. Reducing the peak-to-average power ratio of orthogonal frequency division multiplexing signal through bit or symbol interleaving. Electronics Letters. 2000;**36**(13):1161–1163. DOI: 10.1049/el:20000822

[14] Sharif M, Gharavi-Alkhansari M, Khalaj BH. On the peak-to-average power of OFDM signals based on oversampling. IEEE Transactions on Communications. 2003;**51**(1):72–78. DOI: 10.1109/TCOMM.2002.807619

[15] Tang L, Zuo Q, Cui W. Synchronization scheme using four-dimensional chaotic system for OFDM. Journal of Communication. 2010;**31**(1):73–84 (in Chinese).

[16] Alvarez G, Li S. Some basic cryptographic requirements for chaos-based cryptosystems. International Journal of Bifurcation and Chaos. 2006;**16**(8):2129–2151. DOI: 10.1142/S0218127406015970

[17] Hu X, Yang X, Shen Z, He H, Hu W. Chaos-based partial transmit sequence technique for physical-layer security in OFDM-PON. IEEE Photonics Technology Letters. 2015;**27**(23):2429–2432. DOI: 10.1109/LPT.2015.2466092

[18] Cheng M, et al. Security-enhanced OFDM-PON using hybrid chaotic system. IEEE Photonics Technology Letters. 2015;**27**(3):326–329. DOI: 10.1109/LPT. 2014.2370757

[19] Hu X, Yang X, Hu W. Chaos-based selected mapping scheme for physical-layer security in OFDM-PON. Electronics Letters. 2015;**51**(18):1429–1431. DOI: 10.1049/el.2015.1261

[20] Shen Z, Yang X, He H, Hu W. Secure transmission of optical DFT-S-OFDM data encrypted by digital chaos. IEEE Photonics Journal. 2016;**8**(3). DOI: 10.1109/JPHOT.2016.2564438

[21] Hajomer A A E, Yang X, Hu W. Chaotic Walsh–Hadamard Transform for Physical Layer Security in OFDM-PON. IEEE Photonics Technology Letters. 2017;**29**(6): 527–530. DOI: 10.1109/ LPT. 2017. 2663400

[22] Chen G, Ueta T. Yet another chaotic attractor. International Journal of Bifurcation and Chaos. 1999;**9**(7):1465–1466. DOI: 10.1142/S0218127499001024

[23] Wang X, et al. SSBI mitigation at 60GHz OFDM-ROF system based on optimization of training sequence. Optics Express. 2011;**19**(9):8839–8846. DOI: 10.1364/OE.19.008839

Chapter 17

Impact of Fiber Duplication on Protection Architectures Feasibility for Passive Optical Networks

Waqas Ahmed Imtiaz, Javed Iqbal, Affaq Qamar,
Haider Ali and Sevia M. Idrus

Additional information is available at the end of the chapter

http://dx.doi.org/10.5772/intechopen.68237

Abstract

Adaptability of high capacity passive optical network (PON) requires the provision of an efficient fault detection and restoration mechanism throughout the network at an acceptable cost. The readily adapted pre-planned protection strategy relies on component duplication, which significantly increases the cost of deployment for PON. Therefore, it is imperative to determine a suitable component that requires high redundancy and determine the impact of protection for that component on feasibility of PON. Five protection architecture including ITU-T 983.1 Type C, single ring, dual ring, tree- and ring-based architectures with hybrid star-ring topology at the optical distribution network (ODN), are considered to evaluate the impact of fiber duplication in terms of capital expenditure (CAPEX), operation expenditure (OPEX), reliability, and support for maximum number of subscribers. Reliability block diagram (RBD) based analysis shows that desirable 5 nines connection availability is provided by each protection architecture and utilization of ring topology avoids duplication of the fiber but effects the number of subscribers. Furthermore, it is observed that OF duplication at ODN is the main contributor to CAPEX. Collectively hybrid protection architectures provide efficient performance and proves to be a feasible solution for the deployment of survivable PONs at the access domain.

Keywords: passive optical network, protection, network topology, reliability, CAPEX, OPEX

1. Introduction

Exponential growth in Internet traffic has significantly increased the demand for high bandwidth connectivity at both business and residential premises. Internet service providers are deploying passive optical network (PON) at the access domain to provide the required capacity in terms of reach, bandwidth, and the number of subscribers. In PON, all services are originated from an optical line terminal (OLT) at the central office (CO). End-face of the OLT is connected to a 15–20 km feeder fiber (FF) that extends the network toward the subscriber premises called optical distribution network (ODN). Remote node (RN) receives the FF at ODN, which houses a $1:N$ bidirectional passive optical coupler (POC). N output ports from the POC are fed into short-branched distribution fibers (DFs) that connect the RN to individual optical network unit (ONU) transceiver modules [1].

PON has emerged as a promising candidate to resolve the last-mile bottle, owing to its significant advantages like:

- Support for high network capacity in terms of reach, bandwidth, and the number of subscribers due to a complete optical fiber (OF) path between OLT and ONU modules.
- Minimum capital expenditure (CAPEX) by sharing FF between OLT and multiple ONUs.
- Reduced operational expenditure (OPEX) through passive components at RN, which requires no power, minimum maintenance and planning.
- Smooth service upgradability with existing infrastructures.
- Highly scalable as new subscribers can easily join the network.
- High degree of flexibility, owing to the use of FF between OLT and multiple subscribers.

With the rapid increase in PONs capacity, fault detection and restoration at satisfactory costs have turned the network reliability to a new challenge for Internet service providers. Each subscriber is interested in seamless reception of maximum bandwidth at minimum possible cost. However, the conventional PON architecture has limited protection, which results in significant data loss at the event of failure in optical components including OF medium. Therefore, it is imperative to devise an architecture, which is capable of maintaining a seamless flow of upstream and downstream traffic at required capacity and acceptable costs for a common end subscriber [1, 2].

Two techniques are readily adapted to provide fault detection and restoration in PON, namely pre-planned and dynamic protection. The latter relies on fault detection and restoration through diagnosis at the higher levels and dynamically allocates resources at the event of failure. Such technique requires more time for traffic restoration between OLT and ONU modules, as upper layer recovery techniques usually utilize routing tables, topology recalculations, and slow convergence time. Yet there is no guarantee for fault restoration at the physical layer [1–3]. Therefore, for the facilitation of an effective and prompt fault detection and restoration, it is highly desirable to provide protection measures at the optical layer.

Pre-planned protection utilize an optical-layer approach by providing dedicated backup paths for components including OF medium. This type of protection is planned at the network design phase, owing to the fact that topology of PON remains same, and the proposed solution can address fault restoration at both feeder and ODN. This type of protection provides high reliability at minimum recovery time in the event of failures at both optical components and OF medium. However, path and resources duplication significantly elevates the CAPEX at the network deployment phase [4, 5]. Therefore, it is imperative to encompass the following considerations while designing a pre-planned protection architecture.

2. Consideration for PON protection planning

2.1. Network topology

Network topology significantly effects the design, redundancy, and deployment cost for the PON. Two common network topologies are used for the deployment of PON, namely tree and ring [1]. In tree topology, the optical signal sent from OLT is divided into N equal parts and delivered to designated ONU modules through respective DFs. Such deployment can provide the required bandwidth at desired number of subscribers; however, a single cut or failure at the feeder level can cripple the entire network by disconnecting the working OLT from ONU modules. Moreover, failure at the DF can also result in significant data loss and high customer dissatisfaction. Therefore, such topology requires the provision of redundancy at both levels of PON, which is achieved by duplicating the network components.

Ring topology is adapted to minimize the cost incurred by the provision of redundant paths in the conventional PON. It utilizes a single ring-based fiber that connects the OLT directly to all ONU transceiver modules. This significantly reduces the effect of fiber cuts or failures [6, 7]. Ring topology provides the required reliability at acceptable costs; however, use of the POC between OLT and individual ONU module introduces serious power budget issues, which effects network capacity in terms of the number of subscribers [4, 8, 9]. Besides the commonly used ring- and tree-based network architectures, hybrid topologies are readily adapted for the implementation of survivable PON at the access domain. These architectures utilize a combination of tree- and ring-based architectures with subsequent topologies such as tree-ring, tree-star, ring-star, and bus. Hybrid architectures have proved as a promising candidate to provide the required redundancy at desirable network capacity [10–14].

2.2. Resources to be protected

A typical PON primarily comprises two types of resources that require protection for efficient delivery of information between OLT and ONU modules, namely OF medium and optical components. Both significantly effect the flow of upstream and downstream traffic throughout the network. **Figure 1** shows connection availability for PON components based on **Table 1** [15]. It is observed that active and passive devices, such as OLT, ONU, POC, $X : N$POC, and so on, provide desirable (5 nines) connection availability over the network lifetime, since the rate of failure for these components is significantly low. Furthermore, the mean time to repair (MTTR) for the in-house optical components is minuscule as compared to the on-field components like OF medium that constitutes a major portion of PON architecture and is more prone to failures as shown in **Figure 1**.

Therefore, OF paths require more attention as compared to other components of the networks, in order to ensure seamless transmission of information, minimize the loss of data, service interruption penalty cost, and PON downtime per year [4].

2.3. Number of subscribers

Number of subscribers refer to the amount of users that a PON can accommodate without compromising the reach and provision of nominal bandwidth. It is an important parameter

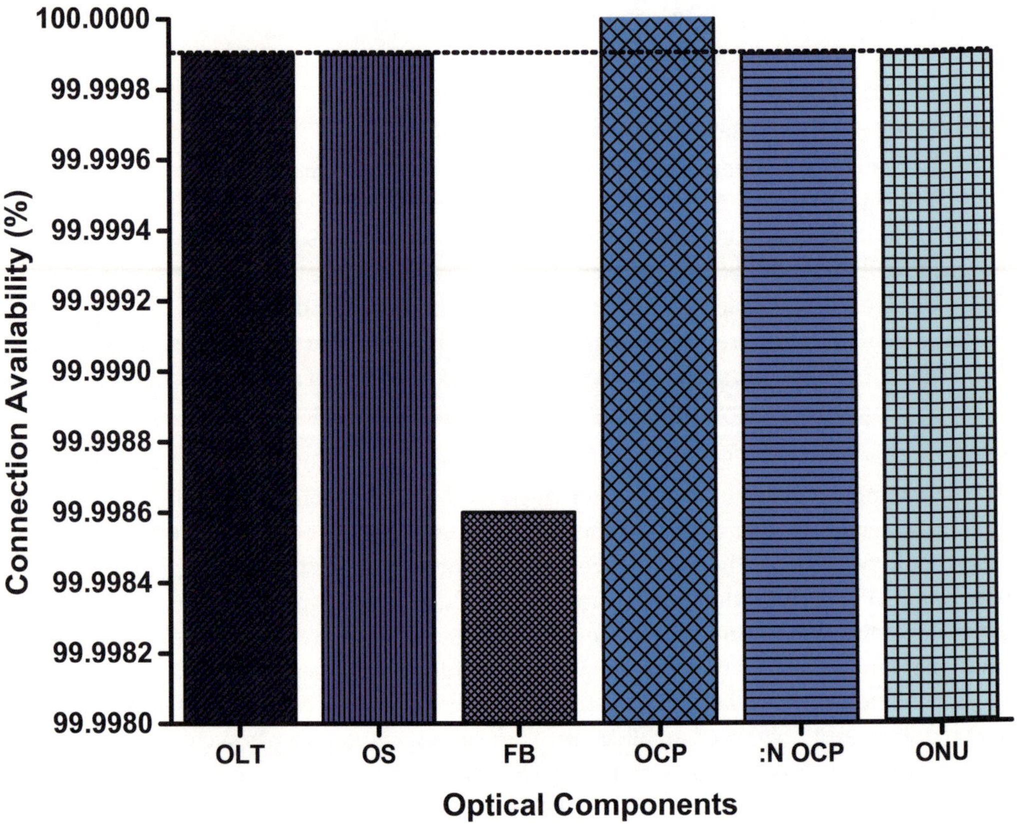

Figure 1. Connection availability of basic optical components in PON.

System components	Unavailability	Cost ($)	Energy consumption
OLT	$5.12e^{-7}$	12100	20 W, 25 W (with EDFA), 2 W (standby)
ONU	$1.54e^{-6}$	350	1 W, 0.25 W (standby)
Optical circulator	$3e^{-7}$	50	
Optical switch	$1.2e^{-6}$	50	
1:2, 2:2 POC	$3e^{-7}$	50	
1:N, 2:N POC	$7.2e^{-7}$	800	
Fiber ($/Km)	$1.3e^{-5}$	160	

Table 1. Components description, connection availability, and cost.

since it is directly associated with the extent and cost of the network. Number of subscribers is primarily effected by the type of topology at both feeder and ODN along with devices at the CO and RN. For example, a typical tree-star topology can accommodate more subscribers as compared to a conventional ring-based architecture due to the use of 1 : N POC. Whereas, the latter utilizes symmetric Y, 1 : 2 or 2 : 2, POC per subscriber, which introduces a power-budget loss of −3 dBm in each symmetric POC, $P_{POC} = 10\text{Log}_{10}\left(\frac{1}{2}\right)$ [6, 16]. This significantly effects capacity of the network since nominal received power is required for high bandwidth communication. Therefore, it is imperative to consider these features at the network planning phase, so that the proposed PON can accommodate maximum number of subscribers at desirable capacity and cost.

2.4. Cost and complexity

Deployment/operational costs and feasibility of PONs primarily depend on complexity of the network architecture. For example, some protection mechanics utilize redundant transceivers at both OLT and ONUs, like ITU-T type C and D, in order to avoid 1 : 1 or 1 + 1 switching [17]. Although such techniques provide an abrupt recovery to maintain a smooth flow of information between OLT and ONU modules, they significantly elevate the deployment cost of the network. Since more CAPEX is spent on OLT duplication as compared to the 1 : 1 or 1 + 1 switching, a trade-off must be made between the cost and recovery time at the event of failure. Therefore, it is desirable to minimize the overall system complexity, without compromising the fault detection and restoration time.

3. Protection architectures for PON

Different protection architectures are proposed to facilitate fault detection and restoration in PON. This section discusses five pre-planned protection architectures, which vary in terms of topology, fiber duplication, and devices at both feeder level and ODN.

3.1. ITU-T 983.1 type C architecture

ITU-T 983.1 type C is a pre-planned protection architecture, which provides fault detection and restoration throughout the network with redundant components at both feeder and the distribution levels [17]. The basic type C PON is shown in **Figure 2**, where each component of the network is duplicated to ensure high connection availability and fast restoration time. OLT is placed at the CO and consists of two transceiver modules, where one acts as primary (OLT_p) and another is set as a secondary (OLT_s) module. Under normal mode of operation, OLT_p is responsible for originating and managing services across the network, whereas OLT_s activates in the event of failure at OLT_p module.

Each transceiver module at OLT is connected to a corresponding FF, namely primary (FF_p) and secondary (FF_s). Both fibers extend the network toward the subscribers' premises. Under normal mode of operation, OLT_p is connected to the FF_p. Two FFs are used to provide maximum connection availability and fast restoration time, such that FF_p is immediately replaced with FF_s in the event of failure. The span of each fiber is about 20 km for a standard PON. Both FFs terminate into RN, which serves as chases for two 1 : N POC modules connected with the corresponding FF at the input port and N DFs at the output port, respectively, where N represents the number of subscribers.

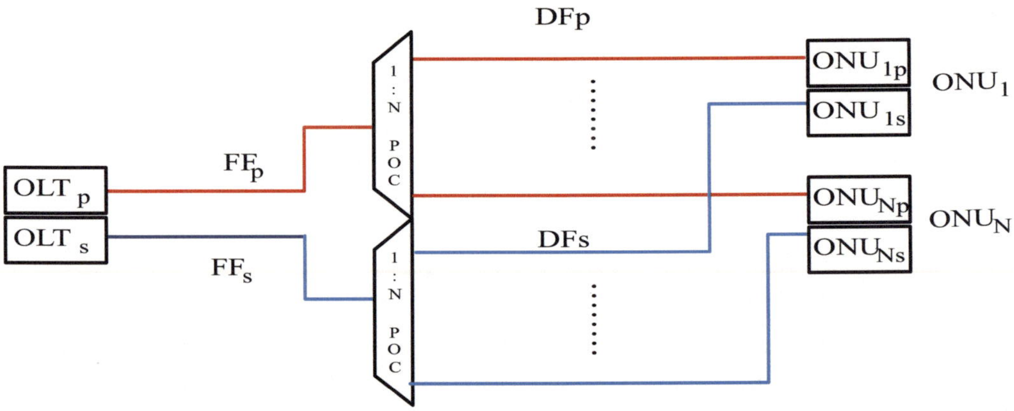

Figure 2. Type C protection architecture.

Consequently, a total of 2*N* DFs are utilized to provide the required protection at the ODN. DFs terminate into ONU with two transceiver modules, (ONU_p,ONU_s), in order to facilitate abrupt fault detection and restoration. Type C protection architecture duplicates every component including OFs to provide abrupt fault detection and restoration. Furthermore, a two-fiber tree-based topology is laid at the feeder level, whereas a star-based topology is adapted to implement the 2*N* DFs at the ODN.

Number of subscribers in type C protection architecture can be determined through power budget analysis from OLT toward ONU module. Power budgeting in PON ensures an efficient communication between the transmitting and receiving modules. Moreover, it also determines the POC splitting ratio, which translates the number of subscribers in PON, respectively. If P_i represents the power loss across each component i, α is attenuation/km in the OF medium, and R_{sen} represents the receiver sensitivity, then the number of subscribers for type C protection architecture can be determined as:

$$P_T - P_{OLT} - \alpha P_{FF} - P_{POC} - \alpha P_{DF} - P_{ONU} - P_{mis} \geq R_{sen} \tag{1}$$

Equation (1) shows that major power loss occurs across the POC, which is determined by $P_{OPC} = 10\text{Log}_{10}(N)$, where N represents the number of subscribers or splitting ration of $1:N$ POC. Therefore, to maintain received power $P_{re} \geq R_{sen}$, the value of N must be adjusted to facilitate extended reach, fault detection, and restoration along with high bandwidth connectivity. For example, when the transmitter power is 10 dBm, $\alpha = -0.25\ \frac{dB}{km}$, $N = 128$, 25 km fiber, and $P_{mis} = -3$ dBm, the approximate power received at the PIN photo-diode will be $P_{re} = -20$ dBm. Consequently, for $R_{sen} = -25$ dBm, type C architecture can efficiently support 128 subscribers simultaneously accessing the medium.

3.2. Single ring architecture

In order to avoid extensive duplication of the OF medium at both feeder and ODN, ring-based topology is utilized to implement PON with desired connection availability and fault detection/

http://dx.doi.org/10.5772/intechopen.68237

restoration between OLT and ONU modules [16]. The basic ring-based PON is shown in **Figure 3**, which contains OLT module at the CO. In order to avoid the high deployment cost of OLT module, this architecture employs a single unit, owing to the fact that failure per year of OLT module is minuscule [4, 18]. End-face of the OLT module is connected to a switching arrangement (SA_{co}) that extends the OF medium in both clockwise (CW) and counter clockwise (CCW) directions. SA_{co} serves as chases for a 1 : 2 POC_{sa}, and 1 : 2 OS_{sa}. POC_{sa} sends and receives the traffic for OLT module. Port 1 splits the optical signal toward the clockwise feeder ring (FR_{cw}), whereas port 2 extends the flow of traffic toward the counter clockwise FR (FR_{ccw}) through a 1 : 2 OS_{sa}. This arrangement recovers the flow of information in case of failures at the FR. Under normal working conditions, OS_{sa} is at port *a* and both upstream and downstream traffic are carried on the FR_{cw}.

FR contains multiple ONU modules $\{ONU_n;\ n = 1, 2, 3 \ldots N\}$, which are placed directly over the FR through individual RNs (RN_x) housing a 2 : 2 $POC_{x,1}$ and a 1 : 2 $POC_{x,2}$, respectively. Consequently, the total number of ONU modules is equal to RNs on the FR $X = N$. $POC_{x,1}$ is used to extend the FR to the neighboring ONU modules. Furthermore, it also connects individual ONU_n with FR through $POC_{x,2}$ and controls the flow of traffic in and out of the ONU module.

Network capacity in single ring-based architecture is significantly effected by the utilization of a single POC per subscriber. If N represents the number of RNs, then the total subscribers accessing the medium simultaneously can be determined by:

$$P_T - P_{OLT} - P_{POC_{sa}} - P_{OS_{sa}} - 2\,XP_{POC_{onu}} - P_{ONU} - P_{mis} \geq R_{sen} \tag{2}$$

X represents the number of RNs in Eq. (2), which is equal to ONUs N that can access the medium simultaneously. Consequently, for the transmitter power of 10, −3 dBm loss across the POC, $P_{mis} = -3$ dBm, 25 km fiber, and $R_{sen} = -25$ dBm, this architecture can scarcely support 16 subscribers simultaneously accessing the medium [6]. It is observed that fiber duplication is avoided at both levels through a ring-based structure; however, power drop across each POC

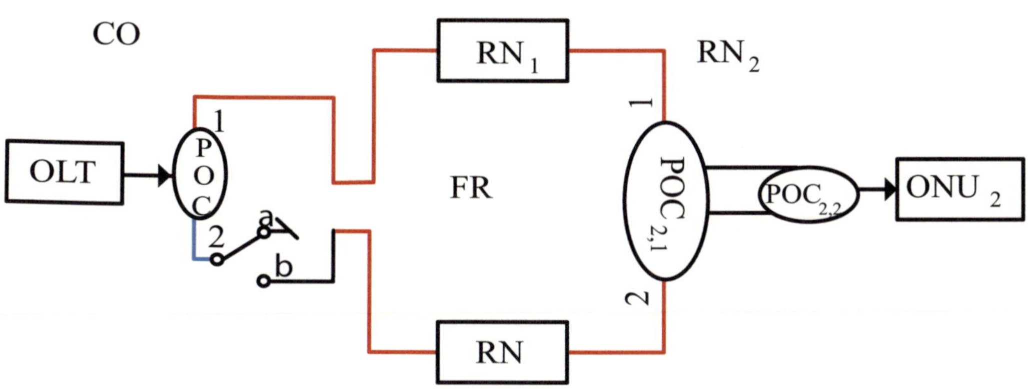

Figure 3. Single ring protection architecture.

introduces serious power budget issues. Consequently, capacity of the PON is effected. It is therefore imperative to utilize a topology that is capable of providing fault detection and restoration without compromising the network capacity and total cost.

3.3. Dual ring protection architecture

Dual ring PON avoids fiber duplication at the feeder level through a single ring-based fiber (FR) as shown in **Figure 4** [9]. Furthermore, several small rings are deployed at ODN to negate the power budget issue. CO contains a single OLT module, which is fed into an optical circulator (OC_{co}) and erbium-doped fiber amplifier (EDFA) module. To maintain passive nature of the PON, EDFA is placed inside the CO as shown in **Figure 4**. Switching arrangement SA_{co} is placed at end-face of the EDFA, which serves as chases for 1 : 2 POC_{sa} and 1 : 2 OS_{sa}. Port 1 of POC_{sa} splits the down-stream traffic towars FR_{cw}, whereas port 2 extends the flow of traffic towars the FR_{ccw} through a 1 : 2 OS_{sa}. In normal mode of operation, all traffic is carried through the FR_{cw}. However, OLT medium access control (MAC) layer flips the switch position when a failure occurs at the FR. This new position of the switch converts the ring-based fiber in two trees-based fibers carrying traffic in clock- and counter clockwise direction between the PoF.

Both FR_{cw} and FR_{ccw} are fed into multiple remote nodes forming several distribution rings (DRs) at the ODN. Each remote node $\{RN_x;\ x = 1, 2, 3 \ldots X\}$ houses two 2 : 2 bidirectional POCs, namely ($POC_{x,1}$,$POC_{x,2}$) as shown in **Figure 4**. The former extends the FR from CW toward CCW direction and the latter carries the traffic from feeder toward the ODN and vice versa. End-face of $POC_{x,2}$ splits the incoming traffic into clock- and counter clockwise DRs (DR_{cw}) and (DR_{ccw}), respectively. Both DR_{cw} and DR_{ccw} are connected to individual ONU

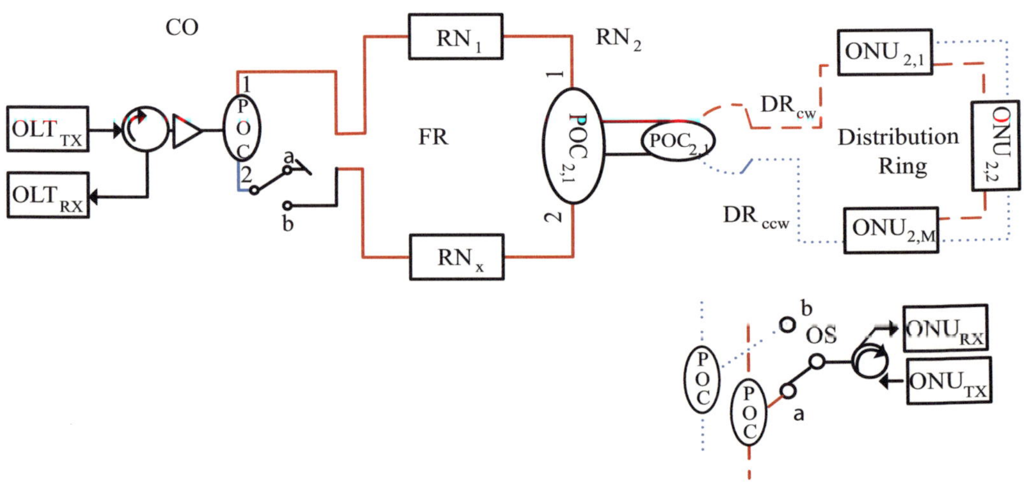

Figure 4. Dual ring protection architecture.

http://dx.doi.org/10.5772/intechopen.68237

$\{ONU_{x,m};\ m = 1, 2, 3...M\}$ module through two 1:2 $POC_{onu_{x,m}}$ and $OS_{onu_{x,m}}$. Under normal working conditions, $OS_{onu_{x,m}}$ is at position a and all traffic is handled through DR_{cw}.

Dual ring architecture utilizes multiple rings at the distribution level to compensate the excessive power drop. Furthermore, EDFA is also employed at the CO to support high capacity transmission. Consequently, the power budget equation from OLT toward ONU module can be written as:

$$P_T - P_{OLT} - P_{OC_{co}} + P_{EDFA} - P_{POC_{sa}} - P_{OS_{sa}} - \alpha P_{FR} - 2XP_{POC_{RN}} - \alpha P_{DR} - MP_{POC_{onu}} - P_{ONU} \geq R_{sen} \quad (3)$$

Where X, in Eq. (3), represents the total number of RNs deployed at the FR and M represents the number ONUs per RN. Consequently, the total ONU modules accessing the medium simultaneously become $N = X \times M$. Now, at $P_T = 10$ dBm, 25 km fiber, and $P_{edfa} = 25$ dB, the maximum value of N that can be achieved is 72 with $X = 9$ and $M = 8$ [9]. It is observed that the dual ring architecture supports more ONU modules in comparison with the single ring architecture. Nevertheless, power drops in the ring topology at the distribution level limits the overall capacity of the network.

3.4. Tree-based hybrid protection architecture

This architecture employs a hybrid topology through combination of tree and star-ring architectures at the feeder and ODN, respectively. OLT is placed at the CO, which is fed into an OC_{co}, EDFA, and SA_{co} as shown in **Figure 5** [15]. SA_{co} consists of a $1:2$ OS_{sa}, with port a connected to the FF_p, whereas port b is fed into FF_s. In normal working conditions, OS_{sa} is at port a and all upstream and down-stream traffic is sent and received through the FF_p. However, in case of failures or cuts at the FF_p, OLT MAC layer flips the switch position and resumes the flow of traffic through the FF_s.

Since a tree-based topology is adapted at the feeder level, the required protection is provided by duplicating the long-span fiber. End-face of both FF_p and FF_s terminates into RN that serves as chases for a $2:N$ POC. Output of the $2:N$ POC is connected to a series of dedicated DFs connecting N ONU transceiver modules as shown in **Figure 5**. In order to avoid extensive duplication of the DFs for protection, a ring-based topology is adapted by connecting intermediate ONU modules, respectively. Consequently, each ONU_n module, where $\{n = 1, 2, 3..., N\}$, consists of $1:2$ POC_n and 1:2 OS_n. Port 1 of the POC is connected to respective DF, whereas port 2 is fed into a 1:2 OS_n. Port 3 extends the DF toward a DR that provides the necessary protection in case of failure at the DF.

OS_n is used to connect OC_n and ONU_n with transmission media at the distribution level. Port "a" of OS_n connects ONU_n with the DF, whereas port "b" is used to connect ONU_n with the redundant DR in case of failure. Consequently, the required protection is achieved by avoiding the extensive duplication of the DFs. Under normal mode of operation, OS_n is at port "a" and all services are delivered through the DF, whereas in case of failure the affected ONU_n MAC

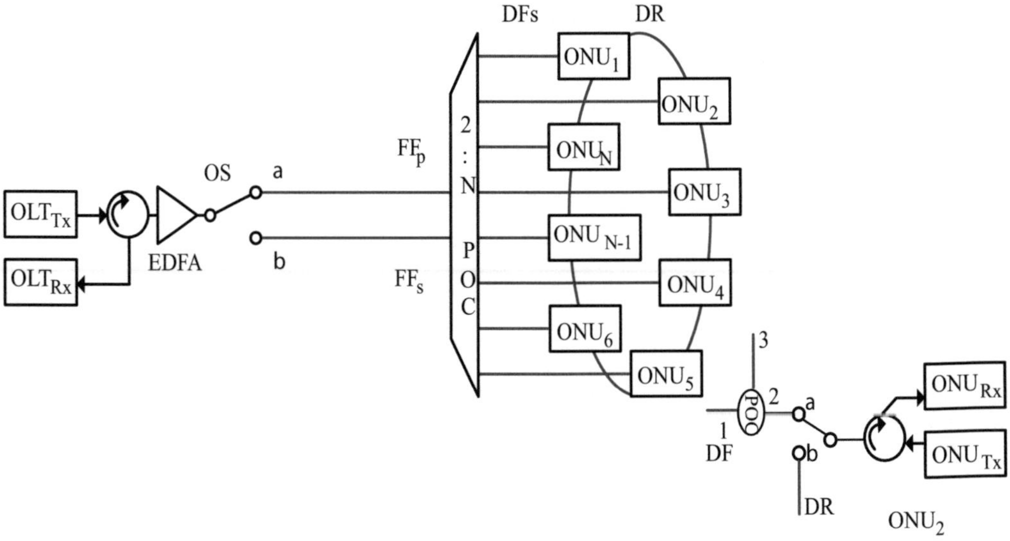

Figure 5. Tree-based hybrid protection architecture.

layer flips the switch position and transfers the flow of traffic to the DR between the adjacent ONU module.

The downstream power budget for tree-based hybrid protection architecture can be written as:

$$P_T - P_{OLT} - P_{OC_{co}} + P_{EDFA} - P_{OS_{co}} - \alpha P_{FF} - 10\text{Log}_{10}N - \alpha P_{DF} \\ - P_{POC_{onu}} - P_{OS_{onu}} - P_{OC_{onu}} - P_{ONU} \geq R_{sen} \tag{4}$$

It is observed from Eq. (4) that for $P_T = 10$ dBm, 25 km fiber, $P_{edfa} = 25$ dB, and $R_{sen} = -25$ dBm, the tree-based hybrid protection architecture can efficiently support 128 subscribers simultaneously accessing the medium [15].

3.5. Ring-based hybrid protection architecture

This architecture is formed by the combination of a single ring topology at the feeder level and a star-ring architecture at the ODN. A single OLT is placed at the CO, which is connected to an OC_{co}, EDFA, and SA_{co} as shown in **Figure 6** [18]. This architecture utilizes a single ring-based fiber to provide the required high bandwidth connectivity and reliability at the feeder level. Port 1 and 2 of the POC_{sa} extend traffic toward the FR_{cw} and FR_{ccw}, respectively. A 1:2 OS_{sa} is connected with port 2 of the POC_{sa} to provide the required fault recovery in the event of failure. Port *a* of the OS_{sa} is connected to ground, whereas port *b* is fed into the FR_{ccw}. Under normal mode of operation, OS_{sa} is at port *a* and all traffic is sent and received through the FR_{cw}.

FR is implemented at each RN by combining FR_{cw} and FR_{ccw} through a special arrangement. This significantly reduces the power budget penalty and cost of the overall architecture. RN

http://dx.doi.org/10.5772/intechopen.68237

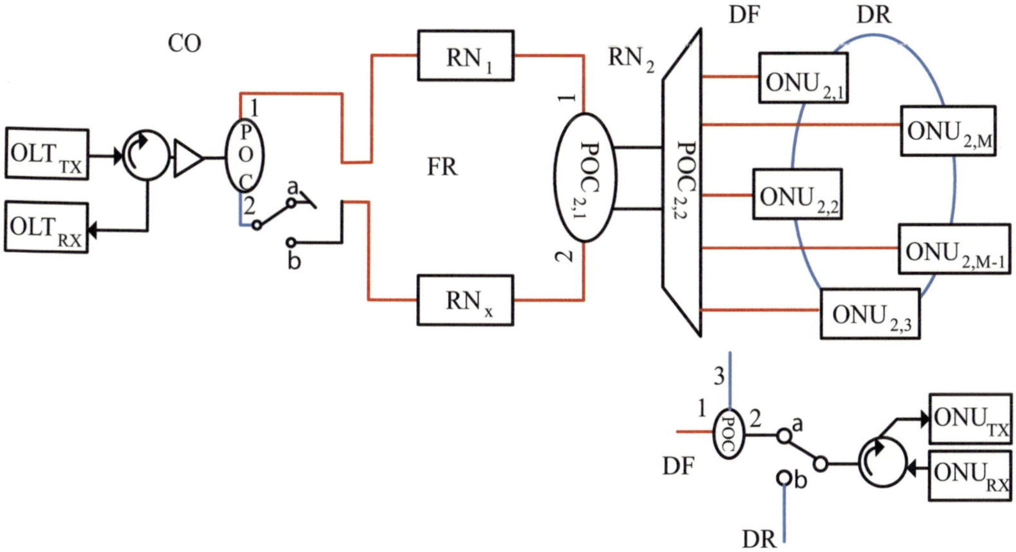

Figure 6. Ring-based hybrid protection architecture.

in the ring-based hybrid network $\{RN_x;\ x = 1, 2, 3 \ldots X\}$ consists of two POCs, namely ($POC_{x,1}$,$POC_{x,2}$). Where $POC_{x,1}$ and $POC_{x,2}$ are bidirectional couplers with 2 : 2 and 2:M ports respectively. FR is formed by connecting both CW and CCW paths through port 1 and 2 of the $POC_{x,1}$ as shown in **Figure 6**. $POC_{x,1}$ further extends the FR toward the ODN through a 2 : M $POC_{x,2}$, where M is the number of ONUs connected to each $POC_{x,2}$ through dedicated DFs. If X represents the total number of RNs deployed over the FR and M is the number ONUs per RN, then the proposed system can support a total of $N = X \times M$ ONUs.

ONU module, $\{ONU_{x,m};\ m = 1, 2, 3 \ldots M\}$, starts with a 1:2 $POC_{x,m}$ with three ports as shown in **Figure 6**. Port 1 is used to connect each ONU to its dedicated DF, whereas port 2 is fed into 1:2 $OS_{x,m}$. Port 3 is used to form a backup DR, which ensures immediate survivability of traffic between ONUs and OLT. Port a of the $OS_{x,m}$ connects $ONU_{x,m}$ with corresponding DF, under normal working conditions, and port "b" is fed into the DR.

The number of subscribers in ring-based hybrid protection architecture can be determined as:

$$P_{\mathrm{T}} - P_{\mathrm{OLT}} - P_{\mathrm{OC_{co}}} + P_{EDFA} - P_{\mathrm{POC_{sa}}} - P_{\mathrm{OS_{sa}}} - \alpha P_{\mathrm{FR}} - XP_{\mathrm{POC}_{x,1}} - 10\mathrm{Log}_{10}(M) - \alpha P_{\mathrm{DF}} \\ - P_{\mathrm{POC_{onu}}} - P_{\mathrm{OS_{onu}}} - P_{\mathrm{OC_{onu}}} - P_{\mathrm{ONU}} \tag{5}$$

If N represents the total number of ONUs, then the value of X can be written as:

$$X \leq \frac{P_{\mathrm{T}} - R_{\mathrm{sen}} - 16.25 - 10\mathrm{Log}_{10}M + P_{\mathrm{EDFA}}}{3} \tag{6}$$

Analysis of Eq. (6) shows that capacity of the network increases when more ONUs are placed per RN. Furthermore, it is observed that for $P_T = 10$ dBm, $P_{edfa} = 25$ dB, and $R_{sen} = -25$ dBm, the ring-based hybrid protection architecture can efficiently support 128 subscribers simultaneously accessing the medium [18].

4. Reliability analysis

Network reliability analysis is an important tool that is used to determine the overall connection availability for a given protection architecture. Reliability block diagrams (RBDs) are commonly used to determine the overall connection availability of PON, owing to its significant advantages like, accuracy, simplicity, visual impact, and flexibility. This section analyzes network reliability in terms of connection availability of the selected protection architectures through RBDs.

RBD represents each component including the OF medium as a functional block connected in series or parallel combination with adjacent blocks. Series and parallel connectivity represent the unprotected and protected components in protection architecture, respectively. The characteristic parameter of each RBD block is the asymptotic unavailability (U_i) of the components that represent their probability of failures. Consequently, if I represents the total number of components in PON, then overall connection availability A is given by:

$$A = 1 - \sum_{i=1}^{I} U_i \tag{7}$$

where U_i is determined by

$$U_i = 1 - \frac{\text{MTBF}}{\text{MTBF} + \text{MTTR}} \tag{8}$$

MTBF in Eq. (8) is the mean time between failures, which is used to represent the number of failures per million hours for an optical component [19]. MTTR represents the mean time to repair, which is the time required for reparation or replacement of faulty hardware modules [5, 20, 21]. In order to determine the overall connection availability for each architecture, RBDs with series and parallel combination of functional blocks are extracted as shown in **Figure 7**.

If $\sum_{i=1}^{I} U_i$ is the summation of i components unavailability in a protection architecture, then connection availability for each survivable PON can be written as:

$$A_{[C]} = 1 - \left[\left(U_{\text{OLT}} \times U_{\text{OLT}_{\text{pt}}}\right) + \left(U_{\text{FF}} \times U_{\text{FF}_{\text{pt}}}\right) + \left(U_{1:N\,\text{POC}} \times U_{1:N\,\text{POC}_{\text{pt}}}\right) + \left(U_{\text{DF}} \times U_{\text{DF}_{\text{pt}}}\right) + \left(U_{\text{ONU}} \times U_{\text{ONU}_{\text{pt}}}\right)\right] \tag{9}$$

$$A_{[SR]} = 1 - \left[U_{\text{OLT}} + U_{\text{POC}} + U_{\text{OS}} + \left(U_{\text{FR}} \times U_{\text{FR}_{\text{pt}}}\right) + U_{\text{POC}} + U_{\text{POC}} + U_{\text{ONU}}\right] \tag{10}$$

http://dx.doi.org/10.5772/intechopen.68237

$$A_{[DR]} = 1 - \Big[U_{OLT} + U_{POC} + U_{OS} + \left(U_{FR} \times U_{FR_{pt}}\right) + U_{POC} + U_{POC} + \left(U_{DR} \times U_{DR_{pt}}\right) + \left(U_{POC} \times U_{POC_{pt}}\right) + U_{OS} + U_{ONU}\Big] \quad (11)$$

$$A_{[HT]} = 1 - \Big[U_{OLT} + U_{OS} + \left(U_{FF} \times U_{FF_{pt}}\right) + U_{2:N\,POC} + \left(U_{DF} \times U_{DR_{pt}}\right) + U_{POC} + U_{OS} + U_{ONU}\Big] \quad (12)$$

$$A_{[HR]} = 1 - \Big[U_{OLT} + U_{POC} + U_{OS} + \left(U_{FR} \times U_{FR_{pt}}\right) + U_{POC} + U_{2:N\,POC} + \left(U_{DF} \times U_{DR_{pt}}\right) + U_{POC} + U_{OS} + U_{ONU}\Big] \quad (13)$$

Figure 8 shows the overall connection availability of the selected protection architectures based on **Table 1**. It is observed that maximum 5 nines connection availability is provided by all architectures, owing to the efficient utilization of redundant components and OF medium throughout the network. Maximum connection availability is provided by type C protection architecture, which duplicates entire PON including OLT and ONU modules. Ring-based architecture at both feeder and ODN, which avoids the duplication of light-wave path, is also observed to provide the required availability. A single failure at the ring-based fiber converts

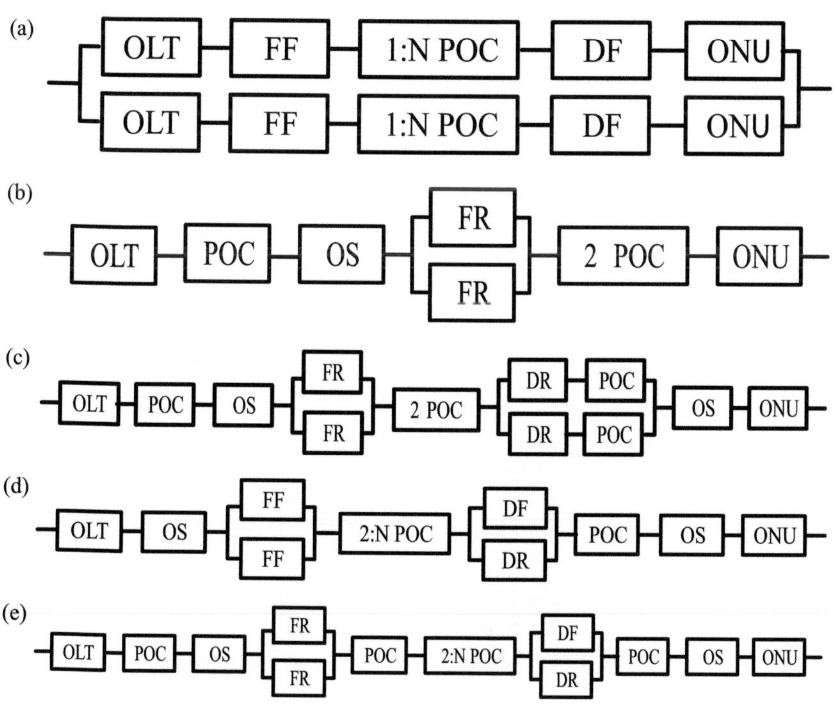

Figure 7. Reliability block diagrams for (a) type C, (b) single ring (SR), (c) dual ring (DR), (d) tree-based hybrid (HT), and (e) ring-based hybrid (HR) protection architectures.

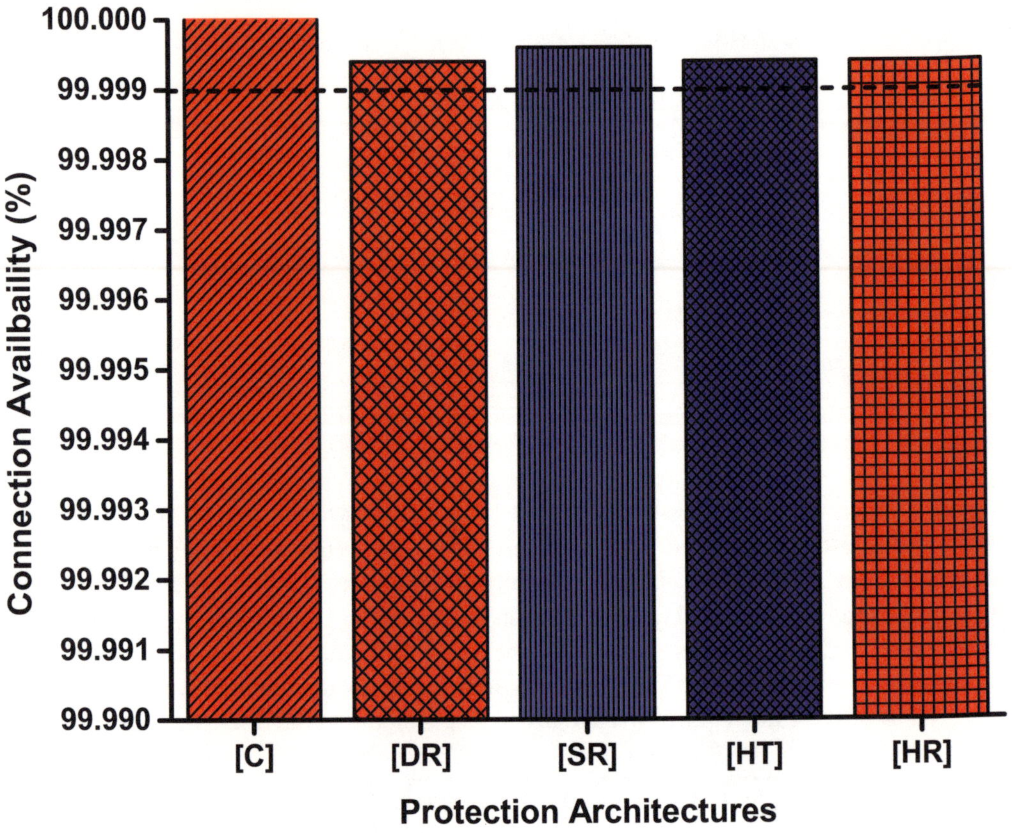

Figure 8. Overall connection availability of selected protection architectures.

the network into two tree-based architectures with the flow of traffic in both clockwise and counter clockwise directions. Thus, redundancy is achieved without duplicating the entire fiber. Furthermore, it is observed that desirable connection availability can be maintained without duplicating the transceiver modules, which further helps in reducing the overall cost of deployment for such architectures.

5. Cost analysis

Cost figures for protection architectures are an important parameter, showing the economic benefits for a common end user at the access domain. This section determines the overall expenditure for the selected architecture while using component costs in **Table 1**. Cost figures for PON can be categorized as CAPEX and OPEX. CAPEX includes the investment utilized for the deployment of PON at access domain and is calculated by computing the cost expenditure on network devices along with the OF infrastructure. OPEX includes the cost incurred on the network operations from the time of deployment till replacement by a new technology [4, 15, 22, 23].

http://dx.doi.org/10.5772/intechopen.68237

OPEX primarily includes the cost required for repairing the faulty, failed equipment (including OF medium), service interruption penalty cost that is commonly applicable for business subscribers, and energy consumed by active components at transceiver modules of OLT and ONU. Reparation cost is determined by multiplying the total downtime/year with resources required to remove the fault, which includes the number of technicians along with their wages and miscellaneous charges. Service interruption penalties include the expenditure that is spent on the fine defined in service level agreement (SLA) between network operators and subscribers. Power consumption by each component is determined by multiplying the unit price of electricity with the sum of energy consumption of all active components over the network life span [15, 23].

Analysis is performed for residential customers only, and life span of each network is taken as 20 years. Furthermore, following specifications are considered for fair analysis of the selected protection architectures

- OLT can support 16, 32, 64, and 128 subscribers based on the type of adapted topology.
- Length of FF = 20 km, DF = 5 km, and DR = 1 km between adjacent nodes in selected architectures.
- EDFA cost is considered in dual ring, tree- and ring-based hybrid architectures.
- Digging cost for OF medium is ignored due to high variation.
- No service interruption penalty cost is considered.
- Repairing cost is 1000 $/h.
- Per hour cost of electricity is taken to be 0.25 $/kWh.

If N is the total number of ONUs in each PON and X represents the number of RNs in ring-based topologies at the feeder level, then the CAPEX equations (based on RBDs) for selected protection architectures can be written as:

$$C_C = (\mathrm{OLT} + \mathrm{OLT}) + (20 \times \mathrm{FF} \times \mathrm{FF}) + (2 \times 1:N\,\mathrm{POC}) + (5 \times N \times \mathrm{DF} \times \mathrm{DF}) + (N \times \mathrm{ONU} \times \mathrm{ONU}) \quad (14)$$

$$C_{SR} = \mathrm{OLT} + \mathrm{POC} + \mathrm{OS} + (25 \times \mathrm{FR}) + (2X \times \mathrm{POC}) + (N \times \mathrm{ONU}) \quad (15)$$

$$C_{DR} = \mathrm{OLT} + \mathrm{POC} + \mathrm{OS} + \mathrm{EDFA} + (20 \times \mathrm{FR}) + (2X \times \mathrm{POC}) + (N \times \mathrm{DR}_{cw}) + (N \times \mathrm{DR}_{ccw}) + (2N \times \mathrm{POC}) + (N \times \mathrm{OS}) + (N \times ONU) \quad (16)$$

$$C_{HT} = \mathrm{OLT} + \mathrm{OS} + \mathrm{EDFA} + (20 \times \mathrm{FF} \times \mathrm{FF}) + (2:N\,\mathrm{POC}) + (5 \times N \times \mathrm{DF}) + (N \times \mathrm{DR}) + (N \times \mathrm{POC}) + (N \times \mathrm{OS}) + (N \times \mathrm{ONU}) \quad (17)$$

$$C_{HR} = \mathrm{OLT} + \mathrm{POC} + \mathrm{OS} + \mathrm{EDFA} + (20 \times \mathrm{FR}) + (X \times \mathrm{POC}) + (X \times 2:N\,\mathrm{POC}) + (5 \times N \times \mathrm{DF}) + (N \times \mathrm{DR}) + (N \times \mathrm{POC}) + (N \times \mathrm{OS}) + (N \times \mathrm{ONU}) \quad (18)$$

Figure 9 shows the CAPEX for each protection architecture at different number of subscribers by referring to **Table 1**. It is observed that ring-based architecture requires minimum CAPEX when compared to conventional tree and hybrid techniques. However, due to the power budget and capacity limitations of single and dual ring-based architectures, their analysis is performed till 16 and 64 number of subscribers, respectively. Overall analysis shows that the deployment cost for each architecture decreases as the number of subscriber increases. It is evident from the fact that total cost incurred is distributed among the number of subscribers, which results in reduced CAPEX per user as the network capacity increases. Furthermore, it is shown that type C architecture requires the highest cost, for all subscribers, in comparison with other schemes due to the extensive duplication of optical components and OF medium throughout the network.

Figure 9 shows that hybrid protection architectures provide nominal performance, for all subscribers, in terms of the deployment cost as they avoid extensive duplication of the OF medium. **Figure 10** further elaborates the CAPEX required for the deployment of light wave path in type C, tree- and ring-based hybrid protection architectures for 128 subscribers in the

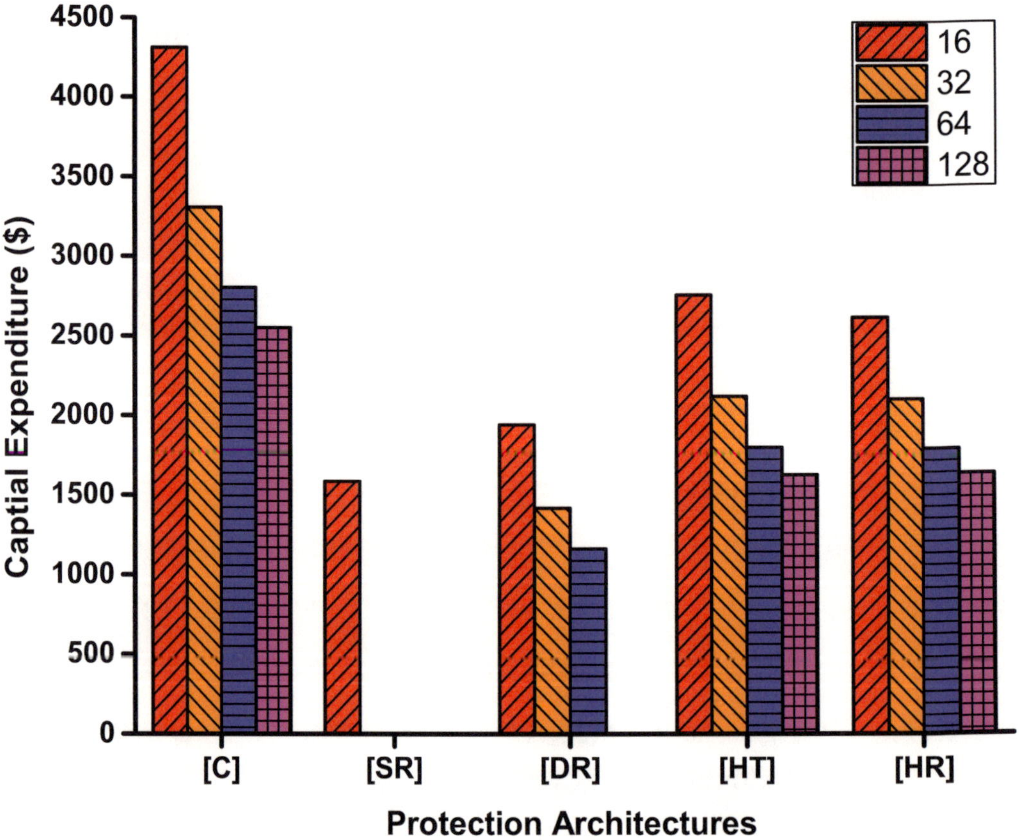

Figure 9. CAPEX at different number of subscribers.

http://dx.doi.org/10.5772/intechopen.68237

network. It is observed that the main contributor for CAPEX, besides OLT module, is the DF as each subscriber requires a dedicated path for high speed connectivity. Furthermore, it is shown in **Figure 10** that type C requires the highest deployment cost for DFs due to the duplication of each fiber between RN and corresponding ONU module. On the other hand, the deployment of star-ring topology at ODN significantly reduces the CAPEX, on redundancy, for hybrid protection architectures.

It is also observed that the CAPEX in hybrid protection architecture varies with the type of topology at the feeder level. Since ring-based architecture requires multiple RNs, the cost of deployment increases with the number of subscribers due to increase in RNs. Consequently, tree-based hybrid architecture requires less cost when compared to ring-based scheme for 64 and 128 subscribers. However, the difference is minuscule.

Figure 11 shows the total expenditure for protection architectures with high network capacity in terms of the number of subscribers. It is evident from the fact that each architecture provides the desired connection availability, which significantly reduces their downtime/year. Hence, minuscule amount is spent on reparation cost over the network life span. Type C protection architecture requires highest OPEX due to the duplication of transceiver modules at both OLT

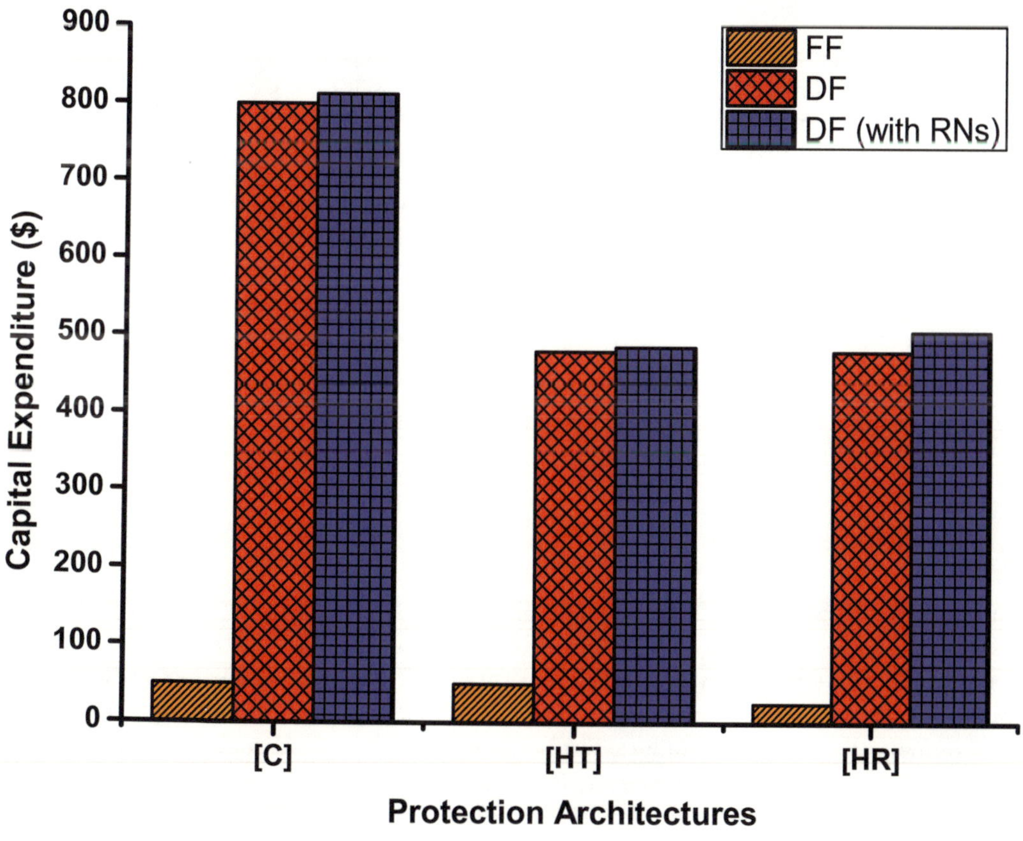

Figure 10. CAPEX for OF infrastructure at 128 subscribers.

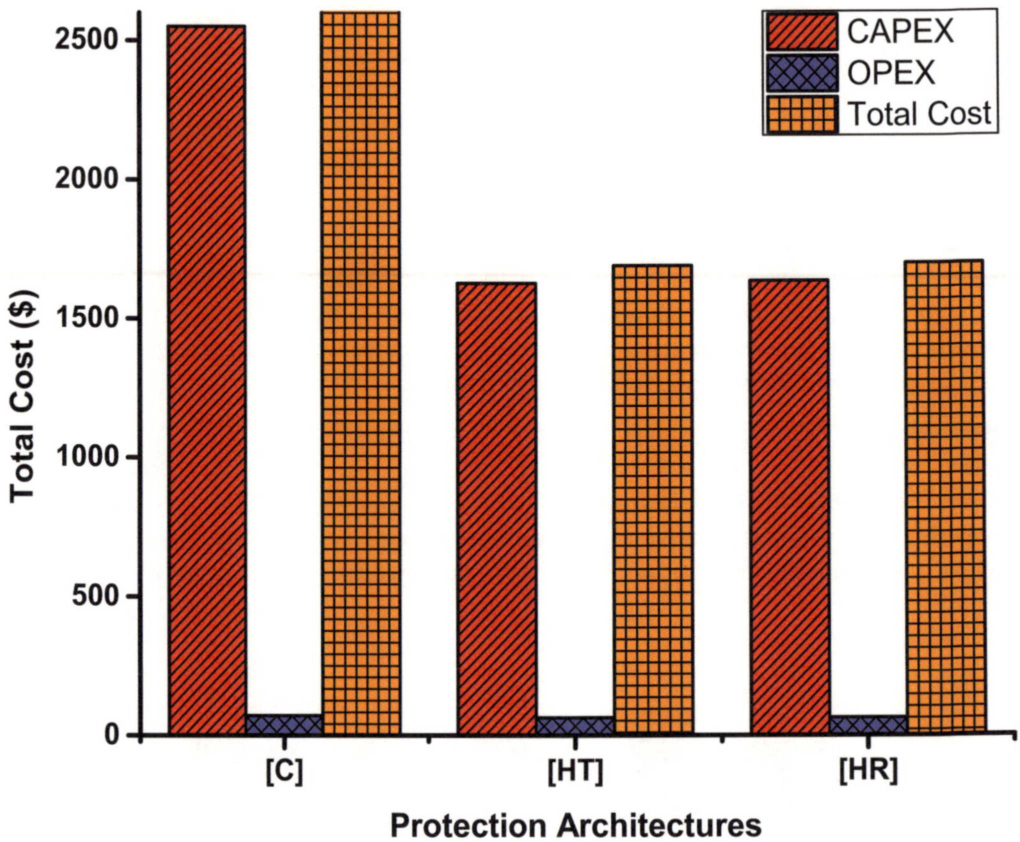

Figure 11. Total cost including CAPEX and OPEX at 128 subscribers.

and ONUs, which consumes more energy as compared to hybrid architectures with single OLT and ONU modules. Consequently, hybrid architecture requires minimum cost, collectively, as compared to the conventional protection scheme.

6. Conclusion

This chapter analyzes the effect of fiber duplication, for protection, on feasibility of PON. Increase in the capacity and adaptability of PON demands for an efficient fault detection and restoration mechanism, which is able to provide the required connection availability at minimum cost. Among all components in the PON, OF medium requires significant protection due to its high rate of failure. Furthermore, it is an on-field component and constitutes a major portion of PON; therefore, it is imperative to provide pre-planned redundant paths in order to ensure swift recovery of failures at the OF medium throughout the network. Five protection architectures are considered, which mainly emphasize on the type of topology laid through the OF medium at both feeder level and ODN. ITU-T 983.1 type C and single ring-based architectures are considered to encompass the conventional tree-star and ring topology, respectively. For further analysis, dual ring architecture and tree- and ring-based hybrid protection architectures for PON are analyzed to study the effect of hybrid topologies on PON feasibility in terms of the number of subscribers, reliability, and cost. It is observed that ring-based architectures provide desirable connection availability without duplicating the OF medium; however, the use of multiple POCs to extend the FR significantly elevates the power budget issues. Consequently, the number of subscribers that can access the medium simultaneously is reduced. Hybrid topologies formed by the variation of tree, ring, and star topology provide nominal performance, in terms of all parameters, as compared to the survivable PONs based on conventional topologies. Furthermore, it is shown that duplication of the OF medium at the ODN network is more critical in comparison with the feeder level duplication. Nevertheless, both tree- and ring-based hybrid pre-planned protection architectures provide desirable network capacity and connection availability at minimum cost.

Author details

Waqas Ahmed Imtiaz[1]*, Javed Iqbal[2], Affaq Qamar[1], Haider Ali[1] and Sevia M. Idrus[3]

*Address all correspondence to: w.imtiaz.1985@ieee.org

1 Abasyn University, Peshawar, Pakistan

2 Sarhad University of Science and Information Technology, Peshawar, Pakistan

3 Faculty of Electrical Engineering, Universiti Teknologi Malaysia, Malaysia

References

[1] Cedric F Lam. *Passive optical networks: principles and practice*. Academic Press, 2007. ISBN 978-0123738530.

[2] Mozhgan Mahloo, Jiajia Chen, Lena Wosinska, Abhishek Dixit, Bart Lannoo, Didier Colle, and Carmen Mas Machuca. Toward reliable hybrid wdm/tdm passive optical networks. *IEEE Communications Magazine*, 52(2):273–280, 2014. doi: 10.1109/MCOM.2014.6736740.

[3] Hussein T Mouftah and Pin-Han Ho. *Optical networks: architecture and survivability*. Springer Science & Business Media, 2012. ISBN 978-1-4615-1169-4.

[4] Lena Wosinska, CHEN Jiajia, and Claus Pop Larsen. Fiber access networks: reliability analysis and swedish broadband market. *IEICE Transactions on Communications*, 92 (10):3006–3014, 2009. doi: 10.1587/transcom.E92.B.3006.

[5] Mozhgan Mahloo. *Reliability versus cost in next generation optical access networks*. Doctoral Dissertation, School of Information and Communication Technology, KTH Royal Institute of Technology, 2013, ISSN 1653-6347 ; 1303.

[6] Pavel Lafata and Jiri Vodrazka. Application of fiber ring for protection of passive optical infrastructure. *Radioengineering*, 22(1):357–362, 2013, ISSN: 1805-9600.

[7] Chien-Hung Yeh and Sien Chi. Self-healing ring-based time-sharing passive optical networks. *IEEE Photonics Technology Letters*, 19(13–16):1139–1141, 2007. doi: 10.1109/LPT.2007.900155.

[8] Byung Tak Lee, Mun Seob Lee, and Ho Young Song. Simple ring-type passive optical network with two-fiber protection scheme and performance analysis. *Optical Engineering*, 46(6), 2007. doi: 10.1117/1.2746929.

[9] Waqas Ahmed Imtiaz, Yousaf Khan, and Khalid Mahmood. Design and analysis of self-healing dual-ring spectral amplitude coding optical code division multiple access system. *Arabian Journal for Science and Engineering*, 41(9):3441–3449, 2016. doi: 10.1007/s13369-015-1988-z.

[10] Haibin Chen, Chaoqin Gan, Maojun Yin, and Cuiping Ni. A single-star multi-ring structure of self-healing wavelength division multiplexing optical access network. *Fiber and Integrated Optics*, 33(1–2):4–16, 2014. doi: 10.1080/01468030.2013.879435.

[11] Yang Qiu and Chun-Kit Chan. A novel survivable architecture for hybrid wdm/tdm passive optical networks. *Optics Communications*, 312:52–56, 2014. doi: 10.1016/j.optcom.2013.09.005.

[12] Yeh C.H., Chow C., and Liu Y.L. Self-protected ring-star-architecture tdm passive optical network with triple-play management. *Optics Communications*, 284:3248–3250, 2011. doi: 10.1016/j.optcom.2011.03.032.

[13] Rastislav Róka. Optimization of traffic protection schemes for utilization in hybrid passive optical networks. *International Journal of Application or Innovation in Engineering and Management*, 5(9):107–116, 2016. ISSN 2319-4847.

[14] Elaine Wong, Carmen Mas Machuca, and Lena Wosinska. Survivable hybrid passive optical converged network architectures based on reflective monitoring. *Journal of Lightwave Technology*, 34(18):4317–4328, 2016. doi: 10.1109/JLT.2016.2593481.

[15] Waqas Ahmed Imtiaz, Yousaf Khan, and Waqas Shah. Cost versus reliability analysis of tree-based hybrid protection architecture for optical code division multiple access system. *Journal of Engineering and Applied Sciences*, 35(1):47–53, 2016.

[16] Chen-Hung Yeh and Sien Chi. Self-protection against fiber fault for ring-based power-splitting passive optical networks. *Optical Engineering*, 47(2): 020501 020501, 2008. doi: 10.1117/1.2841702.

[17] Rec ITU-T. G. 983.1. study group 15: broadband optical access systems based on passive optical networks (pon). *ITU-T, Rec. G*, 983, 1998.

[18] Waqas A Imtiaz, Pir Mehar, Muhammad Waqas, and Yousaf Khan. Self-healing hybrid protection architecture for passive optical networks. *International Journal of Advanced*

Computer Science and Applications (IJACSA), 6(8):144–148, 2015. doi: 10.14569/IJACSA. 2015.060819.

[19] Susan Stanley. Mtbf, mttr, mttf & fit explanation of terms. *IMC Network*, 1–6, 2011.

[20] Patrick O'Connor and Andre Kleyner. *Practical reliability engineering*. John Wiley & Sons, 2012. ISBN 978-0-470-97982-2.

[21] David J Smith. *Reliability, maintainability and risk 8e: Practical methods for engineers including reliability centred maintenance and safety-related systems*. Elsevier, 2011. ISBN 978-00809690 22.

[22] Álvaro Fernández and Norvald Stol. Capex and opex simulation study of cost-efficient protection mechanisms in passive optical networks. *Optical Switching and Networking*, 17:14–24, 2015. doi: j.osn.2015.01.001.

[23] Jiajia Chen and Lena Wosinska Wosinska. Cost vs. reliability performance study of fiber access network architectures. *IEEE Communications Magazine*, 48:56–65, 2010. doi: 10.1109 /MCOM.2010.5402664.